π CHAPMAN & HALL/CRC
Monographs and Surveys in
Pure and Applied Mathematics   **132**

SHOCK WAVES

AND

EXPLOSIONS

CHAPMAN & HALL/CRC
Monographs and Surveys in
Pure and Applied Mathematics   **132**

SHOCK WAVES

AND

EXPLOSIONS

P. L. SACHDEV

CRC Press
Taylor & Francis Group
Boca Raton London New York

CRC Press is an imprint of the
Taylor & Francis Group, an **informa** business

A CHAPMAN & HALL BOOK

CRC Press
Taylor & Francis Group
6000 Broken Sound Parkway NW, Suite 300
Boca Raton, FL 33487-2742

First issued in paperback 2019

© 2004 by Taylor & Francis Group, LLC
CRC Press is an imprint of Taylor & Francis Group, an Informa business

No claim to original U.S. Government works

ISBN-13: 978-1-58488-422-4 (hbk)
ISBN-13: 978-0-367-39417-2 (pbk)
Library of Congress Card Number 2004047803

Library of Congress Cataloging-in-Publication Data

Sachdev, P. L.
    Shock waves and explosions / P.L. Sachdev.
        p. cm. — (Chapman & Hall/CRC monographs and surveys in pure and applied
    mathematics ; 132)
    Includes bibliographical references and index.
    ISBN 1-58488-422-3 (alk. paper)
        1. Differential equations, Hyperbolic—Numerical solutions. 2. Shock
    waves—Mathematics. I. Title. II. Series.

QA377.S24 2004
518'.64—dc22   2004047803

Visit the Taylor & Francis Web site at
http://www.taylorandfrancis.com

and the CRC Press Web site at
http://www.crcpress.com

TO THE MEMORY OF MY PARENTS

Contents

Preface

I have been interested in the theory of shock waves for a long time. It started some three decades ago when I published some work on explosions. Since then my interests have diversified but nonlinearity has remained the focus of all my exertions. I returned to the explosion phenomenon a few years ago when the Defence Research and Development Organisation (DRDO), Ministry of Defence, suggested that I write the present monograph.

I have now looked at the phenomenon of explosions and shock waves in the larger context of hyperbolic systems of partial differential equations and their solutions. Application to explosion phenomenon is the major motivation. My approach is entirely constructive; both analytic and numerical methods have been discussed in some detail. The historical evolution of the subject has dictated the sequence and contents of the material included in the present monograph. A reader with a basic knowledge of fluid mechanics and partial differential equations should find it quite accessible.

I have been much helped in the present venture by many collaborators and students. I may mention Prof. K.T. Joseph, Prof. Veerappa Gowda, Dr. B. Sri Padmavati, Manoj Yadav and Ejan-ul-Haque. Dr. Eric Lord went through the manuscript with much care. I must particularly thank Ms. Srividya for her perseverence in preparing the camera ready copy as I wrote the manuscript, introduced many changes and struggled to make the material meet my expectations. I am also grateful to Prof. Renuka Ravindran and Prof. G. Rangarajan for their support.

I wish to thank Dr. Sunil Nair and Ms. Jasmin Naim, Chapman & Hall, CRC Press for their immense co-operation as I accomplished this most fascinating piece of work.

Finally, I must thank my wife, Rita, who has permitted me to indulge in my writing pursuits over the last two decades. She has during this period borne the major part of family responsibilities with great fortitude.

Financial support through the extra-mural research programme of DRDO and from the Indian National Science Academy is gratefully acknowledged.

Acknowledgements

The following illustrations and Tables are reproduced, with permission, from the sources listed:

Figure 2.1: Rogers, M.H., 1958, Similarity flows behind strong shock waves, *Quart. J. Mech. Appl. Math.*, 11, 411.

Figures 2.2 and 2.3: Kochina, N.N. and Melnikova, N.S., 1958, On the unsteady motion of gas driven outward by a piston, neglecting the counter pressure, *PMM*, 22, 622.

Figure 2.4: Grigorian, S.S., 1958a, Cauchys problem and the problem of a piston for one-dimensional, non-steady motions of a gas (automodel motion), *PMM*, 22, 244; Grigorian, S.S., 1958b, Limiting self-similar, one-dimensional non-steady motions of a gas (Cauchys problem and the piston problem),*PMM*, 22, 417.

Figures 2.5-2.7: Sachdev, P.L. and Venkataswamy Reddy, A., 1982, Some exact solutions describing unsteady plane gas flows with shocks, *Quart. Appl. Math.*, 40, 249.

Tables 2.1-2.3: Rogers, M.H., 1958, Similarity flows behind strong shock waves, *Quart. J. Mech. Appl. Math.*, 11, 411.

Figure 3.1: Sakurai, A., 1954, On the propagation and structure of the blast wave II, *J. Phys. Soc. Japan*, 9, 256.

Figures 3.2-3.5: Bach, G.G. and Lee, J.H.S., 1970, An analytical solution for blast waves, *AIAA J.*, 8, 271.

Figures 3.6-3.9: Laumbach, D.D. and Probstein, R.F., 1969, A point explosion in a cold exponential atmosphere, *J. Fluid Mech.*, 35, 53-75.

Figures 3.10 and 3.11: Raizer, Yu.P., 1964, Motion produced in an inhomogeneous atmosphere by a plane shock of short duration, Sov. Phys. Dokl., 8, 1056.

Figure 3.12: Waxman, E. and Shvarts, D., 1993, Second type self-similar solutions to the strong explosion problem, *Phys. Fluids A*, 5, 1035.

Figures 3.13-3.15: Reinicke, P. and Meyer-ter-Vehn, J., 1991, The point explosion with heat conduction, *Phys. Fluids A*, 3, 1807.

Figure 3.16: Whitham, G.B., 1950, The propagation of spherical blast, *Proc. Roy. Soc. A*, 203, 571.

Table 3.1: Taylor, G.I., 1950, The formation of a blast wave by a very intense explosion, I, *Proc. Roy. Soc. A*, 201, 159.

Tables 3.2-3.4: Sakurai, A., 1953, On the propagation and structure of the blast wave I, *J. Phys. Soc. Japan*, 8, 662.

Figures 6.1-6.3: Chisnell, R.F., 1998, An analytic description of converging shock waves, *J. Fluid Mech.*, 354, 357.

Figures 6.4 and 6.5: Van Dyke, M. and Guttmann, A.J., 1982, The converging shock wave from a spherical or cylindrical piston, *J. Fluid Mech.*, 120, 451-462.

Table 6.1: Fujimoto, Y. and Mishkin, E.A., 1978, Analysis of spherically imploding shocks, *Phys. Fluids*, 21, 1933.

Tables 6.2 and 6.3: Van Dyke, M. and Guttmann, A.J., 1982, The converging shock wave from a spherical or cylindrical piston, *J. Fluid Mech.*, 120, 451-462.

Figures 7.1-7.3: McFadden, J.A., 1952, Initial behaviour of a spherical blast, *J. Appl. Phys.*, 23, 1269.

Figures 8.1-8.4: Brode, H.L., 1955, Numerical solutions of spherical blast waves, *J. Appl. Phys.*, 26, 766.

Figures 8.5-8.12: Payne, R.B., 1957, A numerical method for a converging cylindrical shock, *J. Fluid Mech.*, 2, 185.

Figure 8.13: Sod., G.A., 1977, A numerical study of a converging cylindrical shock, *J. Fluid Mech.*, 83, 785-794.

Figure 8.14: Liu, T.G., Khoo, B.C., and Yeo, K.S., 1999, The numerical simulations of explosion and implosion in air: Use of a modified Hartens TVD scheme, *Int. J. Numer. Meth. Fluids*, 31, 661.

Chapter 1

Introduction

The development of the nuclear bomb during World War II changed the way people thought about power and the way nations conducted themselves in peace and in war. The strategies and methods which were employed in the scientific effort to develop the bomb in each of the countries involved—the U.S., the U.K., the (then) U.S.S.R., and Germany—would themselves make a most fascinating study. That such a bomb went much beyond what was envisioned may be gauged from the way R.J. Oppenheimer, the man at the helm of the Manhattan Project in the U.S., exclaimed when he first saw the blazing light of the bomb. He was so stunned and dazzled that he could only murmur and quote from The Bhagwad Gita, "Brighter than a Thousand Suns"—one of the attributes of the Virat Roopa of God as delineated in this much venerated book.

The mathematical formulation of the problem of the nuclear explosion and the estimation of its mechanical effects on the surroundings was itself a challenging task. There was hardly any literature on this subject. So, some of the best minds in applied mathematics and physics were made to put their heads together to unravel this topic. This gave a great fillip to nonlinear science, which has since made great strides and which now permeates and influences all sciences—pure and applied. The people who initiated the nonlinear studies in this context include G.I. Taylor, John Von Neumann, L.I. Sedov, L.D. Landau, H. Bethe and many others. There were several centers in each advanced country, which devoted their entire effort to the study of blast waves from nuclear explosions, both intense and not so intense. The war time work continued until the 1970s and engaged some other bright minds—M.J. Lighthill, G.B. Whitham, M. Holt, R.F. Probestein and A.K. Oppenheim. There has been some lull in this activity in the last two decades; other concepts such as solitons and chaos have overtaken to carry the study of nonlinearity to a more sophisticated and complex level.

1

We shall first discuss the original contributions of Taylor (1946, 1950) and Von Neumann (1941); the original work of Sedov (1946) is not readily available in English. Sir Geoffrey Taylor published in the Proceedings of the Royal Society (1950) the work on the formation of a blast wave by a very intense explosion in two parts. This work was circulated in early 1941 for the Civil Defense Research Committee in the U.K.; it had been undertaken at their request and was intended to investigate what effect a bomb, with a very large amount of energy released by nuclear fission, would have on the surroundings. It was still not called an atom bomb. How would it differ from a conventional explosive bomb, which is produced by the sudden generation of a large amount of gas at a high temperature in a confined space? It is remarkable that the only reference made by G.I. Taylor is to the work of Rankine (1870) on the so-called Rankine-Hugoniot conditions which hold at a shock, a surface of discontinuity across which the flow variables—pressure, particle velocity, density and entropy—suffer a jump. Taylor (1950) idealised the problem as follows. He envisioned that a finite amount of nuclear energy is suddenly released in an infinitely concentrated form. A high pressure gas, headed by an (infinitely) strong shock, propagates outwards and engulfs the undisturbed gas, suddenly raising its velocity (from zero), temperature, pressure and density. It may be observed that only the front surface of discontinuity is called a shock, while the entire disturbed flow of the gas behind the shock is called blast wave.

As the volume of this high-pressure gas increases, its density decreases and so does pressure, hence the changes it brings about in the surrounding air weaken, that is, it begins to decay. This high pressure gas, however, is always headed by a shock wave; the flow parameters behind it decrease as the center of the blast is approached.

Mathematically, the continuous flow behind the shock is governed by the nonisentropic equations of gas dynamics which must be solved subject to the so-called Rankine-Hugoniot conditions at the shock and the symmetry condition at the center requiring that the particle velocity there is zero. Along the shock trajectory, the theory of shocks imposes more boundary conditions than are appropriate to the given system. This over-determined data, however, leads to the finding of the shock trajectory, which itself is unknown apriori. This, in this sense, constitutes a free boundary value problem. In this simplest model the role of heat conduction is ignored.

Taylor (1950) made some highly intuitive physical statements about this phenomenon. For example, he observed that the explosion forces most of the air within the shock front into a thin shell just inside the front. This, as we shall discuss later, forms the basis of an analytic theory of blast waves in an exponential atmosphere by Laumbach and Probestein (1969). Taylor (1950) also observed that, as the front expands, the maximum pressure decreases till at about 10 atmospheres the analysis under the assumption of an infinitely strong shock ceases to hold.

Taylor (1950) also noted that, at 20 atmospheres, 45% of the energy had degraded into heat which was then unavailable for doing work and was used up in expanding against atmospheric pressure, indicating the rather inefficient nature of a nuclear bomb as a blast producer in comparison with the high explosive bomb. This argument regarding the degradation of mechanical energy into heat was later used by Brinkley and Kirkwood (1947) and Sachdev (1971, 1972) to formulate a (local) analytic shock theory, which determined the trajectory of the leading shock from its inception to final decay at infinity.

Taylor's formulation of the similarity solution was derived entirely from physical arguments. By using dimensional arguments, he wrote the similarity form of the solution in Eulerian co-ordinates in terms of the similarity variables r/R, where R, the radius of the shock, was found to be proportional to $t^{2/5}$; he did not use any sophisticated transformation theory of nonlinear PDEs. Taylor reduced the system of nonlinear PDEs to nonlinear ODEs and numerically solved the latter, subject to the strong shock conditions (appropriately transformed) and the requirement of spherical symmetry, namely, that the particle velocity at the center of the explosion must be zero. He also used the conservation of total energy, E, behind the shock to derive the shock trajectory. The constant $B = E/\rho_0 A^2$, which appears in the shock law $R = Bt^{2/5}$, involves the nondimensional form of energy and was found from the numerical solution; it varies with γ, the ratio of specific heats.

In a typically applied mathematical approach, Taylor (1950) carefully analysed the numerical solution and noticed that the particle velocity distribution behind the shock as a function of the similarity variable was quite close to linear, particularly near the center of the blast. He assumed for particle velocity a form of the solution which is the sum of a linear term and a nonlinear correction term in the similarity variable; he was able to explicitly determine this term by making use of the governing equations and the Rankine-Hugoniot conditions. This enabled him to find an (approximate) closed form solution of the entire problem which was in error in comparison with the numerical solution by less than five percent!

In the second part of his paper, Taylor (1950a) checked the $R\sim t^{2/5}$ law by comparing it with the shock trajectory obtained experimentally from the New Mexico explosion. The agreement of the two for various values of γ, the ratio of specific heats, was uncannily good. In this comparison, photographs were used to measure the velocity of the rise of the slowing center of the heated volume. This velocity was found to be 35 m/sec. The (hemispherical) explosive ball behaves like a large bubble in water until the hot air suffers turbulent mixing with the surrounding cold air. The vertical velocity of this 'equivalent' bubble was computed from this analysis and was found to be 35 m/sec.

While Taylor (1950) was quite aware of the advantages of a Lagrangian approach to the problem, he was rather sceptical of its practicality since, as he remarked, that would "introduce great complexity, and, in general, solutions can only be derived by using step by step numerical integration" of the full system of nonlinear PDEs. Actually, as a particle crosses the shock, it has an adiabatic relationship between pressure and density corresponding with the entropy which is endowed upon it by the shock wave during its passage past it. This naturally suggests a Lagrangian approach wherein the Lagrangian co-ordinate is defined as one which retains its value along the particle path. Indeed, this matter was raised much later again by Hayes (1968) who tried to contradict the suggestion by Zeldovich and Raizer (1967) that the Lagrangian formulation is as convenient as the Eulerian, even more so for the problems of blast wave type. He argued that the basic differential equation to be solved numerically is in a nonanalytic form in the Lagrangian formulation and would therefore pose difficulties, a view in agreement with Taylor's apprehension.

However, Von Neumann (1941), independently and contemporaneously, tackled the point explosion problem in Lagrangian co-ordinates and obtained an analytic solution in a form more explicit than that of Taylor (1950) or Sedov (1946). The solution was expressed in terms of a parameter, which was later physically interpreted. Von Neumann (1941) also found approximate form of his exact solution when $(\gamma - 1)$ is small, which holds for air or heated gas, and confirmed his results by comparison with those of Bethe (1942) under the same approximation.

Much later, Laumbach and Probestein (1969) considered a point explosion in a cold exponential atmosphere. They found an explicit analytic solution by assuming that the flow field was 'locally radial,' implying that the flow gradients in the θ direction were negligible; here θ is the polar angle measured from the vertical direction. The basic assumption in their analysis is that the shock is sufficiently strong so that the counter pressure may be neglected. They also made use of Lagrangian co-ordinates and exploited the physical observation proffered first by Taylor (1950) that most of the mass in the strong shock regime of the blast wave is concentrated in a thin shell immediately behind the front. They used a perturbation analysis, exploiting the above observation, and employed an integral method with an energy constraint. Their results, in the limit of uniform density, agree remarkably with those of Taylor (1950). Laumbach and Probestein (1969) also checked their results for ascending and descending parts of the flow with the known numerical results and found excellent agreement. The far field behaviour of the shock wave in the upward and downward direction, respectively, was found to be of the same form as the self-similar asymptotic solutions for the plane shock found earlier by Raizer (1964). The important point here is that the physically motivated perturbation analysis gave excellent analytic answers for this non-symmetrically stratified problem in both upward and downward directions.

To look at the problem of explosion from a different perspective, Taylor (1946) considered the disturbance of the air wave surrounding an expanding sphere. In a remarkable lead for nonlinear problems, he considered first the (spherical) sound waves produced by the vibration of a spherical piston. From his analysis of such waves, Taylor was led to the form of the solution for the nonlinear waves. He postulated that the solution must depend on the combination $r/a_0 t$ of the independent variables (a_0 being the speed of sound in the undisturbed medium), the so-called progressive wave form. Since the (strong) shock, under this assumption, moves with a constant speed, the flow behind it is of an isentropic character. Thus, with the assumption that the physical variables p, ρ, and u are constant along the line $dr/dt = r/t$, it was possible to reduce the problem to one of solving two coupled nonlinear ODEs with appropriate boundary conditions at the shock; the piston boundary was located such that the particle velocity there is equal to the piston velocity. This system was solved numerically and the results were depicted graphically. It must be stated, however, that this problem is rather artificial since the shock must ultimately decay. Taylor (1946) also confirmed that the sound wave solution here fails even when the piston motion is relatively small.

In an attempt to generalize the work of Taylor (1946, 1950), Rogers (1958) derived similarity solutions which describe flow of a perfect gas behind strong shocks for spherical, cylindrical and plane symmetries. The expanding piston, causing the motion, is now allowed to increase the total energy of the flow behind the shock, $E = E_0 t^s$, where E_0 and $s > 0$ are constant. Interestingly, this class of flows, again called similarity solutions or progressive waves, includes both the problems considered by Taylor (1946, 1950) as special cases, namely, the strong blast wave with $s = 0$ and the uniformly expanding sphere with $s = 3$. We may observe that, in practice, the total energy of the flow—the sum of kinetic energy and internal heat energy of the gas—will suffer losses due to dissipative effects while there will also be gain in the internal energy as the shock advances and encompasses more of the quiescent gas. The latter increase is ignored on the assumption that the shock is of infinite strength so that the pressure (and hence the internal energy of the gas ahead of the shock) are negligible. Rogers (1958), therefore, considered flows for which the total energy behind the shock increases with time, $E = E_0 t^s$. Since, in the present case, it is assumed that the flow behind the shock is adiabatic (no generation of heat) and the shock itself is of infinite strength, the increase in the energy is brought about by the pressure exerted by the piston on the surrounding gas. The mathematical problem is solved numerically in the manner of Taylor (1946), starting from the shock inward and locating the position of the piston as described earlier. Rogers (1958) concluded that, in the case of blast wave ($s = 0$), 80% of the energy of the blast is in the form of internal heat energy, as predicted by Taylor (1950). This ratio decreases as s increases until for the case of the piston moving with uniform speed with $s = 3$ (Taylor (1946)), there is an equi-partition of energy.

The Russian scientists were particularly active in the theory of explosions during the fifties of the last century and did some excellent analysis in the context of self-similar flows. For example, Kochina and Melnikova (1958) considered the general piston motion with the piston moving like $r \propto t^{m+1}$, where m is real. The self-similar form of the solution was considered in the (reduced) sound speed square–particle velocity plane. All the singularities of the DE in these variables were carefully identified, their nature denominated, and local solutions in their neighbourhoods found. Different ranges of the parameter m for a given γ were identified for which the piston motion headed by a shock, peripheral explosion with the neglect of the motion of the products of explosion, and converging shocks were described. Some typical numerical results for specific piston motions were presented.

In a slightly different context, Grigorian (1958a, 1958b) showed how self-similar motions arose from a class of initial conditions at $t = 0$. Again, self-similar form of the solution was assumed and the governing set of one-dimensional gasdynamic equations were reduced to ODEs. Local analysis of the solutions of the ODEs in the neighbourhood of singular points helped to identify the initial conditions, and hence Cauchy problems for various piston motions were posed. It was explained how a flow relating to given initial conditions (in the present context) could be caused by a piston motion. As an example, the piston motion $r \propto e^{t/\tau}$ (where τ is a dimensional parameter) was discussed in detail and the solution with a shock boundary analysed.

A curious work relating to strong explosion into a nonuniform medium is due to Waxman and Shvarts (1993). They considered the nonuniform medium with the density distribution, $\rho_0(r) = K r^{-\omega}$, where ω is a positive number and K is a dimensional parameter. This provides an additional freedom in the discussion of self-similar solutions. In these motions, strong shocks arise from the release of a large amount of energy, E_0, at $r = 0$. It was shown by Waxman and Shvarts (1993) that if $0 < \omega < 3$, blast wave solutions with strong shocks exist. These simply extend the well-known Taylor-Sedov solution to an inhomogenous medium with $\rho_0(r) = K r^{-\omega}$. These are solutions of the first kind which are fully determined by the dimensional parameters appearing in the problem, namely, K and E_0. These are asymptotic solutions which hold for short distance and time. It may also be observed that the front shock decelerates in the present case. When $\omega \geq 3$, it is shown that the short time asymptotic similarity breaks down. A qualitative analysis in the reduced particle velocity–sound speed plane shows that the solutions in the range $\omega \geq 3$ belong to the class called self-similar solutions of the second kind. Here, the explosion energy is not a relevant parameter. The solution must be found by solving the reduced ordinary differential equation from the shock to a 'new' singular point in the phase plane. This constitutes an eigenvalue problem for each γ. These solutions describe asymptotic (large distance and time) flows both when $3 < \omega < 5$ and when $\omega \geq 5$. The shock waves that head these solutions, in contrast to those for $0 < \omega < 3$, are accelarating. For a given γ, these solutions exist only for $\omega_g < \omega < \omega_c$, where ω_g and ω_c are functions of γ. They describe actual flows in some region $D(t) \leq r \leq R(t)$, where $R(t)$ is the shock radius, $D(t)$ diverges linearly with time and $D(t)/R(t)$ tends to zero as R diverges. The mass and energy contained in the self-similar flow region are finite. The distinguishing feature of this work is that self-similar solutions of both kinds exist for the same set of equations, and are demarcated by the parameter ω appearing in the undisturbed density distribution. The nature of each class of solutions, detailed above, is quite different. For decelerating shocks with $\omega < 3$, the explosion energy is divided in some time–dependent manner into internal and kinetic energies. For accelerating shocks with $\omega > 3$, the explosion energy is fully transformed into kinetic energy as the shock radius diverges.

Lighthill (1948) considered the solution of spherical and cylindrical piston motions in a more analytic manner. He introduced velocity potential into the analysis and sought for it a similarity form of the solution. The second order ordinary differential equation resulting from this assumption turns out to be quite complicated. Lighthill 'simulated' it by changing the nonlinear terms in an approximate way such that the simplified 'simulating' equation could be analytically integrated, satisfying the boundary conditions at the shock and at the piston, namely, the velocity potential is continuous across the shock front and the particle velocity is equal to the piston velocity at the piston. The Rankine-Hugoniot condition for the velocity at the shock was also satisfied. This posed a boundary value problem for the ODE, referred to above, which was approximately solved. The solution required a certain relation between the shock Mach number and the piston velocity. Lighthill (1948) determined an approximate form of this relation, which he confirmed later by a more rigorous, yet order of magnitude, argument. This problem however still remains to be solved in its generality (see section 2.5).

As we remarked earlier, while Taylor's solution is neat and gives an excellent description of the early stages of nuclear explosion, it begins to fail as the shock decays and the conditions of constancy of energy behind the shock do not apply. Sakurai (1953) devised a perturbation scheme in which Taylor's solution appears as the zeroth order term. He introduced the variable $x = r/R$ (where r is the spatial co-ordinate and $R(t)$ is the radius of the shock) and time t as the new independent variables and sought solutions in the form of power series in t, with coefficients depending on x. The zeroth order solution is just the Taylor's solution for spherical symmetry. This work was also generalized to cylindrical and plane symmetries. Sakurai (1954) first obtained local (analytic) solution for the first order terms. However, to obtain the unknown parameter in the expansion for the shock radius, he suitably separated the first order equations in two parts—one without the parameter and the other with the parameter—and hence solved the shock boundary value problem. Sakurai (1954) obtained shock velocity versus shock radius curves, the distance-time curves for the shock, and the distributions of velocity, pressure and density behind the shock front. The solution departed considerably from that given by the first term alone in the expansion.

Before we consider a more realistic model of the blast wave where the 'finite' initial gas sphere is allowed to expand in a natural way, we analyse the other asymptotic limit when the shock has decayed and has become weak, or alternatively, when the explosion itself is weak and the shock it produces is of small magnitude. This naturally suggests exploiting the sound wave solution by appropriately nonlinearising it. In fact this is what was accomplished by Whitham (1950). He investigated the attenuation of a spherical blast at large distances from the origin. He assumed the flow to be isentropic since this is a good approximation to the correct (nonisentropic) one when the shocks considered are weak and the entropy changes across them are of third order in shock strength. The main idea is to correct the linear theory so that exact equations of motion are solved for large distances by using certain expansions for the particle velocity and pressure. For the case for which the disturbances are small from the outset, the general theory is used to modify the linearised approximation to yield results which are uniformly valid at all distances from the origin. The shock loci are found both when there is only a leading shock and when a secondary shock is also produced in the flow behind, as is the case for a moderate explosion. The basic idea in Whitham's approach is to replace the linear characteristics by the exact nonlinear ones. The expansions involve powers of the reciprocal of distance, requiring also the introduction of some logarithmic terms. Whitham (1952) later used this theory for a variety of other gasdynamic problems (see Whitham (1974) for a detailed discussion). Reference may also be made to the book of Sedov (1959) who rederived the same results in a different manner.

While the point explosion is a very useful model for initial description of a strong nuclear explosion and Sakurai's (1953, 1954) perturbation scheme extends the validity of this model to greater distances, there is a need of an analytic approach where more realistic initial conditions are assumed. Either of the simplifying assumptions must be discarded: it is a strong point explosion or it starts as a weak explosion. This would require going beyond the Taylor-Sakurai approach as well as the weak shock theory of Whitham (1950). Such an attempt was first made by McFadden (1952). He envisioned that a (nondimensional) unit sphere containing a perfect gas at a uniform high pressure is allowed to expand suddenly at $t = 0$ into a homogenous atmosphere at lower pressure—an equivalent of a spherical shock tube problem. The inner medium may be referred to as gas while the outer one is air. It is required to find the behaviour of the ensuing flow. This flow in the (x, t) plane may be divided into five regions (see section 7.2). The region

A refers to the undisturbed gas at a high pressure; that ahead of the main shock (which is instantaneously formed) is also undisturbed and is called region E. Region B is a rarefaction wave which is bounded on the left by its head, a straight negative characteristic, and on the right by its tail, another characteristic. (In the present model the explosion is not so strong that a secondary shock is formed; in that case, the characteristic at the tail of the rarefaction wave is replaced by a secondary shock). The region C is rarefied gas which moves outward. The region D consists of the air overtaken by the main shock and is bounded by the shock on the right and a contact surface on the left; the latter separates the compressed air from the rarefied gas. In one of the first attempts of its kind, McFadden (1952) proposed a series solution in a variable q, which is proportional to the distance moved by the head of the rarefaction wave in time t, with coefficients which depend on a slope co-ordinate. More explicitly, $q = (1/2N)[(2N - 1) + (1 - x)/y]$, where x is the radial co-ordinate, y is nondimensional time, $N = (\gamma + 1)/2(\gamma - 1)$, and γ is the ratio of specific heats. The zeroth order terms come simply from the plane shock tube solution; higher order terms represent the effect of the spherical geometry. McFadden (1952) found first order correction in time describing the effect of geometry in various regions. He also derived boundary conditions on the curves separating various regions to higher orders but limited his solution to the first order only. His purpose was merely to obtain correct starting conditions for a full numerical treatment, since discontinuous initial data introduced serious numerical errors in the solution. Now, more sophisticated numerical methods are available and this difficulty can be easily circumvented.

The approach proposed by McFadden (1952) is highly promising. It is possible to generate an arbitrary number of terms for the series solution in each region, but matching them across the boundaries poses some difficulties. The solution however can be found iteratively. One may then sum the series directly or by using Padé approximation (see Nageswara Yogi (1995)) and study their numerical convergence. Once such a series solution is constructed, it constitutes a genuine analytic solution for the more realistic model describing a moderately strong explosion which, however, does not involve a secondary shock. This approach may be adapted for explosions in other media such as water. McFadden (1952) fully derived first order solution for a spherical blast wave problem and showed graphically how it differed from the corresponding plane problem; the pressure distribution behind the shock showed some significant changes, resulting in a different location of the tail of the rarefaction wave and the contact surface.

Chisnell (1957) considered a shock wave moving into a channel with varying cross-sectional area and derived an analytic expression which gives a relation between the channel area and the shock strength. In his highly intuitive approach, Whitham (1958) simplified Chisnell's approach by proposing his so-called shock-area rule, which may be stated as follows. Write the differential relation which is to be satisfied by the flow quantities along a characteristic coming into a shock and apply to it the flow quantities just behind the shock as given by the Rankine-Hugoniot conditions. If these quantities are expressed in terms of the shock strength, a differential equation is obtained which relates the shock strength and the radius of the shock. This equation is solved analytically or otherwise, using the initial strength of the shock at some initial radius. This rule may be applied both when the flow ahead of the shock is uniform and when it is nonuniform. It agrees with the results of Chisnell (1957) for the former case.

This intuitive rule has been much used and criticized in literature. The consensus however is that it is a useful simple approach which has been found helpful in the so-called shock-dynamics problems (see Whitham (1974)) where the geometry of the shock surface is also analysed, and it provides some check on numerical results and experiments. However, this approach is clearly not rigorous. It seems reasonable when the shock is weak; paradoxically, it works even for strong shocks. It does not give accurate results for shocks of intermediate strength. Hayes (1968) analysed this approach in the context of explosion and implosion problems. While it has shown phenomenal accuracy with respect to the implosion problems, it is inappropriate for shocks descending in an exponential atmosphere. For rising shocks it was found useful, giving an inaccuracy of about 15%. Hayes (1968a) first considered exact self-similar motions for a strong explosion in an atmosphere at a moderately low altitude. This explosion generally reaches a stage when the shock becomes weak and the propagation is close to acoustic. This is largely due to the geometric divergence of the propagating rays with the attendant increase in the area of the ray tube. On the other hand, the parts of the shock waves that propagate upward travel into regions of ever-decreasing density; this results in strengthening of the shock. As the shock propagates upwards a couple of scale heights, the strengthening effect over-rides the geometrical spreading. It can therefore become strong asymptotically. In this limit the analytic results of Raizer (1964) for the plane shock begin to apply. Assuming the shock to be self-propagating (when the effects from the region on the back of the shock do not catch up with it), Hayes (1968) used the results of self-similar motions to modify the coefficients in the approximate equation found by Chisnell (1957) and Whitham (1958) and rendered it more accurate. He employed this improvised Chisnell-Whitham approach to the study of an axisymmetric shock of self-propagating type moving upwards into the atmosphere.

The high pressure gas model, studied first by McFadden (1952), was taken up again by Friedman (1961); the latter was apparently not aware of this work. The major change in Friedman's model is that the initial compressed gas in the sphere (or cylinder) is at a much higher pressure and, therefore, leads to the phenomenon of secondary shock. The analysis of the problem is rather approximate and makes an extensive use of Whitham's shock area rule. At $t = 0$, a gas sphere of (dimensional) radius x_0 under high pressure is suddenly allowed to expand (or equalize) into the surrounding air at atmospheric pressure. For $x \geq 0$, $t > 0$, five domains in the (x, t) plane may be distinguished (see section 7.3): (0) undisturbed air, (1) air which has been overtaken by the main shock and compressed, (2) a nearly uniform region outside the main expansion, (3) gas in the main expansion region, and (4) the gas which has not been disturbed by the rarefaction wave. The separating boundaries are: the shock between the regions (0) and (1), the contact surface between the compressed air in (1) and the gas in (2), and the secondary shock which separates regions (3) and (2). The head of the rarefaction separates regions (3) and (4).

If the initial compressed gas is at a high pressure, a secondary shock develops for both cylindrical and spherical geometries; it does not occur in the planar case. For the latter, the main shock and the expansion region come into instantaneous equilibrium, being separated by a region of uniform pressure and velocity. For the former cases, the high pressure gas upon passing through a spherical rarefaction wave expands to lower pressures than those reached through an equivalent one-dimensional expansion, clearly due to increase in volume. This leads to pressure at the tail of the rarefaction wave lower than the pressure transmitted back by the main shock. To match these phases a compressive secondary shock must be inserted. It can also be shown mathematically that the negative characteristics in the centered expansion, after first pointing to decreasing x-direction, fan out in the increasing x-direction. These are met by the reflected negative characteristics from the main shock as the latter decays. These characteristics of the same family, arising from different sources, tend to intersect and form what is called a secondary shock. Friedman (1961) found a solution in the expansion wave by a perturbation of the known exact solution for the planar geometry; he made rather drastic approximations in his analysis leading to some gross errors. For the description of the main shock and secondary shock (with disturbed condition ahead of it), he made an extensive use of Whitham's (1958) shock area rule. The latter, as we have observed, is rather approximate unless it is appropriately improvised, as was done by Hayes (1968). We have carefully investigated these points in section 7.3. This approach, after suitable changes, gives a good qualitative picture of the flow analytically. However, the series approach, initiated by McFadden (1952) and considerably modified by us, is much more rigorous. It has the potential of accommodating the secondary shock effects. It may also be extended to explosions in water and other media.

Converging spherical and cylindrical shock waves have wide applications in controlled thermonuclear fusion and cavitation. In contrast to the usual explosion problem, here the shocks converge to the center or axis of implosion and strengthen to become infinitely large. There has been considerable activity on this topic since Guderley first initiated it in 1942. He showed that the similarity solutions that describe these flows are of a different type, now referred to as belonging to the second kind. In contrast to self-similar solutions of the first kind, characterised by Taylor-Sedov solution for strong explosions, the exponent in converging shock propagation law cannot be found from dimensional considerations alone. Instead, one must solve a boundary value problem in the phase plane (now called Guderley map) of the reduced system of ODEs, starting from the shock and hitting an appropriate singular point which often turns out to be a saddle point. For each γ, there is one value of the exponent in the similarity variable (or shock law) which solves this boundary value problem. The importance of the self-similar solutions of the second kind derives also from the fact that they represent what are called intermediate asymptotes, to which a large class of solutions of initial/boundary value problems converge in the limit $t \to 0$ or $t \to \infty$. For a complete description of the motion of imploding shock waves, Payne (1957) and Brode (1955) carried out detailed numerical investigations and verified the values of the similarity exponent obtained from the analysis in the phase plane. As was pointed out earlier, the approximate intuitive approach of Chester (1954), Chisnell (1957) and Whitham (1958)(CCW) seems to give uncannily accurate values for the similarity exponent for the converging shock problem. This is because in the close proximity of the center or axis of implosion the shock flow essentially 'forgets' details of its initial conditions and converges to the similarity solution of the second kind, a circumstance highly conducive to the applicability of CCW approach. The proportionality constant A in the shock law $r = At^\alpha$ must, however, be found from the numerical solution of the problem (Payne (1957)). In a more recent study, Chisnell (1998) gave an analytic description of the flow behind the converging shock which he claimed gave extremely good, though approximate, values of the similarity exponent, as well as a simple analytic description of the entire flow behind the converging shock. The basic idea is to study the singular points in the reduced sound speed square—particle velocity plane, and choose an appropriate trial function in this plane so as to remove the singular behaviour at one of these points. An iterative process is still needed for the accurate evaluation of the exponent but just a couple of iterations give a good value of the similarity exponent as well as the entire flow behind the shock wave.

It is known from the numerical solution of the converging shock problem that, in the reduced variables, the pressure behind the shock first increases and then decreases monotonically (see Zeldovich and Raizer (1967)). Making use of this observation, Fujimoto and Mishkin (1978) gave an analytic approach for the spherically imploding shock problem and obtained good values of the self-similarity exponent. This work was severely criticized by Lazarus (1980), but was appropriately responded to by Mishkin (1980). The matter remains to be resolved satisfactorily.

We conclude the discussion of converging shocks by referring to the work of Van Dyke and Guttman (1982) who considered the global problem of spherical and cylindrical shock waves. They envisioned that the flow was generated by a spherical or cylindrical piston which collapsed inward with a constant speed so that the basic approximation for small time is just the flow produced by the impulsive motion of a plane piston. The piston motion is assumed to be so large that the shock thus produced has an infinite Mach number, that is, the pressure ahead of the shock may be taken to be zero. The flow field is found by expanding the solution in powers of time with coefficients functions of a similarity variable which appears in the plane solution. As a consequence, the zeroth order terms constitute the solution of the plane problem while higher order terms give the effect of geometry, spherical or cylindrical. A coupled system of ODEs of infinite order results, which is solved subject to conditions at the shock and the piston. The shock locus, expressed as a power series in time, is compared to that given by the phase plane analysis of Guderley (1942) and subsequent workers. The tedious computation of the series solution is delegated to the computer and is carried out to 40 terms. This solution describes the whole flow field accurately almost to the time of collapse; it does not hold in the immediate vicinity of the collapse. Thus the global problem was solved and the shock pursued to its limiting self-similar form at the focus. The series solution gives excellent values of the similarity exponent and compares very well with the more recently computed accurate values. It also confirms Gelfand's conjecture, quoted by Fujimoto and Mishkin (1978), that, in the range of the adiabatic exponent γ where Guderley's solution has been shown to be not unique, the smallest admissible similarity exponent is realized. We may refer to an exhaustive review on converging shocks and cavities by Lazarus (1981). It deals, in particular, with the rich variety of previously neglected nonanalytic solutions and a full exploration of the relevant parameter space. New solutions are described which contain additional shocks, arriving at the origin concurrently with the initial shock. Some of these solutions are entirely analytic except at the shocks themselves and some are not. Previously rejected partial solutions are also discussed. Lazarus (1981) suggested that more extensive numerical integration of the original PDEs must be carried out to confirm the evidence for the approach of a class of solutions to a unique self-similar solution.

Chapter 2 treats the piston problem in its several manifestations. The similarity solutions are discussed in the physical plane with strong or weak shocks both when the medium ahead is uniform and when it is nonuniform; the blast wave solution comes out as a special case in this formulation. Other artifices such as new sets of dependent and independent variables help the discovery of new solutions with strong or weak shocks. Chapter 3 considers the blast wave problem directly when the medium ahead is stratified or when it is uniform; both Eulerian and Lagrangian co-ordinates are used.

Attempts at extending the Taylor-Sedov-Von Neumann solution to the time
regime when the shock becomes weak are discussed. More realistic exponentially stratified atmosphere is also analysed in this context. The effect of heat
conduction in the propagation of the blast wave is then considered. Here,
some interesting new results emerge. This chapter ends with the aymptotic
analysis of the blast wave, far away from the point of explosion. Chapter
4 discusses some important war-time shock wave theories due to Bethe and
Kirkwood (see Cole (1948)) and Brinkley and Kirkwood (1947) and their ex-
tensions for both uniform and nonuniform media. Chapter 5 contains exact
solutions of one-dimensional gasdynamic equations with shock boundaries
and vacuum fronts. The work of McVittie (1953) and Keller (1956) is discussed and generalised. Both Eulerian and Lagrangian forms of equations
of motion are treated. Chapter 6 briefly investigates the important phenomenon of converging spherical and cylindrical shocks—their genesis and
asymptotic focusing to the center or axis of symmetry. The effects of heat
transfer in this context are also considered. Chapter 7 deals with the more
realistic explosion model where the point explosion hypothesis is dispensed
with; it is now simulated as release of a high pressure gas sphere or cylinder into the ambient medium at atmospheric pressure, resulting in weak or
moderately strong shocks. The concluding chapter details computational
methods for the study of strong or moderately strong explosions; these are
related to analytic results discussed in earlier chapters.

It is clear that most of the book deals with the mathematical analysis of
explosions but the computational results are also included where ever they
are available. The concluding chapter deals with this matter in some detail.

The mathematics of explosions has spawned many original ideas in the
theory of nonlinear PDEs; in that respect too, it serves as a very fruitful
topic of study and research.

Chapter 2

The Piston Problem

2.1   Introduction

A simple way to simulate an explosion is to view it as a spherical piston motion, pushing out undisturbed air/gas ahead of it. To motivate stronger disturbances it is convenient to view small changes that a 'small' motion of the piston will bring about (see Taylor (1946)). In the spherically symmetric case, small motions are governed by the linear wave equation

$$\phi_{tt} = a_0^2 \left(\phi_{rr} + \frac{2\phi_r}{r} \right), \tag{2.1.1}$$

where ϕ is the velocity potential. The forward moving wave as solution of (2.1.1) is

$$\phi = r^{-1} f(r - a_0 t), \tag{2.1.2}$$

where a_0 is the (constant) speed of sound. The velocity and pressure disturbances from (2.1.2), therefore, are found to be

$$u = \frac{\partial \phi}{\partial r} = r^{-2} f(r - a_0 t) - r^{-1} f'(r - a_0 t), \tag{2.1.3}$$

$$p - p_0 = -\rho_0 \frac{\partial \phi}{\partial t} = \rho_0 a_0 r^{-1} f'(r - a_0 t). \tag{2.1.4}$$

If the piston motion is given by $R = R(t)$, then using the kinematic condition at the piston $u(R(t), t) = dR/dt$, we find from (2.1.3) that

$$\dot{R} = R^{-2} f(R - a_0 t) - R^{-1} f'(R - a_0 t). \tag{2.1.5}$$

If we know the piston motion $R = R(t)$, we may find the form of the function f by solving the first order ODE (2.1.5). This can be done easily for the

15

special case for which the spherical piston moves with a constant speed, starting from the position $R = 0$ at $t = 0$. Thus,

$$R = \alpha a_0 t, \tag{2.1.6}$$

where α is a nondimensional constant. Since we consider small motions, $\alpha << 1$. Writing $w = R - a_0 t = (\alpha - 1)a_0 t < 0$ in (2.1.5), we get an ODE for $f(w)$:

$$\frac{\alpha - 1}{\alpha w} f'(w) - \left(\frac{\alpha - 1}{\alpha w}\right)^2 f(w) + a_0 \alpha = 0. \tag{2.1.7}$$

The solution of (2.1.7) for negative values of w is

$$f(w) = \frac{a_0 \alpha^3}{1 - \alpha^2} w^2 + c(-w)^{(\alpha - 1)/\alpha}, \tag{2.1.8}$$

where c is the constant of integration. Since $0 < \alpha << 1$, $f(w)$ in (2.1.8) is finite only if we choose c equal to zero. Thus, the solution $\phi = r^{-1}f(r - at)$ and hence other physical variables become

$$\phi = \frac{a_0 \alpha^3}{1 - \alpha^2} \frac{(r - a_0 t)^2}{r} = \frac{a_0 \alpha^3}{1 - \alpha^2} r\left(1 - \frac{a_0 t}{r}\right)^2, \tag{2.1.9}$$

$$u = \frac{a_0 \alpha^3}{1 - \alpha^2}\left(\frac{a_0^2 t^2}{r^2} - 1\right), \tag{2.1.10}$$

$$p - p_0 = 2\rho_0 \frac{a_0^2 \alpha^3}{1 - \alpha^2}\left(\frac{a_0 t}{r} - 1\right). \tag{2.1.11}$$

The undisturbed medium ($u = 0$, $p = p_0$), outside the moving sphere $r = at$, remains undisturbed for subsequent time while the motion inside this sphere is governed by (2.1.9)–(2.1.11). The form (2.1.9)–(2.1.11) of the linear solution, namely, the spherically symmetric sound wave, suggests that all physical variables are functions of the nondimensional combination $r/a_0 t$ of the independent variables. The air wave produced by a uniformly expanding sphere expands at a uniform rate and the velocity and pressure disturbances are constant along the lines $r/a_0 t = $ constant. We may observe that, although we started with a damped travelling wave (2.1.2), the actual solution dictated by the boundary condition is a 'progressive' wave or a similarity solution (2.1.9)–(2.1.11). We shall see that the same form holds even when we analyse the full nonlinear problem.

Taylor (1946) assumed that all the variables u, p, and ρ are functions of $x = r/a_0 t$ alone, or equivalently, that they are constant along the direction $dr/dt = r/a_0 t$, and reduced the nonlinear PDEs governing one-dimensional gasdynamic equations to ODEs. He proved that these equations do not have a solution bounded by the sound front $r = a_0 t$. Instead they have solutions bounded by a shock of finite or infinite strength. The problem was solved numerically by integrating the reduced system of ODEs from the

shock and determining the position of the piston with respect to the shock
for each α such that the kinematic condition at the piston was satisfied. It
was concluded that the thickness of the layer of the disturbed air decreases
as the velocity of the expanding sphere increases until at an infinite rate of
expansion it is only 6% of the radius of this sphere.

An approximate analysis based on the theory of sound for small radial
velocity yields results which are inaccurate both near the piston and the
shock.

We discuss a more general piston motion in the next section. It includes
Taylor's (1946) solution for a uniformly moving piston as a special case.

2.2   The Piston Problem: Its Connection with the Blast Wave

That the piston problem does simulate the point explosion as a special case
will now be brought out by introducing energy considerations. The present
approach is due to Rogers (1958). It assumes that, as a result of the piston
motion, the total energy—the sum of the kinetic and the internal energies
of the gas—varies with time according to the law

$$E = E_0 t^s, \tag{2.2.1}$$

where E_0 and s are constants. The case $s = 0$ corresponds to the blast wave
where the energy behind the shock is constant, as we shall detail in the next
chapter. Here, $s \geq 0$ so that the total energy of the flow increases with
time (or at most remains constant). The flow is assumed to be produced
by a large piston motion and is headed by an infinitely strong shock. As
in the problem of Taylor (1946) discussed in section 2.1, the position of the
inner surface of the piston will be found by the numerical integration of the
system of ODEs resulting from the assumption of self-similarity of the flow,
starting the integration at the shock front and locating the piston where the
kinematic condition is satisfied. Here, the dissipation effects are ignored.

We consider one dimensional unsteady flow of a perfect gas described by
the system

$$u_t + uu_r + \frac{1}{\rho}p_r = 0, \tag{2.2.2}$$

$$\rho_t + u\rho_r + \rho u_r + \frac{ku\rho}{r} = 0, \tag{2.2.3}$$

$$(p\rho^{-\gamma})_t + u(p\rho^{-\gamma})_r = 0, \tag{2.2.4}$$

where u, p, and ρ are particle velocity, pressure and density, respectively, of
the gas at the radial distance r and time t; $k = 0, 1, 2$ for plane, cylindrical
and spherical symmetry, respectively.

It is convenient to replace (2.2.4) by the equivalent form

$$\left(\frac{1}{2}\rho u^2 + \frac{p}{\gamma - 1}\right)_t + \frac{1}{r^k}\left\{r^k u\left(\frac{1}{2}\rho u^2 + \frac{\gamma p}{\gamma - 1}\right)\right\}_r = 0. \qquad (2.2.5)$$

The system of nonlinear PDEs (2.2.2), (2.2.3), and (2.2.5) together with (2.2.1) must be solved subject to the boundary conditions at the piston and the shock produced by it. It is pertinent to introduce the variable

$$x = r/R, \qquad (2.2.6)$$

where $R = R(t)$ is the radius of the shock; thus, $x = 1$ represents the shock locus. The velocity of the shock is denoted by

$$V = \frac{dR}{dt}. \qquad (2.2.7)$$

The solution is sought in the self-similar form

$$u = Vf(x), \qquad (2.2.8)$$
$$p = \frac{\rho_0 V^2}{\gamma}g(x), \qquad (2.2.9)$$
$$\rho = \rho_0 h(x), \qquad (2.2.10)$$

where ρ_0 is the uniform density of the gas ahead of the shock. The total energy of the gas is

$$E = \int \frac{1}{2}\rho u^2 d\tau + \int \frac{p}{\gamma - 1}d\tau, \qquad (2.2.11)$$

where $d\tau$ is the volume element. The first integral in (2.2.11) is the total kinetic energy while the second is the total internal energy, contained in the space between the piston surface and the shock surface. If we introduce the form (2.2.8)–(2.2.10) into (2.2.11) we get

$$E = \rho_0\epsilon_k V^2 R^{k+1}\int_{x_0}^{1}\left\{\frac{1}{2}hf^2 + \frac{g}{\gamma(\gamma - 1)}\right\}x^k dx, \qquad (2.2.12)$$

where $\epsilon_k = 2^k\pi^{\frac{1}{2}k(3-k)}$, and x_0 is the co-ordinate of the expanding surface.

The integral in (2.2.12) will involve the parameters γ, k and s (see(2.2.1)). From (2.2.1) and (2.2.12) we have

$$R^{\frac{1}{2}(k+1)}\frac{dR}{dt} = \left[\frac{E_0}{\epsilon_k\rho_0 J}\right]^{1/2}t^{s/2}, \qquad (2.2.13)$$

where

$$J = \int_{x_0}^{1}\left\{\frac{1}{2}hf^2 + \frac{g}{\gamma(\gamma - 1)}\right\}x^k dx. \qquad (2.2.14)$$

On integration, (2.2.13) yields

$$R = \left(\frac{k+3}{s+2}\right)^{2/(k+3)} \left(\frac{E_0}{\epsilon_k \rho_0 J}\right)^{1/(k+3)} t^{(s+2)/(k+3)}, \qquad (2.2.15)$$

where we have used the condition $R = 0$ at $t = 0$. Two important cases may first be identified from (2.2.15): (i) $s = 0$, $R \propto t^{2/(k+3)}$ yielding the shock radius for the point explosion for different geometries corresponding to $k = 0, 1, 2$, and (ii) $s = k+1$, $R \propto t$, the case of uniform expansion of the plane, cylindrical or spherical piston. These special cases also determine the ends of the interval of physical interest for s, as we shall see. The transformation (2.2.8)–(2.2.10) changes (2.2.2),(2.2.3) and (2.2.5) to

$$(x - f)f' = \frac{g'}{\gamma h} + \frac{s - k - 1}{s + 2}f, \qquad (2.2.16)$$

$$(x - f)h' = h\left(f' + \frac{kf}{x}\right), \qquad (2.2.17)$$

$$\left\{x^k f\left(\frac{1}{2}hf^2 + \frac{g}{\gamma - 1}\right)\right\}' = x^{k+1}E_1'(x) + \frac{2(k+1-s)}{s+2}x^k E_1(x), \qquad (2.2.18)$$

where

$$E_1 = \frac{1}{2}hf^2 + \frac{g}{\gamma(\gamma - 1)}. \qquad (2.2.19)$$

The prime in the above denotes differentiation with respect to x. The Rankine-Hugoniot conditions for the strong shock are

$$u_1 = \frac{2}{\gamma + 1}V, \qquad (2.2.20)$$

$$\frac{\rho_1}{\rho_0} = \frac{\gamma + 1}{\gamma - 1}, \qquad (2.2.21)$$

$$\frac{p}{\rho_0 V^2} = \frac{2}{\gamma + 1}. \qquad (2.2.22)$$

In view of (2.2.8)–(2.2.10), they become

$$f(1) = \frac{2}{\gamma + 1}, \qquad (2.2.23)$$

$$h(1) = \frac{\gamma + 1}{\gamma - 1}, \qquad (2.2.24)$$

$$g(1) = \frac{2\gamma}{\gamma + 1}. \qquad (2.2.25)$$

The piston motion is given by $x = x_0$, that is, $r = x_0 R(t)$. The kinematic conditions at the piston requires that the piston velocity is equal to the particle velocity there. Therefore, from (2.2.8) and $r = x_0 R(t)$, we have $f(x_0) = x_0$. The problem thus reduces to finding x_0 such that, for a given γ, the system (2.2.16)–(2.2.18) with the initial conditions (2.2.23)–(2.2.25) when integrated from $x = 1$ backward leads to the value $x = x_0$ for which $f(x_0) = x_0$. The similarity assumption implies that the total mass of the gas between the shock and the piston is the same for all time. This fact was used as a check on the numerical solution of the problem. Tables 2.1–2.3 give x_0, the point at which the expanding surface occurs, the pressure $g(x_0)$ there and the integral J defined by (2.2.14) for plane, cylindrical and spherical geometries, respectively, for different values of s. The function J defines the total energy carried by the flow (see (2.2.12)).

Rogers (1958) also rederived the analytic solution for the blast wave problem, which we shall discuss in great detail in chapter 3. First we observe that equations (2.2.16)–(2.2.18) possess a constant solution for plane symmetry with $k = 0$. This solution is simply

$$f(x) = \frac{2}{\gamma + 1}, \tag{2.2.26}$$

$$g(x) = \frac{2\gamma}{\gamma + 1}, \tag{2.2.27}$$

$$h(x) = \frac{\gamma + 1}{\gamma - 1}. \tag{2.2.28}$$

It is clear that the piston position in this case is given by

$$x_0 = f(x_0) = \frac{2}{\gamma + 1}. \tag{2.2.29}$$

In terms of physical variables, the solution is

$$u = \frac{2}{\gamma + 1} V, \tag{2.2.30}$$

$$\rho = \rho_0 \frac{\gamma + 1}{\gamma - 1}, \tag{2.2.31}$$

$$p = \frac{2\rho_0 V^2}{\gamma + 1}, \tag{2.2.32}$$

where V is the (constant) velocity of the shock front. Clearly, the piston advances with the constant speed $2V/(\gamma + 1)$ so that the volume engulfed by the shock steadily increases with time. The total (kinetic and internal) energy of the flow in a volume having unit cross-sectional area (see (2.2.12)) is

$$\begin{aligned}
\mathcal{E} &= \rho_0 V^2 R \int_{2/(\gamma+1)}^{1} \left(\frac{1}{2} h f^2 + \frac{g}{\gamma(\gamma - 1)} \right) dx \\
&= \rho_0 V^2 R \left(\frac{4}{\gamma^2 - 1} \right) \int_{2/(\gamma+1)}^{1} dx = \frac{4\rho_0 V^2 R}{(\gamma + 1)^2}.
\end{aligned} \tag{2.2.33}$$

Table 2.1. The position of the piston $x_0 = f(x_0)$, pressure $g(x_0)$ and total energy J for different values of s for spherically symmetric piston motions with $\gamma = 1.4$, $k = 2$ (Rogers, 1958).

s	x_0	$g(x_0)$	J
0	0	0.4264	0.4264
0.07	0.578	0.4854	0.3960
0.2	0.736	0.5748	0.3610
0.5	0.846	0.7302	0.3173
1.00	0.900	0.9079	0.2832
2.00	0.932	1.1215	0.2556
3.00	0.942	1.2450	0.2526

Table 2.2. The position of the piston x_0, pressure $g(x_0)$ and total energy J for different values of s for cylindrically symmetric piston motions with $\gamma = 1.4$, $k = 1$ (Rogers, 1958).

s	x_0	$g(x_0)$	J
0	0	0.4351	0.6414
0.05	0.408	0.4814	0.5900
0.07	0.468	0.4986	0.5773
0.1	0.534	0.5232	0.5607
0.2	0.664	0.5975	0.5178
0.4	0.776	0.7206	0.4660
0.6	0.828	0.8211	0.4344
1.0	0.875	0.9776	0.3992
2.0	0.915	1.2246	0.3702

It is interesting to observe that the value of each of the integral terms in (2.2.33) is the same, showing that the kinetic and internal energies are equal in the present case. Moreover, it follows from differentiating (2.2.33) that

$$\frac{d\mathcal{E}}{dt} = \frac{4\rho_0 V^3}{(\gamma + 1)^2}$$

$$= \frac{2\rho_0 V^2}{\gamma + 1} \frac{2V}{\gamma + 1}, \tag{2.2.34}$$

which is simply the product of the surface pressure and the velocity of expansion of the plane piston (see (2.2.30) and (2.2.32)).

In the general case, the rate at which the piston motion feeds energy into the flow is again equal to the product of the surface pressure, the area and the velocity of the expanding surface. This may be written as

$$\Re = \epsilon_k (x_0 R)^k \frac{\rho_0 V^2}{\gamma} g(x_0) V x_0. \tag{2.2.35}$$

From (2.2.1) we also get

$$\Re = s E_0 t^{s-1}. \tag{2.2.36}$$

Table 2.3. The position of the piston x_0, pressure $g(x_0)$ and total energy J for different values of s for plane symmetric piston motions with $\gamma = 1.4$, $k = 0$ (Rogers, 1958).

s	x_0	$g(x_0)$	J
0	0	0.4550	1.2174
0.05	0.221	0.5086	1.1001
0.1	0.365	0.5591	1.0206
0.2	0.536	0.6523	0.9177
0.4	0.694	0.8139	0.8094
0.5	0.736	0.8846	0.7774
0.6	0.768	0.9500	0.7500
0.8	0.808	1.0662	0.7164
1.0	0.833	1.1667	0.6958

Equating (2.2.35) and (2.2.36) and using (2.2.15) we get

$$J = \frac{(s+2)}{\gamma s(k+3)} x_0^{k+1} g(x_0). \tag{2.2.37}$$

The relation (2.2.37) provides an excellent check on the computation of flows for all cases except $s = 0$. Figure 2.1 shows the kinetic energy of the flow expressed as a percentage of the total energy for the spherically symmetric case for $\gamma = 1.4, 1.2$. It is found that as s increases from 0, the value corresponding to a blast wave, to its limiting value 3, this fraction approaches $1/2$, indicating the equipartition of energy for $s = 3$. We shall discuss this matter in greater detail in the context of the blast wave problem in chapter 3. The non-self-similar piston motions with nonzero pressure ahead of the shock were treated by Chernii (1960).

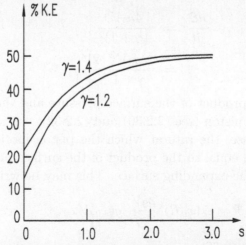

Figure 2.1 Ratio of the kinetic energy to total energy as a function of s, see (2.2.1) (Rogers, 1958).

2.3   Piston Problem in the Phase Plane

Now we consider a general piston motion with the speed $v = ct^m$ giving rise to a strong shock, which propagates into a medium with pressure $p_1 \simeq 0$. In the manner of section 2.2, a strong explosion can be simulated as the motion of a mass of gas driven by an expanding cylinder or sphere with speed $v = ct^m$. This study includes the situation when a peripheral explosion takes place, giving rise to an imploding shock. In the sequel, we follow the work of Kochina and Mel'nikova (1958).

We start again with the equations of motion

$$v_t + vv_r + \frac{1}{\rho}p_r = 0, \qquad (2.3.1)$$

$$\rho_t + (\rho v)_r + (\nu - 1)\frac{\rho v}{r} = 0, \qquad (2.3.2)$$

$$(p\rho^{-\gamma})_t + v(p\rho^{-\gamma})_r = 0, \qquad (2.3.3)$$

where $\nu = 1, 2, 3$ for plane and spherical symmetry, respectively, and $\gamma = c_p/c_v$. We assume that $1 < \gamma < 7$. The characteristic parameters of the problem, $[r] = L, [t] = T, [\rho] = ML^{-3}, [c] = LT^{-m-1}$, lead to the similarity variable

$$\lambda = \frac{ct^{m+1}}{r}. \qquad (2.3.4)$$

The solution may be sought in the self-similar form

$$v = \frac{r}{t}V(\lambda), \quad \rho = \rho_1 R(\lambda), \quad p = \rho_1\frac{r^2}{t^2}P(\lambda), \qquad (2.3.5)$$

where ρ_1 is the uniform density in the medium ahead of the shock.

We recall that the undisturbed pressure p_1 is assumed to be zero, hence the form for p in (2.3.5). Introducing (2.3.5) into (2.3.1)–(2.3.3) and introducing the variable $z = \gamma P/R$, the nondimensional sound speed square, we obtain

$$\frac{dz}{dV} = \frac{z[2(V-1)+\nu(\gamma-1)V](V-m-1)^2}{(V-m-1)[V(V-1)(V-m-1)-(2m/\gamma+\nu V)z]}$$
$$- \frac{z\{(\gamma-1)V(V-1)(V-m-1)+2z(V-m-1+m/\gamma)\}}{(V-m-1)[V(V-1)(V-m-1)-(2m/\gamma+\nu V)z]}, \quad (2.3.6)$$

$$\frac{dR}{dV} = \frac{R}{(V-m-1)}\left\{\frac{[(V-m-1)^2-z]\nu V}{V(V-1)(V-m-1)-(2m/\gamma+\nu V)z} - 1\right\}, \qquad (2.3.7)$$

$$\frac{d\lambda}{dV} = \lambda\frac{(V-m-1)^2-z}{V(V-1)(V-m-1)-(2m/\gamma+\nu V)z}. \qquad (2.3.8)$$

The strong shock conditions (see (2.2.20)–(2.2.22)), in view of (2.3.5), become

$$V_2 = \frac{2(m+1)}{\gamma+1}, \quad z_2 = \frac{2\gamma(\gamma-1)(m+1)^2}{(\gamma+1)^2}, \quad R_2 = \frac{\gamma+1}{\gamma-1}, \tag{2.3.9}$$

where the subscript 2 denotes conditions immediately behind the shock.

At the piston $r_* = r_*(t)$, we have

$$v_* = \frac{dr_*}{dt} = \frac{r_*}{t}V_* = ct^m, \tag{2.3.10}$$

or

$$r_* = \frac{c}{m+1}t^{m+1}, \tag{2.3.11}$$

where we have used (2.3.5) and the law of motion of the piston, $v_* = ct^m$. Thus, from (2.3.4), (2.3.5) and (2.3.11), we have the following conditions at the piston:

$$V_* = m+1, \lambda_* = m+1. \tag{2.3.12}$$

We must solve (2.3.6)–(2.3.8), subject to boundary conditions (2.3.9) and (2.3.12) at the shock and at the piston, respectively. The integral curve, starting from the shock must, for increasing λ, reach either of the points C and D (see cases 2 and 3 below) where $V = m+1$.

It is easy to check from (2.3.6) and its reciprocal that $V = m+1$ and $z = 0$ are members of integral curves.

Kochina and Mel'nikova (1958) carried out a very exhaustive study of the above boundary value problem. We shall discuss the divergent flows with shocks arising from piston motions in some detail, and summarize the results for other possible flows described by (2.3.6)–(2.3.8), (2.3.9) and (2.3.12).

First we discuss the singularities of (2.3.6) in the domain $0 < V < m+1$, $z > 0$. $\widehat{C}$ in the following denotes a constant.

1. Point O $(V = 0, z = 0)$ is a node. The integral curves in its neighbourhood have the form

$$V + \frac{2m}{(m+1)\gamma}z = \widehat{C}z^{1/2}. \tag{2.3.13}$$

2. Point C $(V = m+1, z = 0)$ is a complicated singular point with the asymptotic solution

$$(V - m - 1)^{2m} = \widehat{C}z^{\nu(m+1)}[\nu z - m\gamma(V - m - 1)]^{2m+\nu(\gamma-1)(m+1)}. \tag{2.3.14}$$

3. Point D $(V = m+1, z = \infty)$ has a local solution

$$z = \widehat{C}(m + 1 - V)^{\frac{2m}{2m+\nu\gamma(m+1)}}. \tag{2.3.15}$$

4. Point E $(V = -2m/\nu\gamma, z = \infty)$.

5. Point G $(V = 1, z = 0)$.

6. Point F_i is intersection of the curves

$$z = \{[2(V - 1) + \nu(\gamma - 1)V](V - m - 1) - (\gamma - 1)V(V - 1)\}$$
$$\times (V - m - 1)/2(V - m - 1 + m/\gamma),$$

$$z = \frac{V(V - 1)(V - m - 1)}{2m/\gamma + \nu V}. \tag{2.3.16}$$

The parameter m in the piston motion $v = ct^m$ ranges from -1 to $+\infty$, leading to change in the character of the above singularities.

Case 1. $m > 0$. In this case the points O, C, G are nodes while D and F are saddle points. The point E is not singular in the present case. The integral curves starting from the shock (V_2, z_2) enter the point C where they have the local behaviour

$$z = C_0(m + 1 - V)^{\frac{2m}{2m + \nu\gamma(m+1)}}, \tag{2.3.17}$$

where C_0 is a constant. The solution exists for all $m > 0$. For the special case $m = 0$, describing a uniform piston motion (see section 2.2), points O, C and A $(V = 0, z = 1)$ are nodes while F and D are saddle points. In this case $V = 1$ is not an integral curve.

Here the case with the pressure $p_1 \neq 0$ ahead of the shock can also be treated. The solution of this problem is obtained by joining any point $(1, z)$ to the shock point (V_2, z_2), where V_2 and z_2 are connected by the relation

$$z_2 = (1 - V_2)\left(1 + \frac{\gamma - 1}{2}V_2\right). \tag{2.3.18}$$

Case 2. $m'' < m < 0$, where

$$m'' = -\frac{\nu(\gamma - 1)}{2 + \nu(\gamma - 1)} \tag{2.3.19}$$

is greater than -1 (see (2.3.14)). In this case the points O and D are nodes while E is a saddle point. Three integral curves pass through the point C: the straight lines $V = m + 1$ and $z = 0$, and a certain dividing curve entering this point at an angle $-m\gamma/\nu$. Therefore, this point corresponds to two saddle points. There is an odd number (greater than 1) of points F_i; the point nearest to C is a node.

For the particular value

$$m = m' = -\frac{\nu}{2 + \nu}, \tag{2.3.20}$$

the field of integral curves coincides with those for a strong explosion (for a spherical explosion $m = -2/5$). In this case, the integral curve coincides with 'the dividing curve' entering the point C, referred to above. It has the form

$$z = \left(\frac{\gamma - 1}{2}\right) \frac{V^2[V - 2/(2 + \nu)]}{[2/(2 + \nu)\gamma - V]}. \qquad (2.3.21)$$

It may be checked that the shock point (V_2, z_2) given by (2.3.9) lies on (2.3.21). For $m > m'$, this point lies between the straight line $V = m + 1$ and the dividing curve, while for $m < m'$, it lies between the dividing curve and the z-axis.

It may be checked from (2.3.19) and (2.3.20) that, for $\gamma < 2$, $m = m'$ does not belong to the interval $m'' < m < 0$. The solution of the piston problem exists and the integral curve passes through the point D.

For $\gamma > 2$ and $m < m'$, the integral curve starting from (V_2, z_2) in the direction of increasing λ does not reach either C or D where $V = m + 1$. Therefore, the solution of the piston problem does not exist in this case.

Returning to the case $m = m'$, the point F now has the co-ordinates

$$V = \frac{2}{2 + \nu(\gamma - 1)}, \quad z = \frac{2\nu\gamma(\gamma - 1)(\gamma - 2)}{[2 + \nu(\gamma - 1)]^2[\nu + 2(\gamma - 1)]}. \qquad (2.3.22)$$

It may be verified that for $\nu = 1, 2$ and for $\nu = 3$ with $\gamma < 7$, the point (V_2, z_2) lies between points E and F. If $\nu = 3$ and $\gamma = 7$, the point (V_2, z_2) coincides with the point F. For $\nu = 3$ and $\gamma > 7$, it lies between F and C. Therefore, the integral curve describing the piston motion exists for $\nu = 3$, $\gamma > 7$.

For $\gamma = 7$, the solution of the strong explosion problem is given simply by the point $V_2 = 0.1$, $z_2 = 0.21$, while the solution of the piston problem is given by the segment of the integral curve (2.3.21) between this point and the point C. At the shock point (V_2, z_2), $\lambda = 0$, implying that the shock wave moves instantaneously to infinity.

Summarizing the results so far, the solution of the piston problem exists for all m in the range $m'' < m < 0$ if $\gamma < 2$. For $\gamma > 2$ and $\nu = 1$ or 2, or for $2 < \gamma < 7$ and $\nu = 3$, it does not exist for $m'' < m < m'$.

There are other cases involving peripheral explosion with neglect of the motion of the products of explosion or converging shock waves, which arise for other values of m and different geometries. We refer the reader to the original paper of Kochina and Mel'nikova (1958) for details for these cases. Now we present some of the results of numerical integration of (2.3.6)–(2.3.8) subject to the conditions (2.3.9) and (2.3.12) at the shock and at the piston,

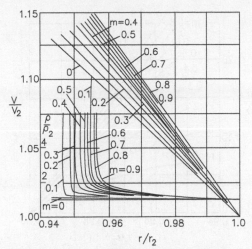

Figure 2.2 Particle velocity v/v_2 and density ρ/ρ_2 as a function of r/r_2 for different values of m, see (2.3.4) and (2.3.23) (Kochina and Mel'nikova, 1958).

respectively. We shall restrict our attention to diverging piston motions for which the shock is ahead of the piston. The integration is carried out, starting from the shock conditions (2.3.9), for increasing values of λ corresponding to decreasing values of r (see (2.3.4)) until (2.3.12) is satisfied.

The solutions may be expressed as

$$\frac{v}{v_2} = \frac{\lambda_2}{\lambda} \frac{V}{V_2}, \quad \frac{\rho}{\rho_2} = \frac{R}{R_2}, \quad \frac{p}{p_2} = \left(\frac{\lambda_2}{\lambda}\right)^2 \frac{zR}{z_2 R_2},$$

$$\frac{T}{T_2} = \left(\frac{\lambda_2}{\lambda}\right)^2 \frac{z}{z_2}, \quad \frac{r}{r_2} = \frac{\lambda_2}{\lambda}. \tag{2.3.23}$$

Figures 2.2–2.3 show $v/v_2, \rho/\rho_2$, and p/p_2 for varying values of the parameter m. The conditions immediately behind the shock are given by

$$p_2 = \frac{2\rho_1}{\gamma + 1} D^2, \quad r_2 = \frac{ct^m}{\lambda_2},$$

$$D = \frac{dr_2}{dt} = \frac{(m+1)ct^m}{\lambda_2}. \tag{2.3.24}$$

Thus, for $m > 0$, the shock speed increases with time so that the counterpressure p_1 ahead may be ignored only after a long time. It is clear from Figures 2.2–2.3 that in this case the pressure on the piston at any time exceeds that at the shock; the density is infinite at the piston while the temperature is infinitesimally small there. Correspondingly, the pressure is finite at the piston.

The case $m = 0$ corresponds to the motion of the piston with constant speed (see section 2.1). In this case all the variables—pressure, density and

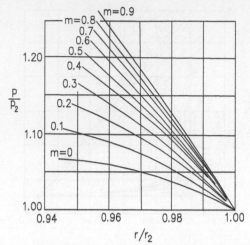

Figure 2.3 Pressure p/p_2 as a function of r/r_2 for varying values of the parameter m, see (2.3.4) and (2.3.23) (Kochina and Mel'nikova, 1958).

temperature—are finite at the piston. Here the motion is self-similar even when $p_1 \neq 0$ (see Sedov (1959)).

The case $m < 0$ corresponds to a decelerating shock from its initial infinite strength so that the flow is realistic only at early times when the shock is of large strength. In these cases, the density at the piston tends to zero, the temperature tends to infinity, while the pressure remains finite.

2.4   Cauchy Problem in Relation to Automodel Solutions of One-Dimensional Nonsteady Gas Flows

In several papers, Grigorian (1958a, 1958b) attempted to show how self-similar (automodel) solutions may arise from a class of initial conditions.

For the one-dimensional gasdynamic equations

$$u_t + uu_r + \frac{1}{\rho}p_r = 0, \tag{2.4.1}$$

$$\rho_t + (\rho u)_r + \frac{(\nu - 1)\rho u}{r} = 0, \tag{2.4.2}$$

$$(p\rho^{-\gamma})_t + u(p\rho^{-\gamma})_r = 0, \tag{2.4.3}$$

we seek self-similar solutions in the form

$$u(r,t) = \frac{r}{t}V(\lambda), \quad p(r,t) = B|r|^{\beta+2}t^{-2}P(\lambda),$$

$$\rho(r,t) = B|r|^{\beta}R(\lambda), \quad z = \frac{\gamma P}{R}, \tag{2.4.4}$$

$$\lambda = \sqrt{\frac{A}{B}} t |r|^{\frac{\alpha-\beta}{2}-1}. \tag{2.4.5}$$

In (2.4.2), $\nu = 1, 2, 3$ for plane, cylindrical and spherical symmetries, respectively. α and β are dimensionless constants while A and B are dimensional constants. Substituting (2.4.4)–(2.4.5) into (2.4.1)–(2.4.3), we get

$$\frac{dz}{dV} = \frac{z}{(V-q)W}\{[2(V-1) + \nu(\gamma-1)V](V-q)^2$$
$$- (\gamma-1)V(V-1)(V-q) - [2(V-1) + \kappa(\gamma-1)]z\}, \tag{2.4.6}$$

$$q\frac{d\ln\lambda}{dV} = \frac{(V-q)^2 - z}{W}, \tag{2.4.7}$$

$$(V-q)\frac{d\ln R}{d\ln\lambda} = q\frac{W}{z - (V-q)^2} + q(\beta+\nu)V, \tag{2.4.8}$$

where

$$W = V(V-1)(V-q) + (\kappa - \nu V)z, \quad q = \frac{2}{2 - (\alpha-\beta)},$$
$$\kappa = -\frac{\alpha}{\gamma}q. \tag{2.4.9}$$

According to the Cauchy-Kowalewskaia theorem, it is possible to write the solution of the system of ODEs (2.4.6)–(2.4.8) in the neighborhood of $\lambda = 0$ in the form

$$V = M_1\lambda + V_2\lambda^2 + \cdots, \tag{2.4.10}$$
$$P = L_1\lambda^2 + P_3\lambda^3 + \cdots, \tag{2.4.11}$$
$$R = N_1 + R_1\lambda + \cdots. \tag{2.4.12}$$

This solution is unique. Here, M_1, L_1, N_1, and the coefficients of higher terms in (2.4.10)–(2.4.12) are dimensionless constants. Equations (2.4.4) and the solution (2.4.10)–(2.4.12) give, in the limit $t \to 0$, the initial distribution of the physical variables as

$$p(r,0) = L_1 A r^\alpha, \quad \rho(r,0) = N_1 B r^\beta,$$
$$u(r,0) = M_1 \left(\frac{A}{B}\right)^{1/2} (r)^{\frac{\alpha-\beta}{2}} \quad (r > 0). \tag{2.4.13}$$

In the (z, V) plane, the corresponding solution in the neighbourhood of $(0, 0)$ has the form

$$z = \frac{\gamma L_2}{N_1 M_1^2} V^2 + \cdots \quad \text{for} \quad M_1 \neq 0, \quad N_1 \neq 0 \tag{2.4.14}$$

and

$$z = -\frac{\gamma}{\alpha}V + \cdots \quad \text{for} \quad M_1 = 0 \quad \text{or} \quad N_1 = 0. \tag{2.4.15}$$

In addition to the above solutions, we have for the planar case $\nu = 1$ another local solution

$$V = -M_2\lambda + \cdots, \tag{2.4.16}$$
$$P = L_2\lambda^2 + \cdots, \tag{2.4.17}$$
$$R = N_2 + \cdots. \tag{2.4.18}$$

Correspondingly, we have

$$z = \frac{\gamma L_2}{N_2 M_2^2}V^2 + \cdots, \quad \text{or} \quad z = -\frac{\gamma}{\alpha}V + \cdots. \tag{2.4.19}$$

Thus, two integral curves in the (z, V) plane correspond to the Cauchy initial value problem. These integral curves have the asymptotic representation (2.4.14) and (2.4.15) for $(\nu = 1, 2, 3)$. Besides, we have (2.4.19) for $\nu = 1$ in the neighbourhood of $V = 0, z = 0$.

Equations similar to (2.4.6) in the (z, V) plane have been much studied in the Russian literature. In the present context, the path of an integral curve originating from (2.4.14), (2.4.15), or (2.4.19) for small values of V and z must be continued for increasing values of λ (see (2.4.5)). This increase should continue to infinity or to some finite value corresponding to the moving boundary of the region occupying the gas if such a boundary exists. An example of this kind is a moving piston pushing the gas ahead. Thus, the Cauchy problem, stated above, will have a solution if the corresponding integral curve is so constructed that, moving along the curve, the parameter λ monotonically goes through the indicated set of values.

The parameters α and β in (2.4.4) and (2.4.5) are uniquely determined in terms of the two parameters q and κ occurring in the present problem (see (2.4.9)):

$$\alpha = -\frac{\gamma\kappa}{q}, \quad \beta = -\frac{\gamma\kappa}{q} - 2\left(1 - \frac{1}{q}\right). \tag{2.4.20}$$

Here, the parameters $\gamma = c_p/c_v > 1$ and $\nu = 1, 2, 3$ are known. From the well-studied integral curves of equations of the type (2.4.6) in the (z, V) plane, describing self-similar or automodel solutions, it is easy to relate to a Cauchy problem. Grigorian (1958a) describes several such problems including strong explosions, uniformly moving piston motion, flame fronts etc.

We may observe that $V = 0, z = 0$ is a singular point of (2.4.6); it is a node in the most general case. The fan of integral curves starting from this point have the asymptotic form (2.4.14), (2.4.15) or (2.4.19) corresponding to varied initial conditions.

To solve a Cauchy problem for these initial distributions one would have to trace each curve of this bundle to a point where either $\lambda = \infty$ or $\lambda = \lambda_\Pi$, where λ_Π is the value corresponding to a moving boundary—a piston, for example. Such piston motions satisfy the kinematic condition $V = q$ (see section 2.3). Thus, if the motions are continuous, the problem reduces to tracing the integral curves of the fan from the rest point $V = 0, z = 0$ to some special points $V = q$ where $z = 0$ or $z = \infty$ or to points $\lambda = \infty$ which may correspond to infinitely distant singular points. In addition, λ must change monotonically along an integral curve.

If we refer to the most general discussion of (z, V) plane in the book of Sedov (1959), we may find that there exists a parabola

$$z = (V - q)^2 \qquad (2.4.21)$$

in the (z, V) plane where λ while moving along an integral curve attains a stationary value. That is, $\partial\lambda/\partial s = 0$ on this curve where s is measured along the integral curve. So λ attains a maximum or minimum there, say, $\lambda = \lambda_0$. This implies that the integral curve on meeting the parabola (2.4.21) folds over back into the region of parameters covered prior to the transition across the parabola (2.4.21). This implies that the solution is discontinuous for all $t > 0$. $\lambda = \lambda_0$ is the limit line demarcating the region where continuous solutions occur. The solution beyond the intersection of the integral curve with this line can be joined to the one before via shock discontinuities. Sometimes neither continuous nor discontinuous solutions of the Cauchy problem may exist (see Grigorian (1958a)).

We discuss now a special piston problem following Grigorian (1958b). We do not discuss power-law piston motions detailed in section 2.3. Instead we turn to piston motions following an exponential law, which can also be obtained from the power law via some transformations and a limiting process requiring the power to tend to infinity. This exponential solution is of interest in its own right (see section 3.4). Let the piston motion be given by

$$r = ae^{t/\tau}, \qquad (2.4.22)$$

where a and τ are dimensional constants. We look for self-similar solutions of the system (2.4.1)–(2.4.3) in the form

$$u(r, t) = \frac{r}{\tau}V(\lambda), \quad p(r, t) = \rho_0 \frac{r^2}{\tau^2}P(\lambda), \quad \rho(r, t) = \rho_0 R(\lambda), \qquad (2.4.23)$$

where

$$\lambda = \frac{a}{r}e^{t/\tau}. \qquad (2.4.24)$$

Thus, the piston path is described by $\lambda = 1$. Substituting (2.4.23)–

(2.4.24) into (2.4.1)–(2.4.3), we get

$$\frac{dZ}{dV} = Z\frac{[2+\nu(\gamma-1)]V(V-1)^2 - (\gamma-1)V^2(V-1) - 2[V-(\gamma-1)\gamma]Z}{(V-1)\{V^2(V-1) - [2/\gamma + \nu V]Z\}},$$

(2.4.25)

$$\frac{d\ln\lambda}{dV} = \frac{(V-1)^2 - Z}{V^2(V-1) - (2/\gamma + \nu V)Z},$$

(2.4.26)

$$(V-1)\frac{d\ln R}{d\ln\lambda} = \frac{V^2(V-1) - (2/\gamma + \nu V)Z}{Z - (V-1)^2} + \nu V,$$

(2.4.27)

where

$$Z = \frac{\gamma P}{R}.$$

(2.4.28)

The field of integral curves of (2.4.25) is shown in Figure 2.4. We assume that the piston motion is so strong that it leads to shocks of infinite strength. These shocks propagate into a stationary gas with zero pressure. We must therefore construct a solution which passes through the point $Z = V = 0$ (see (2.4.23) and (2.4.28)) in the (V, Z) plane.

The strong shock conditions (see section 2.3) in view of (2.4.23) and (2.4.28) become

$$V_2 = \frac{2}{\gamma + 1}, \quad Z_2 = \frac{2\gamma(\gamma - 1)}{(\gamma + 1)^2},$$

(2.4.29)

where the suffix '2' denotes flow immediately behind the shock. These conditions discontinuously connect the point $Z = V = 0$. It follows from (2.4.23$_1$), (2.4.24), and $\dot{R} = RV(\lambda)/\tau$ applied at $\lambda = 1$ that $V = 1$ at the piston. We also find from (2.4.25) that we must have $Z = 0$ when $V = 1$. Thus, the

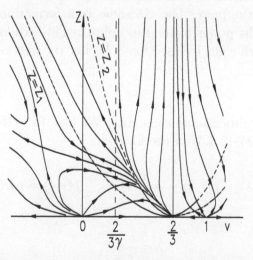

Figure 2.4 The field of integral curves of equation (2.4.25) for $1 < \gamma < 2$ (Grigorian, 1958).

solution starting from the shock point (2.4.29) must pass through the piston point $V = 1, Z = 0$. A local analysis of (2.4.25)–(2.4.27) shows that, in the vicinity of this point, we have

$$Z = C_1(1-V)^{\frac{2}{2+\nu\gamma}}, \quad \lambda = C_2 \left(V + \frac{2}{\nu\gamma}\right)^{\frac{1}{\nu}}, \quad R = C_3 \left(\frac{V + 2/\nu\gamma}{1-V}\right)^{\frac{2}{2+\nu\gamma}},$$

$$(2.4.30)$$

where C_1, C_2, and C_3 are constants. Since the piston motion is assumed to be given by $\lambda = 1$ where $V = 1$, we find from $(2.4.30_2)$ that the constant C_2 must be equal to $\left[\nu\gamma/(2 + \nu\gamma)\right]^{1/\nu}$. The solution of the piston problem may now be obtained as follows. Starting from the shock point (2.4.29) we integrate (2.4.25) until the solution matches with the local solution $(2.4.30_1)$. A numerical fit determines the constant C_1. The parameter λ is found by integrating (2.4.26) with the conditions $\lambda = 1$, $Z = 0$, $V = 1$; the solution is locally known from $(2.4.30_2)$. The density R is found by integrating (2.4.27), starting from the shock where $R_2 = (\gamma + 1)/(\gamma - 1)$. The value of λ at the shock is known from the previous integration. The integration is carried to the piston where $\lambda = 1$. This numerical integration determines the third constant C_3 in (2.4.30). Thus, the approximate solution (2.4.30) near the piston is now fully determined. This solution shows that at the piston the density is infinite, the pressure is finite, and the temperature is zero. This is in agreement with the results obtained in section 2.3 for power law piston motions. The present solution may also be obtained as the power in the law of the piston motion is allowed to tend to infinity appropriately (see Grigorian (1958b)).

2.5   Uniform Expansion of a Cylinder or Sphere into Still Air: An Analytic Solution of the Boundary Value Problem

Lighthill (1948) derived an 'approximate' analytic solution to describe the motion caused by the uniform expansion of a cylinder or sphere into still air. The motion is supposed to be small so that only weak shocks are produced; the changes in entropy are therefore ignored. This is a special case of a kindred class of problems treated by Lighthill (1948). He analysed the flow in terms of velocity potential and introduced the similarity variable to reduce the governing PDE to an ODE. He then wrote appropriate boundary conditions both at the piston and the weak shock. The boundary value problem thus arrived at for the nonlinear second order ODE is still formidable. Lighthill (1948) proceeded to intuitively ignore certain terms in

the equation. This enabled him to solve the resulting ODE in a 'somewhat'
closed form. For the solution of the BVP to exist, the relation between the
shock Mach number, M, and the nondimensional velocity of the piston, α,
was found explicitly.

Consider the uniform expansion of a cylinder, which starts from zero
radius. We summarise the results for the spherical case later. Let r be
the distance measured from the axis of the cylinder. The time t is chosen
to be zero when $r = 0$. For the uniform expansion of the cylinder it is
natural to introduce the nondimensional variable $x = r/(a_0 t)$, where a_0 is
the undisturbed speed of sound. Let $a_0 \alpha$ be the velocity of the cylinder
which leads to the velocity $a_0 M$ of the shock, $M \geq 1$. α is assumed to
be small. The region of flow between the piston and the shock in terms
of x becomes $\alpha < x < M$. Since the flow is assumed to be isentropic, the
equations of continuity and motion may be combined to yield a single second
order PDE for the velocity potential ϕ (see Von Mises (1958)):

$$a^2 \left(\phi_{rr} + \frac{1}{r} \phi_r \right) = \phi_{tt} + 2\phi_r \phi_{rt} + (\phi_r)^2 \phi_{rr}. \qquad (2.5.1)$$

The speed of sound a is found from the Bernoulli's equation

$$\phi_t + \frac{1}{2} q^2 + \frac{a^2}{\gamma - 1} = \frac{a_0^2}{\gamma - 1}, \qquad (2.5.2)$$

where $q = \partial \phi / \partial r$ is the fluid velocity. The boundary conditions to be
satisfied by the solution are as follows.

(i) The particle velocity at the cylinder is equal to that of the cylinder so
   that we have the relation

$$\frac{\partial \phi}{\partial r} = a_0 \alpha, \qquad (2.5.3)$$

   holding at the cylinder.

(ii) The velocity potential ϕ at the shock is continuous; therefore,

$$\phi(M) = 0. \qquad (2.5.4)$$

(iii) From the Rankine-Hugoniot conditions, the velocity behind the shock
   is given by

$$\frac{u}{a_0} = \frac{2}{\gamma + 1} \left(M - M^{-1} \right). \qquad (2.5.5)$$

If we introduce the similarity form of the solution

$$\phi = a_0^2 t f(x), \quad x = r/a_0 t \qquad (2.5.6)$$

into (2.5.1), it reduces to the exact nonlinear ODE

$$\left[1 - (\gamma - 1)\left(f - xf' + \frac{1}{2}f'^2\right)\right](f'' + x^{-1}f') = (x - f')^2 f'' \qquad (2.5.7)$$

while the BCs (i)–(iii) above become

$$f'(\alpha) = \alpha, \qquad\qquad\qquad\qquad (2.5.8)$$
$$f(M) = 0, \qquad\qquad\qquad\qquad (2.5.9)$$
$$f'(M) = 2(M - M^{-1})/(\gamma + 1). \qquad\qquad (2.5.10)$$

Since we have three BCs (2.5.8)–(2.5.10) for the second order ODE (2.5.7), there must exist a relation between the parameters α and M for a solution of this BVP to exist. First we derive a first approximation to this relationship, following the work of Lighthill (1948). He argued that equation (2.5.7) may be approximated by

$$[1 + (\gamma - 1)f']f'' + x^{-1}f' = (x^2 - 2f')f'', \qquad (2.5.11)$$

where some terms have been dropped. This was done by observing that ϕ and, therefore, f are small near $x = 1$ and f'^2 is small compared to f'; besides, x has been put equal to one where ever convenient. Indeed, the terms that have been dropped are assumed to be small even away from $x = 1$. Surprisingly, in spite of these approximations, Lighthill (1948) found a quite accurate relationship between M and α. Equation (2.5.11) can be solved in terms of an integral. Introducing the variables

$$f' = y, \quad x^{-2} = z \qquad\qquad\qquad (2.5.12)$$

into (2.5.11), we get the first order ODE

$$\left[1 + (\gamma + 1)y - \frac{1}{z}\right]\frac{dy}{dz} = \frac{1}{2}\frac{y}{z} \qquad (2.5.13)$$

or

$$\frac{dz}{dy} - 2(\gamma + 1 - y^{-1})z = -2y^{-1}. \qquad (2.5.14)$$

Equation (2.5.14) is linear in z and integrates to give

$$zy^{-2}e^{-2(\gamma+1)y} = \int_y^\infty 2t^{-3}e^{-2(\gamma+1)t}dt + \text{constant}. \qquad (2.5.15)$$

Since $y = f'$ is assumed to be small everywhere, retaining only the first two terms in the asymptotic expansion of the integral in (2.5.15), namely, $-2(\gamma + 1)e^{-2(\gamma+1)y}y^{-1} + e^{-2(\gamma+1)y}y^{-2}$, we get

$$z = 1 - 2(\gamma + 1)y + Ky^2, \qquad\qquad (2.5.16)$$

where K is a constant. The boundary condition (2.5.8) gives $z = \alpha^{-2}$ when $y = \alpha$. Therefore, from (2.5.16), we have

$$K \sim \frac{1}{\alpha^4} \quad \text{as} \quad \alpha \to 0. \tag{2.5.17}$$

The boundary condition (2.5.10) becomes

$$z = M^{-2} \quad \text{when} \quad y = \epsilon = 2(M - M^{-1})/(\gamma + 1). \tag{2.5.18}$$

Therefore, from (2.5.16), we have

$$M^{-2} \approx 1 - 4(M - M^{-1}) + 4K(M - M^{-1})^2/(\gamma + 1)^2 \tag{2.5.19}$$

which, for $M \sim 1$, implies that

$$M - 1 \sim \frac{3(\gamma + 1)^2}{8K} \sim \frac{3}{8}(\gamma + 1)^2 \alpha^4 \tag{2.5.20}$$

as $\alpha \to 0$ (see (2.5.17)).

The corresponding relation for the spherical symmetry was found to be

$$\log(M - 1) \sim -\frac{1}{(\gamma + 1)\alpha^3}. \tag{2.5.21}$$

Lighthill (1948) remarked that the solution of the above 'physical' problem must exist for each α; its uniqueness however is not obvious. He also gave a more rigorous order of magnitude argument to show that the relation (2.5.20) is correct for small values of α.

We may observe that the approximated equation (2.5.11) is analytically quite different from the full equation (2.5.7). The latter, for example, is invariant under the transformation $x \to -x$; this is clearly not true of the former.

We observe that this problem can be numerically treated as an initial value problem. For a given γ and an assumed value of $M = M_1 \sim 1$, one may integrate (2.5.7), starting at $x = M$ with the initial values (2.5.9) and (2.5.10). Since the solution of the problem exists (it should in fact be analytically proved), there would exist a point $x = \alpha << M$, where $f'(\alpha) = \alpha$. Thus, such a value of α may easily be obtained for given values of M_1 and γ; no iteration is needed to find α.

Our numerical solution of the boundary value problems for both spherical and cylindrical piston motions with $\gamma = 1.4$ shows that Lighthill's approximate analysis gives good results for $M \geq 1$. This is in spite of the fact that (2.5.11), approximating the original equation (2.5.7), is qualitatively different.

2.6 Plane Gas Dynamics in Transformed Co-ordinates

Sometimes it is useful to express the basic equations of motion in a new co-ordinate system to discover new forms of solutions. However, the transformation of the governing equations should be such that it is still feasible to impose boundary conditions at both the moving boundaries—the shock and the piston. Hodograph transformations do not generally allow this felicity. On the other hand, if one writes the basic equations in conservation form, it becomes possible to choose new independent variables such that one of them put equal to constant gives particle lines while the other set equal to zero delineates the shock path. Thus, the trajectories of the particles (including that of the piston) and of the shock wave are easily denominated and determined. There are other choices of co-ordinates which we shall summarize at the end of this section. We shall follow the work of Sachdev and Venkataswamy Reddy (1982) in the following.

The gasdynamic equations in planar symmetry for a compressible medium are

$$\rho_t + (\rho u)_x = 0, \tag{2.6.1}$$

$$\rho(u_t + uu_x) + p_x = 0, \tag{2.6.2}$$

$$S_t + uS_x = 0, \tag{2.6.3}$$

where ρ, u, p, and S stand for density, particle velocity, pressure and specific entropy at any point x and time t, respectively. We also have the equation of state, $p = \rho^\gamma \exp\left(\frac{S-S_0}{c_v}\right)$, where $\gamma = c_p/c_v$ is the ratio of specific heats at constant pressure and constant specific volume, respectively. Equation (2.6.1) is already in a conservation form. Equations (2.6.1) and (2.6.2) can be combined to give

$$\frac{\partial}{\partial t}(\rho u) + \frac{\partial}{\partial x}(p + \rho u^2) = 0. \tag{2.6.4}$$

The conservation laws (2.6.1) and (2.6.4) suggest the introduction of the variables τ and ξ:

$$d\tau = \rho dx - (\rho u)dt, \tag{2.6.5}$$

$$d\xi = \rho u dx - (p + \rho u^2)dt. \tag{2.6.6}$$

We can, therefore, write the differential relations

$$dx = -\frac{u}{p}d\xi + \frac{(p + \rho u^2)}{\rho p}d\tau, \tag{2.6.7}$$

$$dt = -\frac{d\xi}{p} + \frac{u}{p}d\tau. \tag{2.6.8}$$

Clearly, $d\tau = 0$ or $dx/dt = u$ gives particle lines. This relation also holds along the piston path. The Rankine-Hugoniot conditions across a shock propagating into a medium with the variable density $\rho_* = \rho_*(x)$, constant pressure $p = p_0$, and particle velocity $u_0 = 0$ are

$$\frac{\rho}{\rho_*} = \frac{(\gamma+1)p + (\gamma-1)p_0}{(\gamma-1)p + (\gamma+1)p_0}, \tag{2.6.9}$$

$$u = \left(\frac{2}{\rho_*}\right)^{1/2}(p - p_0)[(\gamma+1)p + (\gamma-1)p_0]^{-1/2}, \tag{2.6.10}$$

$$U = (2\rho_*)^{-1/2}[(\gamma+1)p + (\gamma-1)p_0]^{1/2}, \tag{2.6.11}$$

where U is the shock velocity. Along the shock, we have

$$dx = U(t)dt, \tag{2.6.12}$$

which, on using (2.6.7) and (2.6.8), can be written as

$$d\xi + \phi(\tau)d\tau = 0 \tag{2.6.13}$$

where we have assumed that we may write

$$\phi(\tau) = \frac{p_0}{\rho_*(x)}U^{-1}. \tag{2.6.14}$$

The relation (2.6.13) suggests introduction of the variable s according to

$$ds = d\xi + \phi(\tau)d\tau \tag{2.6.15}$$

so that the shock is given by some line $s =$ constant. The differential relations (2.6.7)–(2.6.8) can now be written as

$$dx = -\frac{u}{p}ds + \frac{(p + \rho u^2 + \rho u\phi)}{\rho p}d\tau, \tag{2.6.16}$$

$$dt = -\frac{ds}{p} + \frac{(u + \phi)}{p}d\tau. \tag{2.6.17}$$

Since (2.6.15) is invariant under a translation in s, we may choose $s = 0$ as the shock path. As we have observed earlier, the lines $\tau =$ constant give particle paths. Since entropy is constant along particle lines, $\tau =$ constant give lines of constant entropy in the (τ, s) plane. We may, therefore, conveniently drop (2.6.3) and transform (2.6.1) and (2.6.2) according to (2.6.16)–(2.6.17) with τ and s as the new independent variables:

$$w_\tau + ww_s - \frac{f}{\gamma}p^{-1/\gamma}p_s - \phi'(\tau) = 0, \tag{2.6.18}$$

$$p_\tau + wp_s - pw_s = 0, \tag{2.6.19}$$

where

$$f(\tau) = [p\rho^{-\gamma}]^{-1/\gamma}, \quad w = u + \phi \tag{2.6.20}$$

and the function $\phi(\tau)$ is defined by (2.6.14). The functions $\phi(\tau)$ and $f(\tau)$ will be fixed so that the shock conditions meet the requirement of a self-similar solution. Using (2.6.9)–(2.6.11), we may write the following relations along the shock:

$$p_{s=0} = \frac{p_0}{\gamma + 1}\left(\frac{2p_0}{\rho_*\phi^2} - \gamma + 1\right), \tag{2.6.21}$$

$$w_{s=0} = \phi\left(\frac{p_{s=0}}{p_0}\right), \tag{2.6.22}$$

$$f(\tau) = p_{s=0}^{1/\gamma}\left[\frac{(\gamma - 1)p_0 + 2\gamma\rho_*\phi^2}{(\gamma + 1)p_0\rho_*\phi^2}\right]\phi^2. \tag{2.6.23}$$

Since we seek similarity solutions for equations (2.6.18)–(2.6.20) in terms of a combination of variables τ and s, we must choose $\phi(\tau)$, $f(\tau)$, and $\rho_*(\tau)$ at the shock such that $\rho_*\phi^2$ and f/ϕ^2 are constant.

Sachdev and Venkataswamy Reddy (1982) used infinitesimal transformations to discover similarity solutions of the system (2.6.18)–(2.6.20) subject to the conditions (2.6.21)–(2.6.23) with the constraints on the functions ρ_*, ϕ, and f mentioned above. They arrived at two forms of solution—the power law and the exponential form. We discuss the power law form in detail and summarize the results for the exponential form. The power-law similarity solution has the form

$$w = \phi W(\sigma), \tag{2.6.24}$$

$$p = p_0 P(\sigma), \tag{2.6.25}$$

$$\phi = \frac{p_0 t_0}{\rho_0 x_0}\left(1 + \frac{a\tau}{\rho_0 x_0}\right)^\alpha, \tag{2.6.26}$$

$$f = \frac{\gamma p_0^{1/\gamma}}{b\rho_0}C_1\left(1 + \frac{a\tau}{\rho_0 x_0}\right)^{2\alpha}, \tag{2.6.27}$$

where

$$\sigma = \frac{s}{p_0 t_0}\left(1 + \frac{a\tau}{\rho_0 x_0}\right)^{-(1+\alpha)} \tag{2.6.28}$$

is the similarity variable, and x_0, t_0, p_0 and ρ_0 are arbitrary constants with dimensions of distance, time, pressure and density, respectively. α appears in the exponent in the similarity variable. C_1 is a constant. The system of PDEs (2.6.18)–(2.6.19) reduce via (2.6.24)–(2.6.28) to the system of ODEs

$$(W - a(\alpha + 1)\sigma)W' - C_1 P^{-1/\gamma}P' = a\alpha(1 - W), \tag{2.6.29}$$

$$PW' - (W - (\alpha + 1)a\sigma)P' = 0. \tag{2.6.30}$$

The parameter a assumes values $+1$ or -1. The constant C_1 is given by

$$C_1 = P_*^{1/\gamma} \left(\frac{\gamma - 1 + (2\gamma/b)}{\gamma + 1} \right) \frac{b}{\gamma}, \tag{2.6.31}$$

where the pressure P_* behind the shock is given by

$$\frac{p_{s=0}}{p_0} = P_* = \frac{2b - \gamma + 1}{\gamma + 1} \tag{2.6.32}$$

with

$$b = \frac{\rho_0 x_0^2}{p_0 t_0^2}. \tag{2.6.33}$$

Equation (2.6.26), the requirement that $\rho_* \phi^2$ must be constant, and the shock condition (2.6.9) imply that the undisturbed density must have the form

$$\rho_* = \rho_0 \left(1 + \frac{a\tau}{\rho_0 x_0} \right)^{-2\alpha}. \tag{2.6.34}$$

Thus, we have a five parameter family of solutions, the parameters being x_0, t_0, p_0, ρ_0 and $\alpha \neq 0$.

To obtain the initial conditions for the ODEs (2.6.29)–(2.6.30) at the shock $s = 0$, that is, at $\sigma = 0$ (see (2.6.28)), we substitute the similarity form (2.6.24)–(2.6.28) into the shock conditions (2.6.21)–(2.6.22). We obtain

$$W(0) = P(0) = \frac{2b - \gamma + 1}{\gamma + 1} = P_*. \tag{2.6.35}$$

Here we have used the form of the density stratification (2.6.34) in the undisturbed medium ahead of the shock. Choosing $V = (\alpha + 1)a\sigma$ and W as the dependent variables and P as the independent variable, we may write (2.6.29), (2.6.30) and (2.6.35) as

$$\frac{dW}{dP} = (W - V)/P, \tag{2.6.36}$$

$$\frac{dV}{dP} = a_1 (C_1 P^{(\gamma-1)/\gamma} - (W - V)^2)/P(W - 1), \tag{2.6.37}$$

$$V(P_*) = 0, \quad W(P_*) = P_*, \tag{2.6.38}$$

where $a_1 = (\alpha + 1)/\alpha$. From (2.6.16) we have, along the shock $s = 0$, the relation

$$d\tau = \rho_* dx, \tag{2.6.39}$$

where ρ_* is given by (2.6.34). Integrating (2.6.39) we have the x co-ordinate of the shock as

$$\begin{aligned}
\frac{x}{x_0} &= \frac{1}{a(2\alpha + 1)} \left[\left(1 + \frac{a\tau}{\rho_0 x_0} \right)^{2\alpha+1} - 1 \right], \quad \alpha \neq -1/2, \\
&= \frac{1}{a} \ln(1 + a\tau/\rho_0 x_0), \quad \alpha = -1/2. \tag{2.6.40}
\end{aligned}$$

Now, observing that, along the shock, $U = dx/dt$, where U is given in terms of pressure behind the shock by (2.6.11), we may write (2.6.39) as $d\tau = \rho_* U dt$. Integrating this relation etc., we have, along the shock,

$$
\begin{aligned}
\frac{t}{t_0} &= \frac{1}{a(\alpha+1)}\left[\left(1+\frac{a\tau}{\rho_0 x_0}\right)^{\alpha+1} - 1\right], \quad \alpha \neq -1, \\
&= \frac{1}{a}\ln\left(1+\frac{a\tau}{\rho_0 x_0}\right), \quad\quad\quad\quad \alpha = -1, \quad (2.6.41)
\end{aligned}
$$

where we have assumed that $\tau = 0$ when $x = 0, t = 0$. Eliminating τ from (2.6.40) and (2.6.41), we get the shock locus as

$$
\begin{aligned}
\frac{x(t)}{x_0} &= \frac{1}{a(2\alpha+1)}\left\{\left[1 + a(\alpha+1)\frac{t}{t_0}\right]^{(2\alpha+1)/(\alpha+1)} - 1\right\}, \quad \alpha \neq -1, -\frac{1}{2}, \\
&= \frac{1}{a}(1 - \exp(-at/t_0)), \quad\quad\quad\quad\quad\quad\quad\quad\quad\quad \alpha = -1, \\
&= \frac{2}{a}\ln\left(1 + \frac{at}{t_0}\right), \quad\quad\quad\quad\quad\quad\quad\quad\quad\quad\quad \alpha = -\frac{1}{2}.
\end{aligned}
$$
$$(2.6.42)$$

The density distribution ahead of the shock is found from (2.6.34) and (2.6.40) as

$$
\begin{aligned}
\rho_*(x) &= \rho_0\left[1 + a(2\alpha+1)\frac{x}{x_0}\right]^{-(2\alpha/2\alpha+1)}, \quad \alpha \neq -\frac{1}{2}, \\
&= \rho_0 \exp\left(\frac{ax}{x_0}\right), \quad\quad\quad\quad\quad\quad\quad\quad \alpha = -\frac{1}{2}. \quad (2.6.43)
\end{aligned}
$$

The shock velocity $U = dx/dt$ may be found from (2.6.42) as

$$
\begin{aligned}
U &= \frac{x_0}{t_0}\left[1 + a(\alpha+1)\frac{t}{t_0}\right]^{\alpha/\alpha+1}, \quad \alpha \neq -1, \\
&= \frac{x_0}{t_0}\exp\left(-\frac{at}{t_0}\right), \quad\quad\quad\quad\quad \alpha = -1. \quad (2.6.44)
\end{aligned}
$$

It may be checked from the explicit results above that if $a\alpha > 0$, the undisturbed density ρ_* ahead of the shock is a decreasing function of x, the shock velocity U grows with time and the pressure $P = p/p_0$ in the region behind the shock is greater that P_* (see (2.6.32)). Similarly, if $a\alpha < 0$, ρ_* increases with x, U decreases with t and $1 < P < P_*$. Figures 2.5–2.7 give shock locus, pressure behind the shock, and the loci of the particle trajectories including the piston path for some typical values of the parameters.

Now we shall find the loci of particle paths and the piston path explicitly. Substituting

$$
s = p_0 t_0 \sigma\left(1 + \frac{a\tau}{\rho_0 x_0}\right)^{\alpha+1}, \quad u = \phi(W - 1), \quad p = p_0 P \quad (2.6.45)
$$

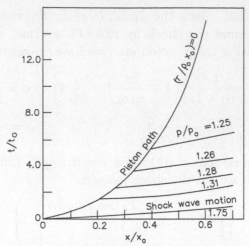

Figure 2.5 Piston path, isobars and shock locus for $\gamma = 5/3$, $b = 8/3$, $\alpha = -1/2$, $a = 1$, see (2.6.54) (Sachdev and Venkataswamy Reddy, 1982).

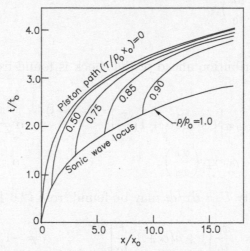

Figure 2.6 Piston path, particle trajectories and sonic wave locus for $\gamma = b = 1.4$, $\alpha = -1$, $a = -1$ (Sachdev and Venkataswamy Reddy, 1982).

into (2.6.16) and (2.6.17), we may obtain the partial derivatives $\partial x/\partial P$ and $\partial t/\partial P$. We recall that $\sigma = \sigma(P)$ under the similarity assumption.

We thus have

$$\frac{\partial x}{\partial P} = -\frac{x_0}{b}\left(\frac{W-1}{P}\right)\left(1 + \frac{a\tau}{\rho_0 x_0}\right)^{2\alpha+1}\frac{d\sigma}{dP}, \qquad (2.6.46)$$

$$\frac{\partial t}{\partial P} = -\frac{t_0}{P}\left(1 + \frac{a\tau}{\rho_0 x_0}\right)^{\alpha+1}\frac{d\sigma}{dP} \qquad (2.6.47)$$

or, using (2.6.36) and (2.6.37), we have

$$\frac{\partial x}{\partial P} = \frac{x_0}{a\alpha b}\left[\left(\frac{dW}{dP}\right)^2 - C_1 P^{-(\gamma+1)/\gamma}\right]\left[1 + \frac{\alpha\tau}{\rho_0 x_0}\right]^{2\alpha+1}, \qquad (2.6.48)$$

$$\frac{\partial t}{\partial P} = \frac{t_0}{a\alpha}\left[\left(\frac{dW}{dP}\right)^2 - C_1 P^{-(\gamma+1/\gamma)}\right]\left[1+\frac{\alpha\tau}{\rho_0 x_0}\right]^{\alpha+1}/(W-1).$$

(2.6.49)

Using the boundary conditions (2.6.40) and (2.6.41) at the shock $P = P_*$, we may write (2.6.48)–(2.6.49) in the integrated form as

$$\frac{x}{x_0} = \frac{\left(1+\frac{a\tau}{\rho_0 x_0}\right)^{2\alpha+1}}{ab\alpha}\int_{P_*}^{P}\left[\left(\frac{dW}{dP}\right)^2 - C_1 P^{-(\gamma+1/\gamma)}\right]dP$$

$$+\frac{1}{a(2\alpha+1)}\left[\left(1+\frac{a\tau}{\rho_0 x_0}\right)^{2\alpha+1}-1\right], \qquad \alpha \neq -\frac{1}{2},$$

$$= -\frac{2}{ab}\int_{P_*}^{P}\left[\left(\frac{dW}{dP}\right)^2 - C_1 P^{-(\gamma+1/\gamma)}\right]dP$$

$$+\frac{1}{a}\ln\left(1+\frac{a\tau}{\rho_0 x_0}\right), \quad \alpha = -\frac{1}{2}.$$

(2.6.50)

$$\frac{t}{t_0} = \frac{\left(1+\frac{a\tau}{\rho_0 x_0}\right)^{\alpha+1}}{a\alpha}\int_{P_*}^{P}\left[\left(\frac{dW}{dP}\right)^2 - C_1 P^{-(\gamma+1/\gamma)}\right]\frac{dP}{(W-1)}$$

$$+\frac{1}{a(\alpha+1)}\left[\left(1+\frac{a\tau}{\rho_0 x_0}\right)^{\alpha+1}-1\right], \qquad \alpha \neq -1,$$

$$= -\frac{1}{a}\int_{P_*}^{P}\left[\left(\frac{dW}{dP}\right)^2 - C_1 P^{-(\gamma+1/\gamma)}\right]\frac{dP}{(W-1)}$$

$$+\frac{1}{a}\ln\left(1+\frac{a\tau}{\rho_0 x_0}\right), \quad \alpha = -1,$$

(2.6.51)

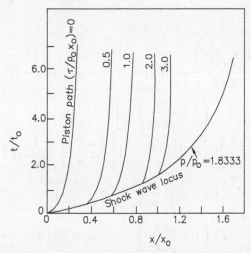

Figure 2.7 Piston path, particle trajectories and shock locus for $\gamma = 1.4$, $b = 2.4$, $\alpha = -0.75$, $a = 1$ (Sachdev and Venkataswamy Reddy, 1982).

where

$$P_* \leq P < \infty \quad \text{for} \quad a\alpha > 0,$$
$$1 < P \leq P_* \quad \text{for} \quad a\alpha < 0.$$

If we set $\tau = \text{constant}$ in (2.6.50)–(2.6.51), we obtain trajectories of the gas particles in a parametric form. In particular, if we put $\tau = 0$, we have the piston path

$$\frac{x(P)}{x_0} = \frac{1}{ab\alpha} \int_{P_*}^{P} \left[\left(\frac{dW}{dP} \right)^2 - C_1 P^{-(\gamma+1/\gamma)} \right] dP,$$

$$\frac{t(P)}{t_0} = \frac{1}{a\alpha} \int_{P_*}^{P} \left[\left(\frac{dW}{dP} \right)^2 - C_1 P^{-(\gamma+1/\gamma)} \right] \frac{dP}{(W-1)}. \quad (2.6.52)$$

This is the piston motion which gives rise to the shock propagation law (2.6.42). We have thus an analytic solution of the following physical problem. At $t = 0$, there is a quiescent gas in the region $x \geq 0$ with constant pressure p_0 and density distribution (2.6.34). The piston begins to move at $t = 0$ according to (2.6.52), giving rise to a flow headed by a shock with the trajectory (2.6.42). The solution, describing this flow, contains five arbitrary constants, namely, x_0, t_0, p_0, ρ_0, and α.

It may be verified from the expression of shock velocity and the undisturbed density and pressure ahead of the shock that the Mach number of the shock is simply $M_0 = (b/\gamma)^{1/2}$. For a compressive shock we must have $M_0 > 1$, that is, $b > \gamma$.

It becomes possible to write the solution in a closed form for $\alpha = -1, -1/2$.

$\alpha = -1$.

$$\frac{x}{x_0} = \frac{\left(1 + \frac{a\tau}{\rho_0 x_0} \right)^{-1}}{ab} [\gamma C_1 (P_*^{-1/\gamma} - P^{-1/\gamma}) - (P - P_*)]$$

$$+ \frac{1}{a} \left[1 - \left(1 + \frac{a\tau}{\rho_0 x_0} \right)^{-1} \right],$$

$$\frac{t}{t_0} = \frac{1}{a} \int_{P_*}^{P} \left[C_1 P^{-(\gamma+1/\gamma)} - 1 \right] \frac{dP}{(P-1)} + \frac{1}{a} \ln \left(1 + \frac{a\tau}{\rho_0 x_0} \right),$$

$$u = \phi(P-1), \quad \phi = \frac{p_0 t_0}{\rho_0 x_0} \left(1 + \frac{a\tau}{\rho_0 x_0} \right)^{-1}, \quad (2.6.53)$$

$$f(\tau) = \frac{\gamma p_0^{1/\gamma}}{b\rho_0} C_1 \left(1 + \frac{a\tau}{\rho_0 x_0} \right)^{-2}, \quad \rho_* = \rho_0 \left(1 - \frac{ax}{x_0} \right)^{-2},$$

$$\frac{x(t)}{x_0} = \frac{1}{a} \left[1 - \exp \left(-\frac{at}{t_0} \right) \right],$$

where $1 < P \leq P_*$ for $a = 1$ and $P_* \leq P < \infty$ for $a = -1$. Ustinov's (1967) special solution corresponds to the case $a = -1$.

$\alpha = -1/2$.

$$\frac{x}{x_0} = \frac{2}{ab} \int_{P_*}^{P} \left[C_1 P^{-(\gamma+1)/\gamma} - \frac{(b - \gamma C_1 P^{-1/\gamma})^2}{Q} \right] dP$$
$$+ \frac{1}{a} \ln\left(1 + \frac{a\tau}{\rho_0 x_0} \right),$$

$$\frac{t}{t_0} = \frac{2\left(1 + \frac{a\tau}{\rho_0 x_0} \right)^{1/2}}{a} \int_{P_*}^{P} \left[C_1 P^{-(\gamma+1/\gamma)} - \frac{(b - \gamma C_1 P^{-1/\gamma})^2}{Q} \right] \frac{dP}{Q^{1/2}}$$
$$+ \frac{2}{a} \left[\left(1 + \frac{a\tau}{\rho_0 x_0} \right)^{1/2} - 1 \right], \qquad (2.6.54)$$

$$u = \phi Q(P)^{1/2}, \quad \phi = \frac{p_0 t_0}{x_0 \rho_0} \left(1 + \frac{a\tau}{\rho_0 x_0} \right)^{-1/2},$$

$$f = \frac{\gamma}{b} - \frac{p_0^{1/\gamma}}{\rho_0} C_1 \left(1 + \frac{a\tau}{\rho_0 x_0} \right)^{-1}, \quad \rho_* = \rho_0 \exp\left(\frac{ax}{x_0} \right),$$

$$\frac{X(t)}{x_0} = \frac{2}{a} \ln\left(1 + \frac{at}{t_0} \right),$$

$$1 < P \leq P_* \quad \text{for} \quad a = 1, \quad \text{and} \quad P_* \leq P < \infty \quad \text{for} \quad a = -1,$$

where

$$Q = 1 + \frac{(\gamma + 1)}{\gamma - 1} P_* + 2bP - \frac{2\gamma^2 C_1 P^{(\gamma-1)/\gamma}}{\gamma - 1}. \qquad (2.6.55)$$

Sachdev and Venkataswamy Reddy (1982) also considered the limiting case when $M_0 \to 1$, corresponding to $b \to \gamma$, so that the shock degenerates into a sonic line. In this case the shock point $P = P_* = 1$ is a singularity. This manifests itself as $W = 1, V = 0$ in the system of ODEs (2.6.36)–(2.6.37). A local analysis about this point was carried out to start the integration from the neighbourhood of the shock. An approximate piston path was also determined, using this local analysis.

We summarize the results for the other class of similarity solutions, namely, the exponential type. Here the solution has the form

$$w = \phi W(\sigma), \qquad (2.6.56)$$

$$p = p_0 P(\sigma), \qquad (2.6.57)$$

$$\sigma = \frac{s}{p_0 t_0} \exp\left(-\frac{a\tau}{\rho_0 x_0} \right), \qquad (2.6.58)$$

$$f = \frac{\gamma p_0^{1/\gamma}}{b \rho_0} C_1 \exp\left(-\frac{a\tau}{\rho_0 x_0} \right), \qquad (2.6.59)$$

$$\phi \;=\; \frac{p_0 t_0}{\rho_0 x_0}\exp\left(\frac{a\tau}{\rho_0 x_0}\right), \tag{2.6.60}$$

where x_0, t_0, p_0, and C_1 are arbitrary constants. The system of PDEs (2.6.18)–(2.6.19) now reduces to

$$(W - a\sigma)W' - C_1 P^{-1/\gamma}P' \;=\; a(1 - W), \tag{2.6.61}$$
$$PW' - (W - a\sigma)P' \;=\; 0, \tag{2.6.62}$$

provided the undisturbed density is chosen in the form

$$\rho_* = \rho_0 \exp\left(-\frac{2a\tau}{\rho x_0}\right). \tag{2.6.63}$$

As before, we may rewrite (2.6.61)–(2.6.62) as

$$\frac{dW}{dP} \;=\; \frac{(W - V)}{P}, \tag{2.6.64}$$

$$\frac{dV}{dP} \;=\; (C_1 P^{\gamma - 1/\gamma} - (W - V)^2)/P(W - 1). \tag{2.6.65}$$

The conditions at the shock are

$$W(P_*) = P_*, \quad V(P_*) = 0, \tag{2.6.66}$$

where $V = a\sigma$. Indeed the above system of ODEs may be recovered from (2.6.36)–(2.6.37) in the limit $\alpha \to \pm\infty$, implying $a_1 = 1$. The degenerate solutions without shocks do not exist for the present class.

As for the power law form, we may derive the parametric form of the shock locus as

$$\frac{x}{x_0} \;=\; \frac{1}{2a}\left[\exp\left(\frac{2a\tau}{\rho_0 x_0}\right) - 1\right], \tag{2.6.67}$$

$$\frac{t}{t_0} \;=\; \frac{1}{a}\left[\exp\left(\frac{a\tau}{\rho_0 x_0}\right) - 1\right]. \tag{2.6.68}$$

The shock locus, the shock velocity and the density of the medium ahead of the shock are therefore given by

$$\frac{x(t)}{x_0} \;=\; \frac{1}{2a}\left(\left(1 + \frac{at}{t_0}\right)^2 - 1\right), \tag{2.6.69}$$

$$U(t) \;=\; \frac{x_0}{t_0}\left(1 + \frac{at}{t_0}\right), \tag{2.6.70}$$

$$\rho_*(x) \;=\; \rho_0/(1 + 2ax/x_0). \tag{2.6.71}$$

The functions x and t may be written as

$$\frac{x}{x_0} \;=\; \frac{\exp\left(\frac{2a\tau}{\rho_0 x_0}\right)}{ab}\int_{P_*}^{P}\left[\left(\frac{dW}{dP}\right)^2 - C_1 P^{-(\gamma + 1/\gamma)}\right]dP$$

$$+\frac{1}{2a}\left[\exp\left(\frac{2a\tau}{\rho_0 x_0}\right) - 1\right], \tag{2.6.72}$$

$$\frac{t}{t_0} = \frac{\exp\left(\frac{a\tau}{\rho_0 x_0}\right)}{a}\int_{P_*}^{P}\left[\left(\frac{dW}{dP}\right)^2 - C_1 P^{-(\gamma+1/\gamma)}\right]\frac{dP}{(W-1)}$$

$$+\frac{1}{a}\left[\exp\left(\frac{a\tau}{\rho_0 x_0}\right) - 1\right], \tag{2.6.73}$$

where $P_* \leq P < \infty$ for $a = 1$ and $1 < P \leq P_*$ for $a = -1$.

The piston path corresponding to $\tau = 0$ is therefore given by

$$\frac{x}{x_0} = \frac{1}{ab}\int_{P_*}^{P}\left[\left(\frac{dW}{dP}\right)^2 - C_1 P^{-(\gamma+1/\gamma)}\right]dP,$$

$$\frac{t}{t_0} = \frac{1}{a}\int_{P_*}^{P}\left[\left(\frac{dW}{dP}\right)^2 - C_1 P^{-(\gamma+1/\gamma)}\right]\frac{dP}{(W-1)}. \tag{2.6.74}$$

Sachdev, Gupta and Ahluwalia (1992) generalised the work of Sachdev and Venkataswamy Reddy (1982) in several ways. The solution of the system (2.6.18)–(2.6.20) was written out as Taylor series in s with coefficients which are functions of τ. The nice thing about this series solution is that no ODEs need to be solved for the coefficients. This is because s does not appear explicitly in (2.6.18)–(2.6.19) and the coefficient functions can be found recursively in an algebraic way. All the self-similar solutions found by Sachdev and Venkataswamy Reddy (1982) were recovered as special cases. Another class of solutions, again in series form, were discovered by introducing a new co-ordinate system which arises from the conservation form of the equations of continuity and energy. For this new formulation, the analysis applied to the previous class was repeated. Both classes of series solutions include solutions of Ustinov (1967, 1986) as special cases. A variety of solutions describing flows, driven by special piston motions and headed by strong shocks or shocks of arbitrary strength, were explicitly written out and graphically depicted. Another class of solutions described flows with characteristic fronts replacing the shocks.

A different formulation, accruing from another conservation form of plane gasdynamic equations, led to a certain nonlinear hyperbolic equation of second order (Ustinov (1967), Sachdev, Dowerah, Mayil Vaganan and Philip (1997)). This equation was thoroughly analysed by Sachdev et al. (1997). New intermediate integrals were found generalising the usual Riemann invariants. This also led to a new class of solutions of gasdynamic equations, which satisfied boundary conditions both at the piston and the shock. The medium ahead of the shock was again assumed to be nonuniform. Some ad hoc approaches yielded additional classes of solutions. The solutions of Ustinov (1967) were recovered as special cases. There are other related studies due to Steketee (1972, 1976, 1977) and Ardavan-Rhad (1970). The former author carried out a detailed analysis of one-dimensional gas-dynamic equations using Lagrangian co-ordinates in plane, cylindrical and spherical symmetries, while the latter, through a set of transformations, obtained an interesting equation for plane geometry wherein particle velocity is the dependent variable and sound speed and entropy are the independent variables. A first integral of this equation was found, generalising the usual Riemann invariant, and was used to study the catching up of a shock by a rarefaction wave. This equation merits further analysis and investigation.

Chapter 3

The Blast Wave

3.1 Introduction

As we remarked in the introductory chapter, in 1941, Sir Geoffrey Taylor was asked by the Defence Ministry in the U.K., much before the atom bomb was actually produced, to attempt to predict the mechanical effects such a nuclear device would bring about—in contrast to those of a common explosive bomb produced by the sudden generation of a large amount of gas at a high temperature in a confined space. The precise question posed was: "Would similar effects be produced if energy could be released in a highly concentrated form unaccompanied by the generation of gas?" It is remarkable how the analysis of Taylor, based on some physical assumptions, led to some precise answers which agreed rather closely with the experimental results made available much later. The self-similar analysis of Taylor gave an accurate answer up to the point where the maximum pressure behind the front decreased to about 10 atmospheres. It was Taylor's special gift to see how apparently complicated phenomena could be expressed in mathematical terms and to introduce a quantitative aspect into their description. He would always idealize the problem so that processes or factors expected to be of little relevance could be ignored. He would also identify all the relevant physical or geometrical parameters occurring in the data of the problem, from which information may be quickly culled.

The physical problem may be succinctly stated as follows. A finite but large amount of energy is suddenly released by nuclear fission in an infinitely concentrated form. This at once leads to the formation of a shock which propagates outward according to the law

$$R(t) = S(\gamma)\rho_0^{-1/5} E^{1/5} t^{2/5}, \qquad (3.1.1)$$

where ρ_0 is the atmospheric density, E is the energy released, and $S(\gamma)$ is a function of γ to be determined. The law (3.1.1) may be derived easily by dimensional considerations. Although a similarity hypothesis is still possible, as in the problem of the expansion of a spherical piston, it ceases to hold soon after the initiation of the bomb, since one of the assumptions underlying the analysis, namely that the intensity of the shock produced is infinitely high breaks down; the strength of the shock becomes finite as it propagates even though the total energy of the blast in the flow behind the shock remains (essentially) constant (see section 2.2 for the piston-driven flows which contain the blast wave as a special case).

We first derive the system of nonlinear ODEs from the basic system of PDEs under the similarity hypothesis, following the original work of G.I. Taylor (1946) and summarize the results of their numerical solution. Later, we shall write out their analytic solution, found by J.L. Taylor (1955) (see also Sedov (1946)). The shock propagation law $\dot{R} \propto t^{-3/5}$ from (3.1.1) suggests that we seek the self-similar solution in the form

$$\frac{p}{p_0} = y = R^{-3} f_1, \tag{3.1.2}$$

$$\frac{\rho}{\rho_0} = \psi, \tag{3.1.3}$$

$$u = R^{-3/2} \phi_1, \tag{3.1.4}$$

where p_0 and ρ_0 are undisturbed pressure and density ahead of the shock, respectively, while f_1, ψ and ϕ_1 are functions of $\eta = r/R(t)$ alone. It will be seen presently that the form (3.1.2)–(3.1.4) is consistent with the equations of motion and the equation of state for a perfect gas.

The equations of motion, continuity, and particle isentropy in spherical geometry are

$$u_t + u u_r + \frac{1}{\rho} p_r = 0, \tag{3.1.5}$$

$$\rho_t + u \rho_r + \rho u_r + \frac{2\rho u}{r} = 0, \tag{3.1.6}$$

$$(p\rho^{-\gamma})_t + u(p\rho^{-\gamma})_r = 0, \tag{3.1.7}$$

respectively, where $u, \rho,$ and p are particle velocity, density and pressure at the point r at time t, and $\gamma = c_p/c_v$ is the ratio of specific heats. Equations (3.1.5)–(3.1.7) reduce via (3.1.2)–(3.1.4) to the ODEs

$$-A(\frac{3}{2}\phi_1 + \eta\phi_1') + \phi_1\phi_1' + \frac{a_0^2}{\gamma}\frac{f_1'}{\psi} = 0, \tag{3.1.8}$$

$$-A\eta\psi' + \psi'\phi_1 + \psi(\phi_1' + \frac{2}{\eta}\phi_1) = 0, \tag{3.1.9}$$

$$A(3f_1 + \eta f_1') + \frac{\gamma f_1}{\psi}\psi'(-A\eta + \phi_1) - \phi_1 f_1' = 0, \tag{3.1.10}$$

provided we assume that

$$\frac{dR}{dt} = AR^{-3/2},$$
(3.1.11)

where A is a constant. The relation (3.1.11) is consistent with the energy relation (3.1.1) which itself follows from dimensional considerations alone. In the above, $a_0^2 = \gamma p_0/\rho_0$ is the square of the (undisturbed) speed of sound. The two constants A and a_0^2 can be eliminated from (3.1.8)–(3.1.10) by the simple scaling

$$f = f_1 a_0^2/A^2,$$
(3.1.12)

$$\phi = \phi_1/A,$$
(3.1.13)

so that we finally obtain the system

$$\phi'(\eta - \phi) = \frac{1}{\gamma}\frac{f'}{\psi} - \frac{3}{2}\phi,$$
(3.1.14)

$$\frac{\psi'}{\psi} = \frac{\phi' + 2\phi/\eta}{\eta - \phi},$$
(3.1.15)

$$3f + \eta f' + \frac{\gamma\psi'}{\psi}f(-\eta + \phi) - \phi f' = 0.$$
(3.1.16)

f' is obtained from (3.1.14)–(3.1.16) as

$$f' = \frac{f\{-3\eta + \phi(3 + \frac{1}{2}\gamma) - 2\gamma\phi^2/\eta\}}{\{(\eta - \phi)^2 - f/\psi\}}.$$
(3.1.17)

ψ' and ϕ' can now be found from (3.1.15) and (3.1.14), respectively. The strong shock conditions

$$\frac{\rho_1}{\rho_0} = \frac{\gamma + 1}{\gamma - 1},$$
(3.1.18)

$$\frac{p_1}{p_0} = \frac{\gamma + 1}{2\gamma}\frac{\dot{R}^2}{a_0^2},$$
(3.1.19)

$$\frac{u_1}{U} = \frac{2}{\gamma + 1},$$
(3.1.20)

in view of (3.1.2)–(3.1.4) and (3.1.11)–(3.1.13), become

$$\psi = \frac{\gamma + 1}{\gamma - 1},$$
(3.1.21)

$$f = \frac{2\gamma}{\gamma + 1},$$
(3.1.22)

$$\phi = \frac{2}{\gamma + 1}.$$
(3.1.23)

It is also relevant to express the total energy of the blast in terms of similarity functions. Thus, the total energy E behind the shock

$$E \;=\; \text{kinetic energy} + \text{heat energy}$$

$$\;=\; 4\pi \int_0^R \left(\frac{1}{2}\rho u^2\right) r^2 dr + 4\pi \int_0^R \frac{p}{\gamma - 1} r^2 dr, \qquad (3.1.24)$$

may be expressed in terms of the similarity functions and the similarity variable $\eta = r/R(t)$ as

$$E \;=\; 4\pi A^2 \left\{ \frac{1}{2}\rho_0 \int_0^1 \psi\phi^2\eta^2 d\eta + \left(\frac{p_0}{a_0^2(\gamma-1)} \int_0^1 f\eta^2 d\eta \right) \right\}$$

$$\;=\; J\rho_0 A^2, \qquad (3.1.25)$$

where

$$J = 2\pi \int_0^1 \psi\phi^2\eta^2 d\eta + \frac{4\pi}{\gamma(\gamma-1)} \int_0^1 f\eta^2 d\eta. \qquad (3.1.26)$$

Since the two integrals in (3.1.26) are functions of γ alone it is clear that the constant A^2, for a given γ, is a function of E/ρ_0.

Taylor (1950) solved the system (3.1.14), (3.1.15), and (3.1.17) for $\gamma=1.4$ numerically starting from the values (3.1.21)–(3.1.23) at the shock $\eta=1$. The numerical solution shows three main features of the flow behind the shock: (1) the velocity curve ϕ rapidly becomes a straight line passing through the origin, (2) the density curve ψ approaches almost zero at $\eta \sim 0.5$ and remains close to it till the center, and (3) the pressure decreases to become a constant and asymptotes to 0.37 times the maximum pressure just behind the shock (see Table 3.1 for the results of the numerical integration). Now we give exact (implicit) solution of this problem due to J.L. Taylor (1955). It is convenient to follow his notation. It is possible by the use of (3.1.5)–(3.1.6) to replace (3.1.7) by the equivalent energy equation

$$E_t + \frac{1}{r^2}(r^2 u I)_r = 0, \qquad (3.1.27)$$

where

$$E = \rho \left\{ \frac{1}{2}u^2 + \frac{p}{(\gamma-1)\rho} \right\} \qquad (3.1.28)$$

and

$$I = \rho \left\{ \frac{1}{2}u^2 + \frac{\gamma p}{(\gamma-1)\rho} \right\} \qquad (3.1.29)$$

are total energy of air per unit mass and total heat of air per unit mass, respectively.

J.L. Taylor, by skipping dimensional constants, sought solutions of (3.1.5), (3.1.6), and (3.1.27) in the form

$$\rho \;=\; f_1(r/t^{2/5}), \qquad (3.1.30)$$

$$u \;=\; t^{-3/5} f_2(r/t^{2/5}), \qquad (3.1.31)$$

$$p \;=\; t^{-6/5} f_3(r/t^{2/5}), \qquad (3.1.32)$$

Table 3.1. The results of numerical integration of spherically symmetric self-similar equations describing the point explosion problem for $\gamma = 1.4$ (see (3.1.14)–(3.1.16) and (3.1.21)–(3.1.23)) (Taylor, 1950).

η	f	ϕ	ψ
1.00	1.167	0.833	6.000
0.98	0.949	0.798	4.000
0.96	0.808	0.767	2.808
0.94	0.711	0.737	2.052
0.92	0.643	0.711	1.534
0.90	0.593	0.687	1.177
0.88	0.556	0.665	0.919
0.86	0.528	0.644	0.727
0.84	0.507	0.625	0.578
0.82	0.491	0.607	0.462
0.80	0.478	0.590	0.370
0.78	0.468	0.573	0.297
0.76	0.461	0.557	0.239
0.74	0.455	0.542	0.191
0.72	0.450	0.527	0.152
0.70	0.447	0.513	0.120
0.68	0.444	0.498	0.095
0.66	0.442	0.484	0.074
0.64	0.440	0.470	0.058
0.62	0.439	0.456	0.044
0.60	0.438	0.443	0.034
0.58	0.438	0.428	0.026
0.56	0.437	0.415	0.019
0.54	0.437	0.402	0.014
0.52	0.437	0.389	0.010
0.50	0.436	0.375	0.007

(cf. (3.1.2)–(3.1.4)); $R = kt^{2/5}$ defines the shock trajectory, as before, so that $\dot{R} = U = \frac{2}{5}kt^{-3/5}$. The similarity form (3.1.30)–(3.1.32) was used to eliminate t derivatives in favour of r derivatives instead of reducing the given system of PDEs to one of ODEs.

We observe that, in view of (3.1.30)–(3.1.32), we have

$$\frac{\partial \rho}{\partial t} = -\frac{2}{5}\frac{r}{t}\frac{\partial \rho}{\partial r} = -\frac{r}{R}U\frac{\partial \rho}{\partial r}, \tag{3.1.33}$$

$$\frac{\partial p}{\partial t} = -\frac{2}{5}\frac{r}{t}\frac{\partial \rho}{\partial r} - \frac{6}{5}\frac{p}{t} = -\frac{r}{R}U\frac{\partial p}{\partial r} - 3\frac{U}{R}p. \tag{3.1.34}$$

Also, since

$$E = t^{-6/5}f(r/t^{2/5}), \tag{3.1.35}$$

we have

$$E_t + \frac{6}{5}\frac{E}{t} + \frac{2}{5}\frac{r}{t}E_r = 0. \tag{3.1.36}$$

Eliminating E_t from (3.1.36) with the help of (3.1.27) and integrating with respect to r, we obtain

$$r^2 u I = \frac{2}{5} r^3 E/t, \tag{3.1.37}$$

or

$$\frac{u}{U} = \frac{r}{R} \frac{E}{I}. \tag{3.1.38}$$

Thus,

$$\bar{u} = \bar{r} \left\{ \frac{1}{2} u^2 + \frac{p}{(\gamma - 1)\rho} \right\} \Big/ \left\{ \frac{1}{2} u^2 + \frac{\gamma p}{(\gamma - 1)\rho} \right\}, \tag{3.1.39}$$

where

$$\bar{u} = \frac{u}{U}, \quad \bar{r} = \frac{r}{R}. \tag{3.1.40}$$

From (3.1.39), we have

$$\frac{p}{\rho} = C_1(t) \bar{u}^2 \frac{\bar{r} - \bar{u}}{\gamma \bar{u} - \bar{r}}, \tag{3.1.41}$$

where $C_1(t)$ is a function of t. From (3.1.5) and (3.1.7), we have

$$\frac{1}{p} \frac{\partial p}{\partial r} - \frac{\gamma - 1}{\rho} \frac{\partial \rho}{\partial r} = -\frac{1}{up} \frac{\partial p}{\partial t} + \frac{\gamma - 1}{u\rho} \frac{\partial \rho}{\partial t} - \frac{1}{u} \frac{\partial u}{\partial r} - \frac{2}{r}. \tag{3.1.42}$$

Eliminating ρ_t and p_t from (3.1.42) with the help of (3.1.33) and (3.1.34), we have

$$\left(\frac{r}{R} - \frac{u}{U} \right) \left(\frac{1}{p} p_r - \frac{\gamma - 1}{\rho} \rho_r \right) = -\frac{3}{R} + \frac{1}{U} u_r + \frac{2}{r} \frac{u}{U}, \tag{3.1.43}$$

or

$$\frac{1}{p} p_r - \frac{\gamma - 1}{\rho} \rho_r = -\frac{2}{r} - \left[\frac{\frac{1}{R} - \frac{1}{U} u_r}{\frac{r}{R} - \frac{u}{U}} \right]. \tag{3.1.44}$$

Equation (3.1.44) can be immediately integrated to yield

$$\frac{p}{\rho^{\gamma - 1}} = C_2(t) (\bar{r} - \bar{u})^{-1} (\bar{r})^{-2}, \tag{3.1.45}$$

where $C_2(t)$ is a function of integration. From (3.1.41) and (3.1.45) we get

$$\rho = C_3(t) [r u (r - u)]^{-2/(2-\gamma)} (\gamma u - r)^{1/(2-\gamma)}, \tag{3.1.46}$$

$$p = C_4(t) r^{-2/(2-\gamma)} u^{2(1-\gamma)/(2-\gamma)} (r - u)^{-\gamma/(2-\gamma)} (\gamma u - r)^{(\gamma - 1)/(2-\gamma)}, \tag{3.1.47}$$

where $\bar{r}$ and $\bar{u}$ have been replaced by r and u (without loss of generality) and $C_3(t)$ and $C_4(t)$ are functions of t. Using the integral (3.1.46) and the expression (3.1.33) for ρ_t in (3.1.6) and integrating with respect to r, we get

$$2 \log u = a \log r + b \log \left(r - \frac{3\gamma - 1}{5} u \right) + h \log(\gamma u - r) + \log C, \tag{3.1.48}$$

where

$$a = \frac{-10(\gamma - 1)}{(3\gamma - 1)},$$

$$b = (-13\gamma^2 + 7\gamma - 12)/(3\gamma - 1)(2\gamma + 1), \qquad (3.1.49)$$

$$h = \frac{5(\gamma - 1)}{2\gamma + 1}.$$

The functions C_i $(i = 1, 2, 3, 4)$ can be obtained by making use of the shock conditions (3.1.18)–(3.1.20) at $r = R$:

$$C_1 = \frac{\gamma - 1}{2} U^2,$$

$$C_2 = 2\rho_0^{2-\gamma}\{(\gamma - 1)^\gamma/(\gamma + 1)^{\gamma+1}\}U^2, \qquad (3.1.50)$$

$$C_3 = 2^{2/(2-\gamma)}\rho_0(\gamma + 1)^{-(\gamma+1)/(2-\gamma)}(\gamma - 1)^{(\gamma-1)/(2-\gamma)},$$

$$C_4 = 2^{\gamma/(2-\gamma)}\rho_0(\gamma + 1)^{-(\gamma+1)/(2-\gamma)}(\gamma - 1)^{1/(2-\gamma)}U^2.$$

The exact solution thus derived is too implicit and is not particularly illuminating. We give in the next section an approximate analytic solution which is more explicit and instructive.

In an interesting paper, Latter (1955) introduced an artificial viscosity term in the inviscid equations to see what effect it would have on the blast wave problem. This concept is originally due to Richtmeyer and Von Neumann (1950) who observed that the addition of a particular viscosity-like term into fluid-dynamic equations could lead to continuous shock-flow fields wherein the discontinuities at the shocks were smeared. The latter were replaced by regions in which physical parameters changed rapidly but smoothly. It was also ensured that the physical variables through such (smooth) shock transitions satisfied the Rankine-Hugoniot conditions. Such an artifice facilitated the numerical solutions of shock flows, where now shocks did not require an explicit fitting. It may be remarked that the choice of the form of the artificial viscosity term is not unique.

Latter (1955) introduced an artificial viscosity term of the form

$$q = \frac{1}{2}K^2\rho r^2 u_r(|u_r| - u_r) \qquad (3.1.51)$$

into the equations of motion and energy

$$u_t + uu_r = -\frac{1}{\rho}(p + q)_r, \qquad (3.1.52)$$

$$p_t + up_r = \frac{\gamma p + (\gamma - 1)q}{\rho}(\rho_t + u\rho_r). \qquad (3.1.53)$$

The equation of continuity in the spherical symmetry remains unaltered:

$$\rho_t + u\rho_r = -\rho\left(u_r + \frac{2u}{r}\right). \qquad (3.1.54)$$

Equations (3.1.52)–(3.1.54) must be supplemented by the adiabatic equation $p/\rho^\gamma = \sigma(s)$. The choice of viscosity term q in (3.1.51), as we remarked, is not unique. The details of the flow in the shock region would depend on the specific choice of q. The choice (3.1.51) admits a similarity form of the solution; this was one of the motivations for writing it in this form. Indeed this similarity solution is the same as Taylor's solution in the smooth region where $q = 0$. It is seen from the form (3.1.51) that the equations (3.1.52)–(3.1.54) in this region are the same as studied by Taylor (1950). This is because $\partial u/\partial r > 0$ for the particle velocity in the entire flow behind the strong shock and so $q = 0$ there. Latter (1955) first rederived Taylor's solution in a closed form in the smooth region where $q = 0$. He also sought a similarity solution in the undisturbed region beyond the shock. For viscosity to enter the analysis, the discontinuities in the slopes of the physical quantities must be admitted. This, however, also indicates a possible deficiency of the viscosity formalism which, in numerical applications, assumes continuity not only of the physical quantities but also of their slopes.

Latter (1955) also found large distance behaviour of the solution. He showed that there exists a solution in $r \geq 0$ which is continuous in $0 \leq r < 1$ and which exhibits a continuous transition region (with slope discontinuities) from the flow conditions at $r = 1$ to an undisturbed state at zero pressure at a large distance.

Latter (1955) computed the viscosity solution in the region $r \geq 1$ for different values of the viscosity coefficient K (see (3.1.51)), starting from the conditions at the shock. The main conclusion of his study is that the ratio of density behind the shock to that at infinity in the undisturbed region changes considerably as the viscosity constant K is increased.

3.2   Approximate Analytic Solution of the Blast Wave Problem Involving Shocks of Moderate Strength

Taylor (1950) first solved numerically the system of nonlinear ODEs governing the self-similar solution of the blast wave, as we have detailed in section 3.1. He observed from the numerical solution that the velocity function ϕ_1 or more accurately ϕ (see (3.1.4) and (3.1.13)) was almost linear. He introduced a correction to this linear behaviour by adding a nonlinear term. This enabled him to get a much closer approximation to the numerical solution. He could, with this approximation, integrate the ODEs in a closed form. This remarkable intuitive idea was generalised to all geometries— planar, cylindrical and spherical—by Sakurai (1953) who also found a more accurate local approximation to the solution, showing how well the intuitive solution of Taylor (1950) compared with the 'rigorous' local solution. In the sequel we follow Sakurai (1953). This approach also helps in finding a first approximation to the more general blast wave problem wherein the shock is not assumed to be infinitely strong; it includes a regime where it is moderately strong.

The equations of flow behind the blast wave may be written as

$$u_t + u u_r + \frac{1}{\rho} p_r = 0, \qquad (3.2.1)$$

$$\rho_t + u \rho_r + \rho \left(u_r + \frac{\alpha u}{r} \right) = 0, \qquad (3.2.2)$$

$$(p \rho^{-\gamma})_t + u (p \rho^{-\gamma})_r = 0, \qquad (3.2.3)$$

where, as before, u, p, and ρ are particle velocity, pressure and density at the position r and time t. The parameter α assumes values 0, 1, 2 for plane, cylindrical, and spherical geometry, respectively. Using (3.2.2) we change (3.2.3) to

$$p_t + u p_r + \gamma p \left(u_r + \frac{\alpha u}{r} \right) = 0. \qquad (3.2.4)$$

Let the position of the shock be given by $R = R(t)$ so that the shock velocity is

$$\frac{dR}{dt} = U. \qquad (3.2.5)$$

The Rankine-Hugoniot conditions holding across a shock of finite strength at $r = R(t)$ are

$$u = \frac{2}{\gamma + 1} U \left(1 - \frac{c_0^2}{U^2} \right), \qquad (3.2.6)$$

$$p = p_0 \left\{ \frac{2\gamma}{\gamma + 1} \left(\frac{U^2}{c_0^2} \right) - \frac{\gamma - 1}{\gamma + 1} \right\}, \qquad (3.2.7)$$

$$\rho = \rho_0 \frac{\gamma + 1}{\gamma - 1} \left\{ \frac{2}{\gamma - 1} \left(\frac{c_0}{U} \right)^2 + 1 \right\}^{-1}, \qquad (3.2.8)$$

where p_0 and ρ_0 are respectively the uniform pressure and density in the undisturbed medium ahead of the shock and $c_0^2 = \gamma p_0 / \rho_0$ is the square of sound speed.

As for the strong shock case, the energy released by the explosive, E_α, is assumed to be constant. Here we define

$$E_\alpha = \int_0^R \left\{ \frac{1}{2} u^2 + \frac{1}{\gamma - 1} \left(\frac{p}{\rho} - \frac{p_0}{\rho_0} \right) \right\} \rho r^\alpha dr, \quad \alpha = 0, 1, 2. \qquad (3.2.9)$$

This is the explosion energy per unit area of the surface of the shock front when R equals unity. If we make use of the Lagrangian form of the equation of conservation of mass, we have

$$\int_0^R \frac{\rho}{\rho_0} r^\alpha dr = \frac{R^{\alpha+1}}{\alpha+1}. \tag{3.2.10}$$

Equation (3.2.9) may now be written as

$$E_\alpha = \int_0^R \left(\frac{1}{2}\rho u^2 + \frac{1}{\gamma-1}p \right) r^\alpha dr - \frac{p_0}{\gamma-1}\frac{R^{\alpha+1}}{\alpha+1}. \tag{3.2.11}$$

Thus, we must solve (3.2.1)–(3.2.3), and (3.2.11), subject to the boundary conditions (3.2.6)–(3.2.8). Moreover, the particle velocity at the center of explosion must be zero. To generalise the self-similar solution of Taylor (1950) reported in section 3.1, we must retain the t-dependence in the solution and write a perturbation scheme in which the Taylor solution comes out as the zeroth order term. The boundary conditions must also assume the strong shock limit at zeroth order.

Thus, we introduce

$$\frac{r}{R} = x, \quad \left(\frac{c_0}{U} \right)^2 = y, \tag{3.2.12}$$

as the new independent variables. The unknown functions are now written in the more general form

$$u = Uf(x,y), \tag{3.2.13}$$

$$p = p_0 \left(\frac{U}{c_0} \right)^2 g(x,y), \tag{3.2.14}$$

$$\rho = \rho_0 h(x,y), \tag{3.2.15}$$

where the functions f, g, and h are nondimensional. It is clear that

$$\frac{\partial}{\partial r} = \frac{1}{R}\frac{\partial}{\partial x}, \tag{3.2.16}$$

$$\left(\frac{\partial}{\partial t} + u\frac{\partial}{\partial r} \right) = \frac{U}{R}\left\{ (f-x)\frac{\partial}{\partial x} + \lambda y\frac{\partial}{\partial y} \right\}. \tag{3.2.17}$$

Substitution of (3.2.13)–(3.2.15) into (3.2.1), (3.2.2) and (3.2.4) leads to the following PDEs for f, g, and h:

$$h\left\{ -\frac{1}{2}\lambda f + (f-x)f_x + \lambda y f_y \right\} = -\frac{1}{\gamma}g_x, \tag{3.2.18}$$

$$(f-x)h_x + \lambda y h_y = -h\left(f_x + \frac{\alpha f}{x} \right), \tag{3.2.19}$$

$$-\lambda g + (f-x)g_x + \lambda y g_y = -\gamma g\left(f_x + \frac{\lambda f}{x} \right), \tag{3.2.20}$$

where $\lambda = (R/y)\,(dy/dR)$ is a function of y alone. Equation (3.2.11) now becomes

$$y\left(\frac{R_0}{R}\right)^{\alpha+1} = \int_0^1 \left(\frac{\gamma}{2}hf^2 + \frac{1}{\gamma-1}g\right) x^\alpha dx - \frac{y}{(\alpha+1)(\gamma-1)}, \qquad (3.2.21)$$

where

$$R_0 = (E_\alpha/p_0)^{\frac{1}{(\alpha+1)}}. \qquad (3.2.22)$$

The shock conditions (3.2.6)–(3.2.8) assume the form

$$f(1,y) = \frac{2}{\gamma+1}(1-y), \qquad (3.2.23)$$

$$g(1,y) = \frac{2\gamma}{\gamma+1} - \frac{\gamma+1}{\gamma-1}y, \qquad (3.2.24)$$

$$h(1,y) = \left\{(\gamma+1)/(\gamma-1)\right\} \Big/ \left\{1 + (2/(\gamma-1))y\right\}. \qquad (3.2.25)$$

Writing

$$\int_0^1 \left(\frac{\gamma}{2}hf^2 + \frac{g}{\gamma-1}\right) x^\alpha dx = J \qquad (3.2.26)$$

in (3.2.21) and differentiating it, we get a relation for λ:

$$\lambda = \frac{R}{y}\left(\frac{dy}{dR}\right) = \frac{(\alpha+1)J - y/(\gamma-1)}{J - y\frac{dJ}{dy}}. \qquad (3.2.27)$$

In the more general setting here, the Taylor's solution should form zeroth order approximation in the limit $y \to 0$ when the shock velocity is large compared to the sound speed in the undisturbed medium ahead. To that end we write the series form of the solution as

$$f = f^{(0)}(x) + yf^{(1)}(x) + y^2 f^{(2)}(x) + \cdots, \qquad (3.2.28)$$

$$g = g^{(0)}(x) + yg^{(1)}(x) + y^2 g^{(2)}(x) + \cdots, \qquad (3.2.29)$$

$$h = h^{(0)}(x) + yh^{(1)}(x) + y^2 h^{(2)}(x) + \cdots. \qquad (3.2.30)$$

The energy integral J in (3.2.26) is also expanded in the form

$$J = J_0(1 + \sigma_1 y + \sigma_2 y^2 + \cdots). \qquad (3.2.31)$$

Substituting the expressions (3.2.28)–(3.2.31) into (3.2.26) and equating coefficients of different powers of y on both sides, we have

$$J_0 = \int_0^1 \left\{\frac{\gamma}{2}h^{(0)}f^{(0)^2} + \frac{1}{\gamma-1}g^{(0)}\right\} x^\alpha dx, \qquad (3.2.32)$$

$$\sigma_1 J_0 = \int_0^1 \left(\gamma f^{(0)}h^{(0)}f^{(1)} + \frac{\gamma}{2}f^{(0)^2}h^{(1)}\right)$$

$$+\frac{1}{\gamma - 1}g^{(1)}\Big)x^{\alpha}dx, \tag{3.2.33}$$

$$\sigma_2 J_0 = \int_0^1 \Big(\gamma f^{(0)}h^{(0)}f^{(2)} + \frac{\gamma}{2}f^{(0)2}h^{(2)} + \frac{1}{\gamma - 1}g^{(2)}\Big)x^{\alpha}dx$$

$$+\frac{\gamma}{2}\int_0^1 (h^{(0)}f^{(1)2} + 2h^{(1)}f^{(1)}f^{(0)})x^{\alpha}dx, \tag{3.2.34}$$

$$\dots\dots\dots\dots$$

Use of (3.2.31) in (3.2.21) leads to

$$y\Big(\frac{R_0}{R}\Big)^{\alpha+1} = J_0\Big[1 + \Big\{\sigma_1 - \frac{1}{J_0(\alpha + 1)(\gamma - 1)}\Big\}y + \sigma_2 y^2 + \cdots\Big] \tag{3.2.35}$$

or more explicitly, in view of the definition $y = (c_0/U)^2$,

$$\Big(\frac{c_0}{U}\Big)^2\Big(\frac{R_0}{R}\Big)^{\alpha+1} = J_0\Big[1 + \Big\{\sigma_1 - \frac{1}{J_0(\alpha + 1)(\gamma - 1)}\Big\}\Big(\frac{c_0}{U}\Big)^2$$

$$+\sigma_2\Big(\frac{c_0}{U}\Big)^4 + \cdots\Big]. \tag{3.2.36}$$

Equation (3.2.36) gives a relation between the shock velocity and R if J_0, σ_i etc. are known. λ defined by (3.2.27) becomes

$$\lambda = (\alpha + 1)\Big[1 + \Big\{\sigma_1 - \frac{1}{J_0(\alpha + 1)(\gamma - 1)}\Big\}y + 2\sigma_2 y^2 + \cdots\Big], \tag{3.2.37}$$

if the expression (3.2.31) for J is introduced.

Equation (3.2.37) may be rewritten more conveniently as

$$\lambda = (\alpha + 1)\Big[1 + \lambda_1 y + \lambda_2 y^2 + \cdots\Big]$$

where

$$\sigma_1 - \frac{1}{J_0(\alpha + 1)(\gamma - 1)} = \lambda_1, \tag{3.2.38}$$

$$2\sigma_2 = \lambda_2,$$

$$\dots\dots\dots\dots$$

To get the ODEs governing the functions $f^{(i)}$, $g^{(i)}$, $h^{(i)}$, we substitute (3.2.28)–(3.2.30) into (3.2.18)–(3.2.20) and compare coefficients of various powers of y on both sides. We obtain

$$(f^{(0)} - x)h^{(0)}f_x^{(0)} + g_x^{(0)}/\gamma = (\alpha + 1)f^{(0)}h^{(0)}/2, \tag{3.2.39}$$

$$h^{(0)}f_x^{(0)} + (f^{(0)} - x)h_x^{(0)} = -\alpha f^{(0)}h^{(0)}/x, \tag{3.2.40}$$

$$\gamma g^{(0)} f_x^{(0)} + (f^{(0)} - x) g_x^{(0)} = g^{(0)} (\alpha + 1 - \alpha \gamma f^{(0)} / x), \qquad (3.2.41)$$

$$
\begin{aligned}
h^{(0)}(f^{(0)} - x) f_x^{(1)} + g_x^{(1)} / \gamma &= -\left\{ (\alpha + 1)/2 + f_x^{(0)} \right\} h^{(0)} f^{(1)} \\
&\quad + \left\{ (\alpha + 1) f^{(0)}/2 + (x - f^{(0)}) f_x^{(0)} \right\} h^{(1)} \\
&\quad + (\alpha + 1) \lambda_1 f^{(0)} h^{(0)} / 2,
\end{aligned}
$$
$$(3.2.42)$$

$$
\begin{aligned}
h^{(0)} f_x^{(1)} + (f^{(0)} - x) h_x^{(1)} &= -(h_x^{(0)} + \alpha h^{(0)}/x) f^{(1)} \\
&\quad - (f_x^{(0)} + \alpha f^{(0)}/x + \alpha + 1) h^{(1)},
\end{aligned}
$$
$$(3.2.43)$$

$$
\begin{aligned}
\gamma g^{(0)} f_x^{(1)} + (f^{(0)} - x) g_x^{(1)} &= -(g_x^{(0)} + \alpha \gamma g^{(0)}/x) f^{(1)} \\
&\quad - \gamma (f_x^{(0)} + \alpha f^{(0)}/x) g^{(1)} \\
&\quad + (\alpha + 1) \lambda_1 g^{(0)},
\end{aligned}
$$
$$(3.2.44)$$

. . . .

The shock conditions (3.2.23)–(3.2.25) via (3.2.28)–(3.2.30) become

$$f^{(0)}(1) = \frac{2}{\gamma + 1}, \quad g^{(0)}(1) = \frac{2\gamma}{\gamma + 1}, \quad h^{(0)}(1) = \frac{\gamma + 1}{\gamma - 1}, \qquad (3.2.45)$$

$$f^{(1)}(1) = -\frac{2}{\gamma + 1}, \quad g^{(1)}(1) = -\frac{\gamma - 1}{\gamma + 1}, \quad h^{(1)}(1) = -2\frac{(\gamma + 1)}{(\gamma - 1)^2},$$
$$(3.2.46)$$

. . . .

The zeroth order solutions for different geometries ($\alpha = 0, 1, 2$) are obtained by solving (3.2.39)–(3.2.41) subject to the BCs (3.2.45) at the shock. This solution is substituted into (3.2.32) to find the integral J_0 and hence the solution of the strong blast wave problem; the solution of Taylor discussed in section 3.1 for spherical symmetry is a special case with $\alpha = 2$. This solution may be written as

$$
\begin{aligned}
u &= U f^{(0)}(x), \quad p = p_0 (U/c_0)^2 g^{(0)}(x), \\
\rho &= \rho_0 h^{(0)}(x), \quad (c_0/U)^2 (R_0/R)^{\alpha+1} = J_0.
\end{aligned}
$$
$$(3.2.47)$$

The shock locus is found from the last of (3.2.47) where use is made of (3.2.32). The first order system (3.2.42)–(3.2.44) which is linear and inhomogeneous involves an unknown parameter λ_1, which is obtained from (3.2.38) in terms of σ_1; the latter is given by (3.2.33). Since (3.2.33) itself involves $f^{(1)}$, $g^{(1)}$ and $h^{(1)}$, the problem must be solved iteratively by assuming some value of λ_1 and hence integrating the system (3.2.42)–(3.2.44) subject to the shock conditions (3.2.46) relevant to this order. The higher order terms $f^{(i)}$, $g^{(i)}$, $h^{(i)}$, $i = 2, 3, \cdots$ are governed by systems similar to those for $i = 1$, and hence may be found in the same manner.

Now we discuss the zeroth order solution in some detail. As we remarked earlier, this analysis was first initiated by G.I. Taylor (1950) for the case of spherical symmetry. It is profitable to rewrite the zeroth order system (3.2.39)–(3.2.41) as

$$(f^{(0)} - x)h^{(0)}f_x^{(0)} + g_x^{(0)}/\gamma = (\alpha + 1)f^{(0)}h^{(0)}/2, \qquad (3.2.48)$$

$$h_x^{(0)}/h^{(0)} = (f_x^{(0)} + \alpha f^{(0)}/x)/(x - f^{(0)}), \qquad (3.2.49)$$

$$g_x^{(0)}/g^{(0)} = (\gamma f_x^{(0)} + \alpha\gamma f^{(0)}/x - \alpha - 1)/(x - f^{(0)}). \qquad (3.2.50)$$

Eliminating $g_x^{(0)}$ from (3.2.48) with the help of (3.2.50), we have

$$f_x^{(0)} = \frac{(\alpha + 1)/\gamma - \alpha f^{(0)}/x + (\alpha + 1)h^{(0)}(x - f^{(0)})f^{(0)}/2g^{(0)}}{1 - h^{(0)}(x - f^{(0)})^2/g^{(0)}}. \qquad (3.2.51)$$

Sakurai (1956) found it more convenient to use an intermediate integral to solve the problem numerically. Writing $(\gamma - 1)\times$ (3.2.49)–(3.2.50), we have

$$(\gamma - 1)h_x^{(0)}/h^{(0)} - g_x^{(0)}/g^{(0)} = (1 - f_x^{(0)})/(x - f^{(0)}) + \alpha/x. \qquad (3.2.52)$$

An integration gives

$$g^{(0)}(x - f^{(0)})h^{(0)-(\gamma-1)}x^\alpha = \{2\gamma/(\gamma + 1)\}\{(\gamma - 1)/(\gamma + 1)\}^\gamma, \qquad (3.2.53)$$

where we have used the shock conditions for $f^{(0)}$, $g^{(0)}$, $h^{(0)}$ from (3.2.45). Making use of (3.2.53), the system (3.2.39)–(3.2.41) can be reduced to

$$f_x^{(0)} = \left(\frac{\alpha + 1}{\gamma} - \frac{\alpha f^{(0)}}{x} + \frac{\alpha + 1}{2}Df^{(0)}\right)\Big/\{1 - (x - f^{(0)})D\}, \qquad (3.2.54)$$

$$\frac{D_x}{D} = \left\{\alpha + 2 - \gamma f_x^{(0)} - \frac{\alpha(\gamma - 1)}{x}f^{(0)}\right\}\Big/(x - f^{(0)}), \qquad (3.2.55)$$

where

$$D = h^{(0)}(x - f^{(0)})/g^{(0)}. \qquad (3.2.56)$$

The boundary conditions at the shock for $f^{(0)}$ and D, derived from (3.2.45), are

$$f^{(0)}(1) = 2/(\gamma + 1), \quad D(1) = (\gamma + 1)/2\gamma. \qquad (3.2.57)$$

The numerical solutions to this order for the plane and cylindrical geometries are shown in Tables 3.2 and 3.3. Now we revert to the matter of the approximate solution of the zeroth order system.

Table 3.2. Numerical and approximate solutions for the planar blast wave problem ($\alpha = 0$) for $\gamma = 1.4$ (Sakurai, 1953).

	numerical			approximate		
x	$f^{(0)}$	$g^{(0)}$	$h^{(0)}$	$f^{(0)}$	$g^{(0)}$	$h^{(0)}$
1.00	0.8333	1.167	6.000	0.8333	1.167	6.000
0.98	0.8086	1.070	5.183	0.8087	1.071	5.186
0.96	0.7844	0.987	4.508	0.7848	0.991	4.518
0.94	0.7607	0.917	3.945	0.7616	0.923	3.961
0.92	0.7376	0.856	3.472	0.7390	0.865	3.494
0.90	0.7151	0.804	3.071	0.7170	0.815	3.096
0.88	0.6931	0.756	2.729	—	—	—
0.86	0.6717	0.720	2.435	—	—	—
0.84	0.6509	0.686	2.181	—	—	—
0.82	0.6307	0.656	1.960	—	—	—
0.80	0.6110	0.631	1.766	0.6151	0.647	1.785
0.78	0.5917	0.608	1.595	—	—	—
0.76	0.5730	0.588	1.443	—	—	—
0.74	0.5547	0.571	1.308	—	—	—
0.72	0.5369	0.555	1.187	—	—	—
0.70	0.5194	0.542	1.079	0.5239	0.556	1.080
0.68	0.5023	0.530	0.980	—	—	—
0.66	0.4855	0.520	0.891	—	—	—
0.64	0.4691	0.511	0.811	—	—	—
0.62	0.4529	0.503	0.736	—	—	—
0.60	0.4370	0.496	0.669	0.4405	0.504	0.658
0.58	0.4213	0.490	0.607	—	—	—
0.56	0.4058	0.484	0.549	—	—	—
0.54	0.3904	0.480	0.497	—	—	—
0.52	0.3753	0.476	0.448	—	—	—
0.50	0.3602	0.472	0.403	0.3624	0.473	0.389
0.48	0.3453	0.469	0.362	—	—	—
0.46	0.3305	0.467	0.323	—	—	—
0.44	0.3158	0.465	0.288	—	—	—
0.42	0.3012	0.463	0.255	—	—	—
0.40	0.2866	0.461	0.225	0.2877	0.456	0.214
0.30	0.215	0.456	0.107	0.2148	0.447	0.102
0.20	0.143	0.455	0.039	0.1430	0.443	0.037
0.10	0.072	0.455	0.006	0.0714	0.442	0.006
0.00	0.000	0.455	0.000	0.0000	0.442	0.000

Taylor (1950) observed that, for the spherical symmetry, the particle velocity near $x = 0$ was linear, with slope $1/\gamma$. So he attempted to improve upon it by assuming that

$$f^{(0)} = x/\gamma + Ax^n. \tag{3.2.58}$$

He determined the constants A and n by computing $f^{(0)}$ and $f_x^{(0)}$ at $x = 1$ from the shock conditions (3.2.45) and the exact slope (3.2.51), respectively. Thus, we may write for $\alpha = 0, 1, 2$,

$$f^{(0)}(1) = \frac{2}{\gamma + 1}, \quad f_x^{(0)}(1) = \frac{(3 - \alpha)\gamma + 3(\alpha + 1)}{(\gamma + 1)^2} \tag{3.2.59}$$

Table 3.3. Numerical and approximate solutions for the cylindrical blast wave problem ($\alpha = 1$) for $\gamma = 1.4$ (Sakurai, 1953).

	numerical			approximate		
x	$f^{(0)}$	$g^{(0)}$	$h^{(0)}$	$f^{(0)}$	$g^{(0)}$	$h^{(0)}$
1.00	0.8333	1.167	6.000	0.8333	1.167	6.000
0.98	0.8035	1.009	4.578	0.8037	1.011	4.584
0.96	0.7750	0.890	3.575	0.7758	0.895	3.590
0.94	0.7479	0.799	2.845	0.7494	0.807	2.862
0.92	0.7223	0.728	2.300	0.7245	0.739	2.317
0.90	0.6980	0.673	1.884	0.7008	0.685	1.898
0.88	0.6749	0.629	1.560	0.6783	0.641	1.570
0.86	0.6531	0.593	1.303	0.6568	0.605	1.309
0.84	0.6322	0.564	1.095	0.6362	0.576	1.098
0.82	0.6124	0.541	0.926	0.6164	0.551	0.925
0.80	0.5934	0.522	0.786	0.5973	0.531	0.783
0.78	0.5751	0.506	0.670	—	—	—
0.76	0.5574	0.493	0.572	—	—	—
0.74	0.5404	0.482	0.488	—	—	—
0.72	0.5238	0.474	0.417	—	—	—
0.70	0.5076	0.466	0.356	0.5104	0.468	0.347
0.68	0.4917	0.460	0.304	—	—	—
0.66	0.4762	0.455	0.258	—	—	—
0.64	0.4608	0.451	0.219	—	—	—
0.62	0.4457	0.448	0.186	—	—	—
0.60	0.4308	0.445	0.157	0.4322	0.441	0.153
0.50	0.360	0.438	0.061	0.3582	0.429	0.058
0.40	0.288	0.435	0.019	0.2859	0.425	0.019
0.30	—	—	—	0.2143	0.424	0.005
0.20	—	—	—	0.1429	0.424	0.001
0.10	—	—	—	0.0714	0.424	0.000
0.00	—	—	—	0.0000	0.424	0.000

and, therefore,

$$A = \frac{1}{\gamma}\frac{\gamma - 1}{\gamma + 1}, \quad n = \frac{(2 - \alpha)\gamma^2 + (3\alpha + 1)\gamma - 1}{\gamma^2 - 1}. \tag{3.2.60}$$

By making use of the expressions for f_0 and $(f_0)_x$ from (3.2.58) in (3.2.49) and (3.2.50), integrating the latter, and using the shock conditions (3.2.45), one obtains

$$g^{(0)} = \frac{2\gamma}{\gamma + 1}\left(\frac{\gamma + 1 - x^{n-1}}{\gamma}\right)^{-\frac{2\gamma^2 + (3\alpha+1)\gamma - (\alpha+1)}{(1-\alpha)\gamma + 3\alpha + 1}}, \tag{3.2.61}$$

$$h^{(0)} = \frac{\gamma + 1}{\gamma - 1}x^{\frac{\alpha+1}{\gamma-1}}\left(\frac{\gamma + 1 - x^{n-1}}{\gamma}\right)^{-\frac{2(2\alpha+1+\gamma)}{(1-\alpha)\gamma + 3\alpha + 1}}. \tag{3.2.62}$$

Table 3.4. The values of J_0 (see (3.2.32)) for $\alpha = 0, 1, 2$ for different values of γ (see Sakurai (1953) and Taylor (1950)).

$\gamma \setminus \alpha$	0	1	2
1.2	3.024	1.547	1.031
1.3	2.147	1.102	0.755
1.4	1.696	0.877	0.596
	(1.701)	(0.880)	
1.667	1.137	0.585	0.404

This approximate solution for $\alpha = 0, 1$ and $\gamma = 1.4$ is given in Tables 3.2 and 3.3 along with the exact numerical solutions (see Table 3.1 for $\alpha = 2$). The accuracy of this approximate solution is remarkably good, the error never exceeding five percent. Table 3.4 gives values of the integral J_0, obtained numerically and by the use of approximate solution (in brackets). Again the agreement is rather close.

Continuing the local analysis of Taylor (1950), Sakurai (1953) developed a similar approach for the first order solution in the neighbourhood of $x = 0$ (see chapter 7 of Sachdev (2000) for a general discussion of local analysis for nonlinear PDEs and ODEs). It may be easily checked that $f^{(0)} \sim x/\gamma$ is the correct zeroth order behaviour of this function as $x \to 0$. It follows from (3.2.55) that $D_x/D \sim [(\gamma+\alpha)/(\gamma-1)x]$ as $x \to 0$, leading immediately to $D \sim D_0 x^{(\gamma+\alpha)/(\gamma-1)}$, where D_0 is the constant of integration. We also observe that, in this limit,

$$x - f^{(0)} \to \frac{\gamma - 1}{\gamma} x, \quad 2f_x^{(0)} + \frac{\alpha - 1}{2} \to \frac{2}{\gamma} + \frac{\alpha + 1}{2},$$

$$f_x^{(0)} + \frac{\alpha + 1}{2} \frac{f^{(0)}}{x - f^{(0)}} \to \frac{1}{\gamma} + \frac{\alpha + 1}{2} \frac{1}{\gamma - 1}, \tag{3.2.63}$$

$$\exp\left(\int_1^x \frac{\alpha + 1}{x - f^{(0)}} dx\right) \to G_0 x^{(\alpha+1)\gamma/(\gamma-1)},$$

where G_0 is a constant. Now we introduce the transformation

$$f^{(1)} = (x - f^{(0)})\phi, \quad g^{(1)} = g^{(0)}\psi, \quad h^{(1)} = h^{(0)}\chi, \tag{3.2.64}$$

into (3.2.42)–(3.2.44), (3.2.46) and (3.2.33), and use (3.2.39)–(3.2.41) for the derivatives of $f^{(0)}$, $g^{(0)}$ and $h^{(0)}$ etc. to arrive at the system

$$-(x - f^{(0)})\phi_x + \frac{1}{\gamma D}\psi_x = -\left(2f_x^{(0)} + \frac{\alpha - 1}{2}\right)\phi$$

$$+ \left(f_x^{(0)} + \frac{\alpha + 1}{2} \frac{f^{(0)}}{x - f^{(0)}}\right)(\chi - \psi)$$

$$+ \frac{\alpha + 1}{2} \frac{f^{(0)}}{x - f^{(0)}}\lambda_1, \tag{3.2.65}$$

$$(x - f^{(0)})(-\phi_x + \chi_x) = (\alpha + 1)(\phi + \chi), \tag{3.2.66}$$

$$(x - f^{(0)})(-\gamma\phi_x + \psi_x) = (\alpha + 1)\{(\gamma - 1)\phi + \psi - \lambda_1\}, \qquad (3.2.67)$$

$$\int_0^1 \{\gamma f^{(0)}(x - f^{(0)})h^{(0)}\phi + \frac{g^{(0)}}{\gamma - 1}\psi + \frac{\gamma}{2}f^{(0)2}h^{(0)}\chi\}x^\alpha dx$$

$$= \lambda_1 J_0 + \frac{1}{(\gamma - 1)(\alpha + 1)}, \qquad (3.2.68)$$

$$\begin{aligned}
\phi(1) &= -2/(\gamma - 1), \quad \psi(1) = -(\gamma - 1)/2\gamma, \\
\chi(1) &= -2/(\gamma - 1).
\end{aligned} \qquad (3.2.69)$$

It is again possible to obtain an intermediate integral for this first order system. If we multiply (3.2.67) by 2, subtract from it $(2\gamma - 1)$ times (3.2.66) and integrate the resulting equation with the conditions (3.2.69), we obtain

$$\phi - 2\psi + (2\gamma - 1)\chi + 2\lambda_1$$
$$= \left(2\lambda_1 - \frac{3\gamma - 1}{\gamma}\frac{\gamma + 1}{\gamma - 1}\right)\exp\left(\int_1^x \frac{\alpha + 1}{x - f^{(0)}}dx\right). \qquad (3.2.70)$$

Sakurai (1954) showed in the appendix to his paper that an integral similar to (3.2.70) exists for each of the higher order systems. It is convenient to use (3.2.70) instead of (3.2.66). Now, if we use the approximation (3.2.63) for $f^{(0)}(x)$ and its derivative etc., we may change the system (3.2.65), (3.2.67) and (3.2.70) to

$$-\frac{\gamma - 1}{\gamma}x\phi_x + \frac{1}{\gamma D_0}x^{-(\gamma + \alpha)/(\gamma - 1)}\psi_x$$
$$= -\left(\frac{2}{\gamma} + \frac{\alpha - 1}{2}\right)\phi + \left(\frac{1}{\gamma} + \frac{\alpha + 1}{2}\frac{1}{\gamma - 1}\right)(\chi - \psi)$$
$$+\frac{\alpha + 1}{2}\frac{\lambda_1}{\gamma - 1}, \qquad (3.2.71)$$

$$\{(\gamma - 1)/\gamma\}x(-\gamma\phi_x + \psi_x) = (\alpha + 1)\{(\gamma - 1)\phi + \psi - \lambda_1\}, \qquad (3.2.72)$$

$$\phi - 2\psi + (2\gamma - 1)\chi = -2\lambda_1 + \left\{2\lambda_1 - (3\gamma - 1)\frac{\gamma + 1}{\gamma(\gamma - 1)}\right\}$$
$$\times G_0 x^{(\alpha + 1)\gamma/(\gamma - 1)}. \qquad (3.2.73)$$

Writing

$$\xi = x^\nu, \quad \nu = \frac{(\gamma + \alpha)}{\gamma - 1} + 1, \qquad (3.2.74)$$

in (3.2.71)–(3.2.73), we arrive at the simpler system

$$\frac{\gamma - 1}{\gamma}\nu\left(-\xi\phi_\xi + \frac{1}{\gamma - 1}\frac{1}{D_0}\psi_\xi\right) = -\left(\frac{2}{\gamma} + \frac{\alpha - 1}{2}\right)\phi + \left(\frac{1}{\gamma} + \frac{\alpha + 1}{2}\frac{1}{\gamma - 1}\right)$$
$$\times(\chi - \psi) + \frac{\alpha + 1}{2}\frac{\lambda_1}{\gamma - 1},$$

$$(3.2.75)$$

$$\left(\frac{\gamma-1}{\gamma}\right)\nu(-\gamma\xi\phi_\xi+\xi\psi_\xi) = (\alpha+1)\{(\gamma-1)\phi+\psi-\lambda_1\}, \tag{3.2.76}$$

$$\phi-2\psi+(2\gamma-1)\chi = -2\lambda_1+\{2\lambda_1-(3\gamma-1)(\gamma+1)/\gamma(\gamma-1)\}$$
$$\times G_0\xi^{(\alpha+1)\gamma/(2\gamma+\alpha-1)}. \tag{3.2.77}$$

We now attempt to solve the system of linear inhomogeneous equations (3.2.75)–(3.2.77). It may be checked by local analysis that, to the lowest order, this system has a constant solution. Proceeding in the manner described in Sachdev (2000), one may determine a particular integral in the form

$$\phi_S = \phi_0+\xi^{(\alpha+1)\gamma/(2\gamma+\alpha-1)}(C_{00}+C_{01}\xi+\cdots), \tag{3.2.78}$$

$$\psi_S = \psi_0+\xi^{(\alpha+1)\gamma/(2\gamma+\alpha-1)}(C_{11}\xi+\cdots), \tag{3.2.79}$$

$$\chi_S = \chi_0+\xi^{(\alpha+1)\gamma/(2\gamma+\alpha-1)}(C_{20}+C_{21}\xi+\cdots), \tag{3.2.80}$$

where

$$\phi_0 = -\chi_0$$
$$= -\left\{2+\frac{(\gamma-1)}{2}\gamma-(\gamma-2)\left(1+\frac{\alpha+1}{2}\frac{\gamma}{\gamma-1}\right)\right\}^{-1}\lambda_1,$$

$$\psi_0 = \lambda_1-(\gamma-1)\phi_0,$$

$$C_{00} = 0, \quad C_{20}=\frac{G_0}{2\gamma-1}\left(2\lambda_1-\frac{3\gamma-1}{\gamma}\frac{\gamma+1}{\gamma-1}\right), \tag{3.2.81}$$

$$C_{11} = \frac{(\gamma-1)D_0}{\gamma(\alpha+3)+\alpha-1}\left(1+\frac{\alpha+1}{2}\frac{\gamma}{\gamma-1}\right)C_{20}.$$

To get the complementary part of the solution, we eliminate χ from (3.2.75)–(3.2.77). We obtain

$$-(\gamma-1)\xi\phi_\xi+\frac{1}{D_0}\psi_\xi = -\frac{1}{\nu}\left\{2+\frac{\alpha-1}{2}\gamma\right.$$
$$+\frac{1}{2\gamma-1}\left(1+\frac{\alpha+1}{2}\frac{\gamma}{\gamma-1}\right)\right\}\phi$$
$$+\frac{1}{\nu}\frac{3-2\gamma}{2\gamma-1}\left(1+\frac{\alpha+1}{2}\frac{\gamma}{\gamma-1}\right)\psi, \tag{3.2.82}$$

$$-\gamma\xi\phi_\xi+\xi\psi_\xi = \{\gamma(\alpha+1)/\nu\}\phi+\{\gamma(\alpha+1)/\nu(\gamma-1)\}\psi. \tag{3.2.83}$$

By the balancing argument etc. (Sachdev (2000)), the solution of (3.2.82)–(3.2.83) may be written out in the form

$$\phi = A\xi^{-(\alpha+1)/\nu}(1+a_{11}\xi+\cdots)+B(a_{20}+a_{21}\xi+\cdots), \tag{3.2.84}$$

$$\psi = A\xi^{-(\alpha+1)/\nu}(b_{10}+b_{11}\xi+\cdots)+B(1+b_{21}\xi+\cdots), \tag{3.2.85}$$

where A and B are arbitrary constants; a_{ij} and b_{ij} are given by

$$a_{11} = -\frac{1}{\gamma^2}\left(\frac{2\gamma-1}{\gamma-1}n-\gamma\right)b_{10}, \quad a_{20} = -\frac{1}{\gamma-1},$$

$$a_{21} = -\frac{n-\gamma+1}{(\gamma-1)(n+\gamma)}b_{21},$$

$$b_{10} = \frac{\gamma l+(\gamma-1)n}{(n-\gamma)}D_0,$$

$$b_{11} = \frac{D_0}{2\gamma-n}[\gamma mb_{10}-\{\gamma l-(\gamma-1)(\gamma-n)\}]a_{11},$$

$$b_{21} = \left(m+\frac{l}{\gamma-1}\right)D_0, \tag{3.2.86}$$

$$l = \frac{1}{\nu}\left\{2+\frac{\alpha-1}{2}\gamma+\frac{1}{2\gamma-1}\left(1+\frac{\alpha+1}{2}\frac{\gamma}{\gamma-1}\right)\right\},$$

$$m = \frac{1}{\nu}\frac{3-2\gamma}{2\gamma-1}\left(1+\frac{\alpha+1}{2}\frac{\gamma}{\gamma-1}\right),$$

$$n = \frac{(\alpha+1)\gamma}{\nu}.$$

The general solution of the linear system (3.2.75)–(3.2.77) in terms of x is obtained by using (3.2.78)–(3.2.80) and (3.2.84)–(3.2.85) appropriately:

$$\begin{aligned}
\phi &= \phi_0 + x^{(\alpha+1)\gamma/(\gamma-1)}(c_{01}x^\nu+\cdots)\\
&\quad+Ax^{-(\alpha+1)}(1+a_{11}x^\nu+\cdots)\\
&\quad+B(a_{20}+a_{21}x^\nu+\cdots),
\end{aligned} \tag{3.2.87}$$

$$\begin{aligned}
\psi &= \psi_0 + x^{(\alpha+1)\gamma/(\gamma-1)}(c_{11}x^\nu+\cdots)\\
&\quad+Ax^{\nu-(\alpha+1)}(b_{10}+b_{11}x^\nu+\cdots)\\
&\quad+B(1+b_{21}x^\nu+\cdots),
\end{aligned} \tag{3.2.88}$$

$$\begin{aligned}
\chi &= \Big[-2\lambda_1+\{2\lambda_1-(3\gamma-1)(\gamma+1)/\gamma(\gamma-1)\}\\
&\quad\times G_0x^{(\alpha+1)\gamma/(\gamma-1)}-\phi+2\psi\Big]\Big/(2\gamma-1).
\end{aligned} \tag{3.2.89}$$

Since ν and $(\alpha+1)\gamma/(\gamma-1)$ are somewhat large and positive, it follows that $\psi \to \psi_0 + B$ as $x \to 0$ while ϕ and χ, for finite A, both tend to infinity as $x \to 0$.

The system (3.2.65)–(3.2.67) must now be solved numerically subject to (3.2.69). The unknown constant λ_1 must be found such that (3.2.68) is satisfied.

Sakurai (1954) separated the solution in the form

$$\phi = \phi_1 + \lambda_1\phi_2, \quad \psi = \psi_1 + \lambda_1\psi_2, \quad \chi = \chi_1 + \lambda_1\chi_2, \tag{3.2.90}$$

and substituted it into the system (3.2.65)–(3.2.67) to get the ODEs for ϕ_i, ψ_i, and χ_i ($i = 1, 2$). He also separated the boundary conditions (3.2.69) at the shock in the form

$$\phi_1(1) = -2/(\gamma - 1), \qquad \psi_1(1) = -(\gamma - 1)/2\gamma,$$
$$\chi_1(1) = -2/(\gamma - 1), \tag{3.2.91}$$

$$\phi_2(1) = \psi_2(1) = \chi_2(1) = 0. \tag{3.2.92}$$

The same separation may be carried through for the 'energy equation' (3.2.68) wherein (3.2.90) is substituted and coefficients with and without λ_1 equated. We obtain

$$I_1 + \lambda_1 I_2 = \lambda_1 J_0 + \frac{1}{(\gamma - 1)(\alpha + 1)} \tag{3.2.93}$$

or

$$\lambda_1 = \left\{ I_1 - \frac{1}{(\gamma - 1)(\alpha + 1)} \right\} \bigg/ (J_0 - I_2) \tag{3.2.94}$$

where

$$\begin{aligned} I_1 &= \int_0^1 \left\{ \gamma f^{(0)}(x - f^{(0)}) h^{(0)} \phi_1 + \frac{g^{(0)}}{\gamma - 1} \psi_1 \right. \\ &\quad \left. + \frac{\gamma}{2} f^{(0)2} h^{(0)} \chi_1 \right\} x^\alpha dx, \end{aligned} \tag{3.2.95}$$

$$\begin{aligned} I_2 &= \int_0^1 \left\{ \gamma f^{(0)}(x - f^{(0)}) h^{(0)} \phi_2 + \frac{g^{(0)}}{\gamma - 1} \psi_2 \right. \\ &\quad \left. + \frac{\gamma}{2} f^{(0)2} h^{(0)} \chi_2 \right\} x^\alpha dx. \end{aligned} \tag{3.2.96}$$

Now the separated systems of ODEs obtained from (3.2.65)–(3.2.67) for ϕ_i, ψ_i, χ_i ($i = 1, 2$) are solved subject to initial conditions (3.2.91) and (3.2.92) at $x = 1$. The integration is carried to $x = 0$. The parameter λ_1 is then determined from (3.2.94)–(3.2.96). Putting together all this, ϕ, ψ, and χ are found from (3.2.90).

The above process is rendered simpler by eliminating χ from (3.2.65) and (3.2.70) and rewriting (3.2.67). We then have the equations governing ψ and ϕ:

$$\psi_x = P_1 \phi + P_2 \psi + P_3 + P_4 \lambda_1, \tag{3.2.97}$$
$$\phi_x = \psi_x/\gamma + P_5\{(\gamma - 1)\phi + \psi - \lambda_1\}, \tag{3.2.98}$$

where

$$\begin{aligned} P_1 &= -\frac{D}{1 - (x - f^{(0)})D} \left\{ 2\gamma f_x^{(0)} + \frac{\alpha - 1}{2} \gamma \right. \\ &\quad \left. + \frac{\gamma}{2\gamma - 1} \left(f_x^{(0)} + \frac{\alpha + 1}{2} \frac{f^{(0)}}{x - f^{(0)}} \right) + (\alpha + 1)(\gamma - 1) \right\}, \end{aligned}$$

$$P_2 = \frac{D}{1 - (x - f^{(0)})D}\left\{\gamma\frac{3 - 2\gamma}{2\gamma - 1}\right.$$

$$\left. \times\left(f_x^{(0)} + \frac{\alpha + 1}{2}\frac{f^{(0)}}{x - f^{(0)}}\right) - \alpha - 1\right\}, \tag{3.2.99}$$

$$P_3 = -\frac{D}{1 - (x - f^{(0)})D}\frac{3\gamma - 1}{2\gamma - 1}\frac{\gamma + 1}{\gamma - 1}$$

$$\times\left(f_x^{(0)} + \frac{\alpha + 1}{2}\frac{f^{(0)}}{x - f^{(0)}}\right)$$

$$\times\exp\left(\int_1^x \frac{\alpha + 1}{x - f^{(0)}}dx\right),$$

$$P_4 = \frac{D}{1 - (x - f^{(0)})D}\left[\frac{-2\gamma}{2\gamma - 1}\left(f_x^{(0)} + \frac{\alpha + 1}{2}\frac{f^{(0)}}{x - f^{(0)}}\right)\right.$$

$$\times\left\{1 - \exp\left(\int_1^x \frac{\alpha + 1}{x - f^{(0)}}dx\right)\right\}$$

$$\left. +\frac{\alpha + 1}{2}\frac{\gamma f^{(0)}}{x - f^{(0)}} + \alpha + 1\right],$$

$$P_5 = -\frac{1}{\gamma}\frac{\alpha + 1}{x - f^{(0)}}.$$

These functions depend on the zeroth order solution via $f^{(0)}$, $f_x^{(0)}$ and D. Separating (3.2.97)–(3.2.98) according to (3.2.90) one obtains

$$\psi_{1x} = P_1\phi_1 + P_2\psi_1 + P_3, \tag{3.2.100}$$

$$\phi_{1x} = \frac{\psi_{1x}}{\gamma} + P_5\{(\gamma - 1)\phi_1 + \psi_1\}, \tag{3.2.101}$$

$$\psi_{2x} = P_1\phi_2 + P_2\psi_2 + P_4, \tag{3.2.102}$$

$$\phi_{2x} = \frac{\psi_{2x}}{\gamma} + P_5\{(\gamma - 1)\phi_2 + \psi_2 - 1\}. \tag{3.2.103}$$

The system (3.2.100)–(3.2.103) was solved numerically by Sakurai (1954) for $\gamma = 1.4$ and $\alpha = 0, 1, 2$, using the conditions (3.2.91)–(3.2.92). Some difficulties due to large gradients in ϕ_1, ϕ_2 near $x = 0$ were encountered. The local solution (3.2.87)–(3.2.88) near $x = 0$ was matched with the numerical solution by evaluating the constants A and B appropriately.

Using ϕ_1, ϕ_2, ψ_1, and ψ_2 from the numerical solution, the integrals (3.2.95) and (3.2.96) were evaluated (χ may be obtained from (3.2.73)) and hence the value of λ_1 computed from (3.2.94). This value of λ_1 was used to obtain ϕ, ψ, χ from (3.2.90). The particle velocity

$$u = f^{(0)} + (x - f^{(0)})\phi \tag{3.2.104}$$

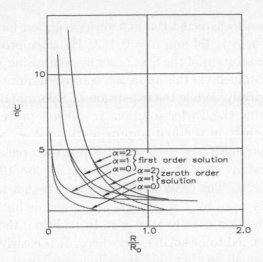

Figure 3.1 Shock velocity–distance curves for zeroth and first order solutions for $\alpha = 0, 1, 2$; see (3.2.105) (Sakurai, 1954).

tends to zero as $x \to 0$. We recall that $f^{(0)} \to x/\gamma$ in this limit. It is also observed that ϕ, ψ, and χ have nearly constant values away from $x = 1$. Using the values of λ_1 and J_0 obtained above, one may find the shock velocity–distance relation as

$$\left(\frac{c_0}{U}\right)^2 \left(\frac{R_0}{R}\right)^{\alpha+1} = J_0 \left\{ 1 + \lambda_1 \left(\frac{c_0}{U}\right)^2 \right\}. \qquad (3.2.105)$$

The relation (3.2.105) is depicted in Figure 3.1 both for zeroth and first order solutions for $\alpha = 0, 1, 2$. This relation, together with the Rankine-Hugoniot condition at the shock

$$p_1/p_0 = \frac{2\gamma}{\gamma+1} \left(\frac{U}{c_0}\right)^2 - \frac{\gamma-1}{\gamma+1}, \qquad (3.2.106)$$

yields the relation between the pressure p_1 behind the shock and the shock radius R:

$$\left(\frac{R_0}{R}\right)^{\alpha+1} = \frac{\gamma+1}{2\gamma} J_0 \left\{ \frac{2\gamma\lambda_1}{\gamma+1} + \frac{\gamma-1}{\gamma+1} + \frac{p_1}{p_0} \right\}. \qquad (3.2.107)$$

The relation (3.2.107) for $\gamma = 1.4$ corresponding to different geometries simplifies to

$$\begin{aligned}
(p_1/p_0 - 2.07)(R/R_0)^3 &= 1.96 \quad \text{for} \quad \alpha = 2, \\
(p_1/p_0 - 2.16)(R/R_0)^2 &= 1.33 \quad \text{for} \quad \alpha = 1, \qquad (3.2.108) \\
(p_1/p_0 - 2.33)(R/R_0) &= 0.69 \quad \text{for} \quad \alpha = 0.
\end{aligned}$$

The comparison of (3.2.108) with some early experimental results showed a mixed agreement.

Sakurai (1954) also depicted the first order solution for Mach numbers $U/c_0 = 2, 3, 5, \infty$ for $\gamma = 1.4$ and $\alpha = 0, 1, 2$. He attempted to find an approximate analytic solution of the first order system, using WKB approach. The values of λ_1 obtained via this approach compared quite favourably with those found numerically. While the expansion of Sakurai (1953) was a good attempt at extending the Taylor solution, it did not prove very successful since it predicted that, in the first approximation, $R \to \infty$ for a value of the inverse shock strength $y = -\frac{1}{\lambda_1}$, which is less than one. Sakurai (1959) attempted to extend his analysis in the manner of Taylor (1950) to cover the entire range $0 < y < 1$. The basic assumption here is that the velocity field behind the shock is given by $f = \frac{2}{\gamma+1}(1-y)x$, which is linear in x and satisfies the shock condition exactly at $x = 1$. In addition, the term $\lambda y \partial h/\partial y$ in (3.2.19) was assumed to be small everywhere. It is clearly small at $y \sim 0$; near $y \sim 1$, λ is small so that this term is again small. Sakurai (1959) argued that it may be assumed to be small in the entire interval $0 < y < 1$. With these two assumptions, Sakurai could solve the system of equations (3.2.18)–(3.2.20) exactly and obtain a relation between the integral J and the shock radius R. This approximate approach, however, led to an asymptotic behaviour for $R \to \infty$ which is at variance with the well-known analytic results obtained in this limit by Whitham (1950) and Landau (1945) (see section 3.9). Thus, although this approximate 'solution' has a simple form, it is rather unsatisfactory from the analytic point of view; it is also not in good agreement with the numerical solution.

In view of the above, Bach and Lee (1970) proceeded in a different manner to obtain an approximate analytic solution which is valid during the entire evolution of the blast—from the strong shock to the acoustic wave. The main assumption in the analysis of Bach and Lee (1970) is that the density behind the shock wave in the blast has a power law behaviour, the exponent being a function of time; this exponent is determined from the mass integral. This is in contrast to Sakurai's assumptions in his approximate analysis, namely, that the particle velocity profile behind the shock is linear and that the derivative of the density with respect to shock Mach number in the continuity equation is small and may be neglected. The latter assumption introduces serious error in the weak shock regime where the shock wave gradually decays to become a sound wave. The analysis of Bach and Lee (1970), in other respects, is quite similar to that of Sakurai (1959).

If we introduce the transformations

$$u(r,t) = \dot{R}\phi(\xi,\eta), \qquad (3.2.109)$$

$$p(r,t) = \rho_0 \dot{R}^2 f(\xi,\eta), \qquad (3.2.110)$$

$$\rho(r,t) = \rho_0 \psi(\xi,\eta), \qquad (3.2.111)$$

with

$$\xi = \frac{r}{R(t)}, \quad \eta = \frac{c_0^2}{\dot{R}^2} = \frac{1}{M^2}, \qquad (3.2.112)$$

$R = R(t)$ being the radius of the shock, into (3.2.1)–(3.2.3), we get

$$(\phi - \xi)\psi_\xi + \psi\phi_\xi + \alpha\phi(\psi/\xi) = 2\theta\eta\psi_\eta, \qquad (3.2.113)$$

$$(\phi - \xi)\phi_\xi + \theta\phi + \frac{1}{\psi}f_\xi = 2\theta\eta\phi_\eta, \qquad (3.2.114)$$

$$(\phi - \xi)\left(f_\xi - \frac{\gamma f}{\psi}\psi_\xi\right) + 2\theta f = 2\theta\eta\left(f_\eta - \frac{\gamma f}{\psi}\psi_\eta\right), \qquad (3.2.115)$$

where

$$\theta(\eta) = \frac{R\ddot{R}}{\dot{R}^2}, \qquad (3.2.116)$$

and $\alpha = 0, 1, 2$ for planar, cylindrical, and spherical symmetry, respectively. It is assumed that the mass and total energy enclosed by the blast wave remain constant so that one may obtain two further relations among the functions f, ψ, ϕ and the vairable

$$y = (R/R_0)^{\alpha+1}, \qquad (3.2.117)$$

where $R_0 = (E_0/\rho_0 c_0^2 k_\alpha)^{1/(\alpha+1)}$ is the characteristic explosion length and $k_\alpha = 1, 2\pi, 4\pi$ for $\alpha = 0, 1, 2$, respectively.

The Lagrangian equation of continuity gives

$$\frac{1}{R^{\alpha+1}}\int_0^R \frac{\rho}{\rho_0}r^\alpha dr = \int_0^1 \psi\xi^\alpha d\xi = \frac{1}{\alpha+1}, \qquad (3.2.118)$$

where we have used (3.2.111) and (3.2.112) (cf. (3.2.10)). The equation for the conservation of energy (3.2.9) in view of (3.2.109)–(3.2.112) becomes

$$1 = y\left(\frac{I}{\eta} - \frac{1}{\gamma(\gamma-1)(\alpha+1)}\right), \qquad (3.2.119)$$

where

$$I = \int_0^1 \left(\frac{f}{\gamma-1} + \frac{\psi\phi^2}{2}\right)\xi^\alpha d\xi \qquad (3.2.120)$$

and y is defined by (3.2.117).

The Rankine-Hugoniot conditions at the shock $\xi = 1(r = R(t))$ are

$$\phi(1, \eta) = [2/(\gamma+1)](1 - \eta), \qquad (3.2.121)$$

$$f(1, \eta) = [2/(\gamma+1)] - [(\gamma-1)/\gamma(\gamma+1)]\eta, \qquad (3.2.122)$$

$$\psi(1, \eta) = (\gamma+1)/(\gamma-1+2\eta). \qquad (3.2.123)$$

The geometrical requirement that the particle velocity at the center (axis) of symmetry must vanish gives $\phi(0, \eta) = 0$. We must therefore solve (3.2.113)–(3.2.115), subject to (3.2.121)–(3.2.123) and the symmetry condition $\phi(0, \eta) = 0$. The integral conditions on mass and energy of the blast further define the dynamics of the blast and its flow structure. The third equation (3.2.115) is therefore not explicitly used.

The basic approximation in the work of Bach and Lee (1970) is that the (nondimensional) density behind the blast is given by the power law

$$\psi(\xi, \eta) = \psi(1, \eta)\xi^{q(\eta)}, \tag{3.2.124}$$

where the boundary condition on $\psi(\xi, \eta)$ at $\xi = 1$ (see (3.2.123)) is automatically satisfied. (Observe the complicated nature of (3.2.124).) The exponent q in (3.2.124) is a function of η and hence time t (see (3.2.112)). Substituting (3.2.124) into the mass integral (3.2.118) leads to the evaluation of this exponent:

$$q(\eta) = (\alpha + 1)(\psi(1, \eta) - 1). \tag{3.2.125}$$

With $\psi(\xi, \eta)$ explicitly known via (3.2.124) and (3.2.125), we can write (3.2.113) as a first order PDE for ϕ alone:

$$\phi_\xi + (q + \alpha)\frac{\phi}{\xi} = q + \frac{2\theta\eta}{\psi(1, \eta)}[1 + (\alpha + 1)\psi(1, \eta)\ln\xi]\frac{d\psi(1, \eta)}{d\eta}. \tag{3.2.126}$$

Equation (3.2.126) can be integrated to yield

$$\phi = \phi(1, \eta)\xi(1 - \Theta\ln\xi), \tag{3.2.127}$$

where

$$\Theta = \frac{-2\theta\eta}{\phi(1, \eta)\psi(1, \eta)}\frac{d}{d\eta}\psi(1, \eta). \tag{3.2.128}$$

The solution (3.2.127)–(3.2.128) involves θ, a function of $R(t)$, which itself must be found as part of the solution (see (3.2.116)).

It may be observed that, in the strong shock limit ($M \to \infty, \eta \to 0$), Θ tends to zero; therefore, the solution ϕ for the velocity becomes linear, the form assumed by Sakurai (1965).

Substituting the expressions for ψ and ϕ thus obtained into the equation of motion (3.2.114) and integrating with respect to ξ, we get

$$\begin{aligned}
f = {} & -\int[-2\theta\eta\xi\left(\frac{d}{d\eta}\phi(1, \eta) - \frac{d}{d\eta}[\phi(1, \eta)\Theta]\ln\xi\right) \\
& + (\phi - \xi)\phi(1, \eta)(1 - \Theta - \Theta\ln\xi) + \theta\phi] \\
& \times \psi(1, \eta)\xi^{q(\eta)}d\xi + C(\eta),
\end{aligned} \tag{3.2.129}$$

where $C(\eta)$ is the function of integration. Using the boundary condition (3.2.122) at the shock and simplifying, (3.2.129) may be written as

$$\begin{aligned}
f(\xi, \eta) = {} & f(1, \eta) + f_2(\xi^{q+2} - 1) \\
& + f_3\{\xi^{q+2}[(q + 2)\ln\xi - 1] + 1\} \\
& + f_4\{2 - \xi^{q+2}[(q + 2)^2\ln^2\xi \\
& - 2(q + 2)\ln\xi + 2]\}.
\end{aligned} \tag{3.2.130}$$

Here,

$$f_2 = \frac{\psi(1,\eta)}{q+2}[(1-\Theta)\{\phi(1,\eta) - \phi^2(1,\eta)\}$$

$$-\theta\{\phi(1,\eta) - 2\eta\frac{d}{d\eta}\phi(1,\eta)\}], \tag{3.2.131}$$

$$f_3 = \frac{\psi(1,\eta)}{(q+2)^2}\left(\theta\left\{\Theta\phi(1,\eta) - 2\eta\frac{d}{d\eta}[\Theta\phi(1,\eta)]\right\}\right.$$

$$\left.-\Theta\phi(1,\eta) - \Theta^2\phi^2(1,\eta) + 2\Theta\phi^2(1,\eta)\right), \tag{3.2.132}$$

$$f_4 = \Theta^2\phi^2(1,\eta)\psi(1,\eta)/(q+2)^3. \tag{3.2.133}$$

In the strong shock limit $\eta \to 0$, the pressure profile (3.2.130) coincides with that obtained earlier by Sakurai (1965):

$$f(\xi) = f(1,0) + \frac{\psi(1,0)\phi(1,0)}{(q+2)}(\xi^{q+2} - 1)[1 - \phi(1,0) - \theta(0)]. \tag{3.2.134}$$

The pressure function (3.2.130) still requires the knowledge of $\theta(\eta)$ and $\theta'(\eta)$ for its explicit evaluation. To find $\theta = \theta(\eta)$, we substitute (nondimensional) velocity, density, and pressure (3.2.127), (3.2.124), and (3.2.130) into the energy relation (3.2.119) and solve for $d\theta/d\eta$:

$$\frac{d\theta}{d\eta} = -\frac{1}{2\eta}\left\{\theta + 1 - 2\phi(1,\eta) - \frac{(D_1 + 4\eta)}{\gamma + 1}\right.$$

$$-(\gamma - 1)(\alpha + 1)\left[\phi(1,\eta) - \frac{(D_1 + 4\eta)^2}{4\theta y(\gamma + 1)}\right]\right\}$$

$$+\frac{(D_1 + 4\eta)}{8\eta^2(\gamma + 1)}\left[\frac{(D_1 + 4\eta)\phi(1,\eta)}{\theta} - \frac{\phi(1,\eta)(\gamma + 1)}{\theta\psi(1,\eta)}\right.$$

$$\left.+2(\eta + 1) + (\gamma - 1)(\alpha + 1)\frac{(\gamma + 1)}{2\theta}\phi^2(1,\eta)\right]$$

$$+\frac{2\theta[2 + (\gamma - 1)(\alpha + 1)]}{D_1 + 4\eta}, \tag{3.2.135}$$

where

$$D_1 = \gamma(\alpha + 3) + (\alpha - 1). \tag{3.2.136}$$

Equation (3.2.135) involves y, which is related to the radius of the shock. We must find another equation involving y to determine both θ and y. This is easily done by differentiating (3.2.117) with respect to η (see (3.2.112)). We thus have

$$\frac{dy}{d\eta} = -(\alpha + 1)y/2\theta\eta, \tag{3.2.137}$$

where we have made use of the identity

$$R\frac{d}{dR} = \theta M\frac{d}{dM} = -2\theta\eta\frac{d}{d\eta} \tag{3.2.138}$$

(see (3.2.116) and (3.2.117)). Equations (3.2.135) and (3.2.137) now can be solved simultaneously for θ and y. The shock radius itself can be found from the definition $dR/dt = \dot{R}$ which can be expressed in terms of y and θ with the help of (3.2.112), (3.2.116) and (3.2.117), and hence integrated. We thus have

$$\frac{c_0 t}{R_0} = -\frac{1}{2} \int_0^\eta \frac{y^{1/(\alpha+1)} d\eta}{\theta \eta^{1/2}}. \tag{3.2.139}$$

Knowing θ and y as function of η from the integration of the coupled system (3.2.135) and (3.2.137), we may evaluate the integral in (3.2.139) to find t as a function of η, and hence the shock locus, in the (y, t) or (R, t) plane (see (3.2.117)).

Since the shock is at $r = 0$ where $\eta = 0$ and since, according to the Taylor-Sedov shock law, $R \propto t^{2/5}$, $\theta(0) = \theta_0$, a constant (see (3.2.116)), we must solve the system (3.2.135) and (3.2.137) subject to the conditions

$$\theta(0) = \theta_0, \quad y(0) = 0. \tag{3.2.140}$$

However, it is easily checked that both these equations assume indeterminate form $(0/0)$ at this initial point; we must therefore carry out a local analysis in the neighbourhood of the point $\eta = 0$. Writing

$$\theta = \theta_0 + \theta_1 \eta + \theta_2 \eta^2 + \cdots, \tag{3.2.141}$$

$$y = y_1 \eta + y_2 \eta^2 + y_3 \eta^3 \cdots, \tag{3.2.142}$$

and substituting into (3.2.135) and (3.2.137) etc., we get the following expressions for the coefficients:

$$
\begin{aligned}
\theta_0 &= -(\alpha+1)/2, \\
y_1 &= -E_1/(C_1 + A_1\theta_0), \\
\theta_1 &= \theta_0^2[A_2 + B_1\theta_0 + C_2/\theta_0 + E_2/(y_1\theta_0)]/C_1, \\
y_2 &= -y_1\theta_1/\theta_0, \\
\theta_2 &= \theta_0^2[A_3 + (B_1 - 1)\theta_1 + B_2\theta_0 + (C_1\theta_1^2/\theta_0^2 - C_2\theta_1/\theta_0 + C_3)/\theta_0 \\
&\quad + E_3/(y_1\theta_0)]/[C_1 + E_1/(2y_1)], \\
y_3 &= y_1[\theta_1^2/\theta_0^2 - \theta_2/(2\theta_0)], \\
\theta_3 &= \theta_0^2[(B_1 - 2)\theta_2 + B_2\theta_1 + B_3\theta_0 + \{C_1[2\theta_1\theta_2/\theta_0^2 - \theta_1^3/\theta_0^3] \\
&\quad + C_2(\theta_1^2/\theta_0^2 - \theta_2/\theta_0) - C_3\theta_1/\theta_0 + C_4\}/\theta_0 \\
&\quad + \{E_1\theta_1\theta_2/(3\theta_0^2) - E_2\theta_2/(2\theta_0)\}/(y_1\theta_0)]/\{C_1 + 2E_1/(3y_1)\}, \\
y_4 &= y_1\{7\theta_1\theta_2/(2\theta_0^2) - 3\theta_1^3/\theta_0^3 - \theta_3/\theta_0\}/3,
\end{aligned}
$$

$$\tag{3.2.143}$$

where

$$A_1 = D_1/[4(\gamma+1)],$$

$$B_1 = -1/2,$$
$$C_1 = D_1[D_1 + \alpha(\gamma+1)]/[4(\gamma+1)^2],$$
$$D_1 = \gamma(\alpha+3) + \alpha - 1,$$
$$E_1 = -(\gamma-1)(\alpha+1)D_1^2/[8(\gamma+1)],$$
$$A_2 = -1/2 + [2 + (\gamma-1)(\alpha+1)]/(\gamma+1) + (3D_1+4)/4(\gamma+1)],$$
$$B_2 = 2[2 + (\gamma-1)(\alpha+1)]/D_1,$$
$$C_2 = \{-D_1^2 + D_1[6 - (\gamma-1)(2\alpha+1)] + 4\alpha(\gamma-1)\}/[4(\gamma+1)^2],$$
$$E_2 = -(\gamma-1)(\alpha+1)D_1/(\gamma+1),$$
$$A_3 = [1 - (\gamma-1)(\alpha+1)]/(\gamma+1),$$
$$B_3 = -8[2 + (\gamma-1)(\alpha+1)]/D_1^2,$$
$$C_3 = \{D_1[-6 + (\gamma-1)(\alpha+1)] + 4[2 - (\gamma-1)(2\alpha+1)]\}/[4(\gamma+1)^2],$$
$$E_3 = -2(\gamma-1)(\alpha+1)/(\gamma+1).$$

It is interesting to note that, in the limit of infinite shock strength, $\eta \to 0$, $\theta_0 \to -(\alpha+1)/2$; this leads via (3.2.116) to the known exponents $N = 2/5, 1/2$, and $2/3$ in the shock law $R \propto t^N$ for spherical, cylindrical and plane symmetries, respectively.

Substituting (3.2.141)–(3.2.142) into (3.2.139) and integrating with respect to η, we can find an (approximate) explicit form of the shock trajectory for $\eta << 1$:

$$\frac{c_0 t}{R_0} = 2y_1^{1/\alpha+1}\eta^{[(\alpha+3)/2(\alpha+1)]}$$

$$\times \left[\frac{1}{\alpha+3} + \frac{T_1\eta}{3\alpha+5} + \frac{T_2\eta^2}{5\alpha+7} + \frac{T_3\eta^3}{7\alpha+9} + \cdots \right],$$

(3.2.144)

where

$$T_1 = y_2(\alpha+2)/[y_1(\alpha+1)],$$
$$T_2 = (2\alpha+3)\{y_3/y_1 - \alpha y_2^2/[2(\alpha+1)y_1^2]\}/(\alpha+1),$$
$$T_3 = (3\alpha+4)\{y_4/y_1 - \alpha y_2 y_3/[y_1^2(\alpha+1)]$$
$$+ \alpha(2\alpha+1)y_2^3/[6y_1^3(\alpha+1)^2]\}/(\alpha+1).$$

Bach and Lee (1970) carried out a numerical solution of the problem by their approach for the spherically symmetric case for $\gamma = 1.4$. Their results along with the exact numerical solution of Goldstine and Von Neumann (1955), the perturbation solution of Sakurai (1954), the linear velocity (approximate) solution of Sakurai (1965) and the quasi-similar solution of Oshima (1960) as computed by Lewis (1961) are shown in Figure 3.2. This figure gives the shock decay coefficient $-\theta$ versus $\eta = 1/M^2$. It is clear that the results of Bach and Lee (1970) are quite close to the numerical solution, although there is a clear departure as the shock Mach number diminishes to become one. It appears (though it has not been explicitly stated) that their analytic solution does not tend exactly to the appropriate weak shock limit as $M \to 1$. The main approximation in their theory lies in the assumption (3.2.124) for the density distribution behind the shock. The solution here does not satisfy the energy equation (3.2.115) exactly; this error manifests itself in the departure from the exact solution. This error is somewhat mitigated by the (imposed) physical conditions that total energy and mass behind the shock are conserved. This leads to a better determination of the trajectory of the shock.

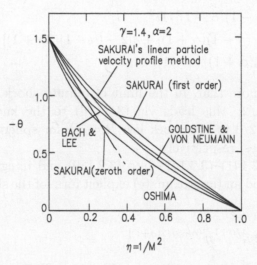

Figure 3.2 Variation of the shock decay coefficient $-\theta$ with shock strength η for spherical blast waves for $\gamma = 1.4$ (Bach and Lee, 1970).

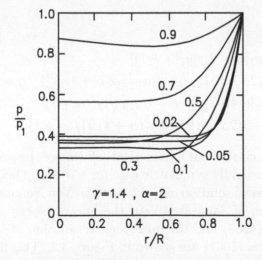

Figure 3.3 Pressure distributions behind spherical blast waves of various shock strengths η for $\gamma = 1.4$ (Bach and Lee, 1970).

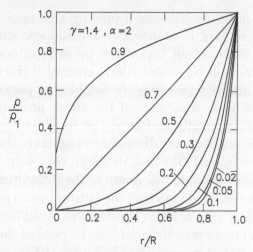

Figure 3.4 Density distributions behind spherical blast waves of various shock strengths η for $\gamma = 1.4$ (Bach and Lee, 1970).

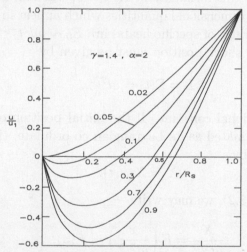

Figure 3.5 Particle velocity distributions behind spherical blast waves of various shock strengths η for $\gamma = 1.4$ (Bach and Lee, 1970).

The distribution of pressure, density, and particle velocity behind the shock for different shock strengths η for $\alpha = 2, \gamma = 1.4$ is shown in Figures 3.3–3.5. The interesting feature here is the development of a negative phase in the velocity distribution as the shock becomes weaker. This accords with the exact numerical results reported in the monograph of Sedov (1959).

3.3 Blast Wave in Lagrangian Co-ordinates

It was Von Neumann (1941) who contemporaneously with G.I. Taylor modelled the blast wave problem in his own independent way. He envisioned the blast wave as arising from the sudden release of energy from a homogeneous gas of much higher pressure. (We shall take up this more realistic model in subsequent chapters, where we discuss what happens when such a high pressure gas sphere expands out into a low pressure atmosphere, sending a shock wave into it). Von Neumann (1941) envisaged the unknown shock boundary as a free surface which must be found as a part of the solution. Thus, the continuous flow of air behind the shock must be found as the solution of the governing nonlinear PDEs; the Rankine-Hugoniot relations holding at the shock impose more boundary conditions than are required for the governing system of PDEs and therefore they help find the trajectory of the shock. Von Neumann was aware of the difficulties that would be encountered when the shock decays and the entropy jump across it changes with decreasing shock strength. Being conscious of the difficulties associated with the general (and more realistic) problem, he posited the simpler model of a point explosion in the manner of G.I. Taylor (1950) (see section 3.1). However, he found it more useful to use Lagrangian co-ordinates and seek a similarity solution in this formulation. At $t = 0$, when an intense explosion takes place, the only dimensional quantities which appear in the data (apart from $\gamma = c_p/c_v$, the ratio of specific heats) are $E_0 \sim ml^2t^{-2}$ and $\rho_0 \sim ml^{-3}$. They imply that the shock position must be given by

$$\Sigma = \Sigma(t) = at^{2/5}, \tag{3.3.1}$$

where a is a dimensional constant. If the initial position of a gas element at $t = 0$ is x, denominated as its Lagrangian co-ordinate, then its Eulerian co-ordinate is given by

$$X = X(x,t). \tag{3.3.2}$$

From (3.3.1) and (3.3.2), we may write

$$\frac{X}{at^{2/5}} = F\left(\frac{x}{at^{2/5}}\right) = F(z), \tag{3.3.3}$$

where

$$z = \frac{x}{\Sigma} = \frac{x}{at^{2/5}}. \tag{3.3.4}$$

Ahead of the shock, pressure p_0 is assumed to be zero, while density, temperature, and particle velocity are ρ_0, T_0, and 0, respectively. After the particle enters the shock, its density, following the conservation of mass becomes

$$\rho = \rho_0 \left[\frac{\partial(x^3)}{\partial(X^3)}\right]_t = \rho_0 \frac{x^2}{X^2}\left(\frac{\partial X}{\partial x}\right)^{-1} = \rho_0 \frac{z^2}{F^2(z)}\frac{1}{F'(z)}. \tag{3.3.5}$$

The velocity of the particle, by definition, is

$$u = \left(\frac{\partial X}{\partial t}\right)_x = \frac{2}{5}at^{-3/5}[F(z) - zF'(z)]. \tag{3.3.6}$$

The shock velocity according to (3.3.1) clearly is

$$U = \frac{d\Sigma}{dt} = \frac{2}{5}at^{-3/5}. \tag{3.3.7}$$

The particle position, immediately after entering the shock, becomes $X = x = \Sigma$. (There is no discontinuous change in the position of the particle, its velocity gets a jump). Therefore, the value of the similarity variable z behind the shock is $z = 1$; thus, $F(z) = 1$ at $z = 1$ (see (3.3.3)–(3.3.4)). The geometrical symmetry requires that the particle at the 'point' of explosion remains there. Therefore, $X(0, t) = 0$. This gives the second boundary condition

$$F(z) = 0 \quad \text{at} \quad z = 0. \tag{3.3.8}$$

The strong shock conditions at $z = 1$ are

$$\rho = \frac{\gamma+1}{\gamma-1}\rho_0, \tag{3.3.9}$$

$$u = \frac{2}{\gamma+1}U, \tag{3.3.10}$$

$$p = \frac{2}{\gamma+1}\rho_0 U^2, \tag{3.3.11}$$

where U is the velocity of the shock. Comparing (3.3.9) and (3.3.5) applied at the shock $z = 1$ and using the condition

$$F(z) = 1 \quad \text{at} \quad z = 1, \tag{3.3.12}$$

(see below (3.3.7)), we get the second condition at the shock, namely,

$$F'(z) = \frac{\gamma-1}{\gamma+1} \quad \text{at} \quad z = 1. \tag{3.3.13}$$

While (3.3.5) gives ρ in terms of $F(z)$ and $F'(z)$ in the flow behind the shock, the expression for p is obtained by using the fact that every particle x, after crossing the shock, retains the value of the entropy it acquired there. The Lagrangian co-ordinate of a particle may be written in terms of t and z as $x = at^{2/5}z$. Since x is continuous across the shock, the time $t = t'$ at which a particle crosses the shock may be obtained from $at'^{2/5} = at^{2/5}z$ or $t' = tz^{5/2}$. The entropy function $p\rho^{-\gamma}$ may be obtained by evaluating it from the conditions immediately behind the shock, given by (3.3.9) and (3.3.11):

$$p\rho^{-\gamma} = \frac{2}{\gamma+1}\rho_0 U^2 \left(\frac{\gamma+1}{\gamma-1}\rho_0\right)^{-\gamma} = \frac{2(\gamma-1)^\gamma}{(\gamma+1)^{\gamma+1}}\rho_0^{-(\gamma-1)}U^2. \tag{3.3.14}$$

On using (3.3.5) and (3.3.7) in (3.3.14), we get

$$p = \frac{8(\gamma-1)^\gamma}{25(\gamma+1)^{\gamma+1}}\rho_0 a^2 t'^{-6/5} \frac{z^{2\gamma}}{[F(z)]^{2\gamma}} \frac{1}{[F'(z)]^\gamma}. \tag{3.3.15}$$

Replacing t' by $tz^{5/2}$ in (3.3.15), we have

$$p = \frac{4}{25}\Phi\rho_0 a^2 t^{-6/5}\frac{z^{2\gamma-3}}{[F(z)]^{2\gamma}}\frac{1}{[F'(z)]^\gamma}, \tag{3.3.16}$$

where

$$\Phi = \frac{2}{\gamma+1}\left(\frac{\gamma-1}{\gamma+1}\right)^\gamma. \tag{3.3.17}$$

Von Neumann (1941), having used the equations of continuity and entropy, employed the energy argument rather than the equation of motion to finally obtain the equation for the unknown function $F(z)$. The functions p, ρ and u are expressed in terms of $F(z)$ and its derivative through (3.3.16), (3.3.5), and (3.3.6), respectively. The equation for $F(z)$ must be solved in the interval $0 < z < 1$ subject to the conditions (3.3.8), (3.3.12) and (3.3.13).

The total energy per unit mass of the gas behind the shock is

$$\epsilon = \frac{1}{\gamma-1}\frac{p}{\rho} + \frac{1}{2}u^2. \tag{3.3.18}$$

The gas in the spherical shell, reaching from the particles x to the particles $x+dx$, is the same for all time and, therefore, may be taken to be $4\pi\rho_0 x^2 dx$, the value at $t = 0$. Thus the total energy inside the sphere of the particles x is

$$\epsilon_1(x) = 4\pi\rho_0\int_0^x \epsilon x^2 dx \tag{3.3.19}$$

$$= 4\pi\rho_0\int_0^x \left(\frac{1}{\gamma-1}\frac{p}{\rho} + \frac{1}{2}u^2\right)x^2 dx. \tag{3.3.20}$$

In terms of z (see (3.3.4)) this becomes $\epsilon_2(z)$:

$$\epsilon_1(x) = \epsilon_2(z)$$

$$= 4\pi\rho_0\int_0^z \left\{\frac{1}{\gamma-1}\frac{(4/25)\Phi\rho_0 a^2 t^{-6/5}\{z^{2\gamma-3}/[F(z)]^{2\gamma}\}[F'(z)]^{-\gamma}}{\rho_0\{z^2/[F(z)]^2\}[F'(z)]^{-1}}\right.$$

$$\left. + \frac{1}{2}\cdot\frac{4}{25}a^2 t^{-6/5}[F(z) - zF'(z)]^2\right\}a^3 t^{6/5}z^2 dz,$$

or equivalently,

$$\epsilon_2(z) = \frac{8\pi}{25}\rho_0 a^5\int_0^z \left\{\frac{2}{\gamma-1}\Phi\frac{z^{2(\gamma-1)-3}}{[F(z)]^{2(\gamma-1)}}\frac{1}{[F'(z)]^{\gamma-1}}\right.$$

$$\left. + [F(z) - zF'(z)]^2\right\}z^2 dz. \tag{3.3.21}$$

Putting $z = 1$ in (3.3.21), we again arrive at the conclusion that the energy behind the shock, $0 < z < 1$, is independent of time. This is the total energy acquired by the gas between $t = 0$ and t finite. Since, by assumption, $p_0 = 0$,

$u = 0$, the initial energy of the gas is zero. The energy given by (3.3.21) with $z = 1$ must be equal to the energy of explosion. Thus,

$$E_0 = \frac{8\pi}{25}\rho_0 a^5 \int_0^1 \left\{ \frac{2}{\gamma-1}\Phi \frac{z^{2(\gamma-1)-3}}{[F(z)]^{2(\gamma-1)}}\frac{1}{[F'(z)]^{\gamma-1}} \right.$$

$$\left. +[F(z)-zF'(z)]^2 \right\} z^2 dz. \tag{3.3.22}$$

It may also be observed that (3.3.21) holds for each z sphere, that is, each x-sphere, $x = at^{2/5}z$, $0 < z < 1$. This constancy of energy in each x-sphere implies that the energy flowing into the sphere with the new material that enters it exactly balances the work which its original surface does by expanding against the surrounding pressure. This, in fact, is the energy principle and is equivalent to the equation of motion. The energy of the material entering the z-sphere, $x = at^{2/5}z$, in the time between t and $t + dt$ is

$$4\pi\rho_0 x^2 (dx)_t \epsilon = 4\pi\rho_0 \left(\frac{1}{\gamma-1}\frac{p}{\rho} + \frac{1}{2}u^2 \right) x^2(dx)_t$$

$$= 4\pi\rho_0 \left\{ \frac{1}{\gamma-1}\frac{(4/25)\Phi\rho_0 a^2 t^{-6/5}\left\{z^{2\gamma-3}/[F(z)]^{2\gamma}\right\}[F'(z)]^{-\gamma}}{\rho_0\left\{z^2/[F(z)]^2\right\}[F'(z)]^{-1}} \right.$$

$$\left. +\frac{1}{2}\frac{4}{25}a^2 t^{-6/5}[F(z)-zF'(z)]^2 \right\} a^2 t^{4/5} z^3 \frac{2}{5}at^{-3/5}dt$$

$$= \frac{16\pi}{125}\rho_0 a^5 t^{-1}\left\{ \frac{2}{\gamma-1}\Phi\frac{z^{2(\gamma-1)-3}}{[F(z)]^{2(\gamma-1)}}\frac{1}{[F'(z)]^{\gamma-1}} \right.$$

$$\left. +[F(z)-zF'(z)]^2 \right\} z^3 dt. \tag{3.3.23}$$

The work done by the original (z) surface against the surrounding pressure is $4\pi p X^2 u dt$. This is equal to

$$4\pi\frac{4}{25}\Phi\rho_0 a^2 t^{-6/5}\frac{z^{2\gamma-3}}{[F(z)]^{2\gamma}}\frac{1}{[F'(z)]^{\gamma}}a^2 t^{4/5}[F(z)]^2\frac{2}{5}at^{-3/5}[F(z)-zF'(z)]dt$$

$$= \frac{32}{125}\rho_0 a^5 t^{-1}\Phi\frac{z^{2\gamma-3}}{[F(z)]^{2(\gamma-1)}}\frac{1}{[F'(z)]^{\gamma}}[F(z)-zF'(z)]dt. \tag{3.3.24}$$

We have made use of (3.3.3), (3.3.4), (3.3.6) and (3.3.15) to arrive at (3.3.23) and (3.3.24). Equating (3.3.23) and (3.3.24) we get an ODE for the function $F(z)$:

$$\frac{2}{\gamma-1}\Phi\frac{z^{2(\gamma-1)}}{[F(z)]^{2(\gamma-1)}}\frac{1}{[F'(z)]^{\gamma-1}} + z^3[F(z)-zF'(z)]^2$$

$$= 2\Phi\frac{z^{2\gamma-3}}{[F(z)]^{2(\gamma-1)}[F'(z)]^{\gamma}}[F(z)-zF'(z)]. \tag{3.3.25}$$

Equation (3.3.25) must be solved subject to the conditions (3.3.8), (3.3.12), and (3.3.13); the constant a in the shock law (3.3.7) is obtained from (3.3.22).

Von Neumann (1941) changed (3.3.25) through the transformation

$$z = e^s, \tag{3.3.26}$$

$$F(z) = e^{\nu s}\phi(s), \tag{3.3.27}$$

to the autonomous form

$$\frac{2}{\gamma-1}\Phi\frac{1}{\phi^{2(\gamma-1)}\left(\frac{d\phi}{ds}+\nu\phi\right)^{\gamma-1}} + \left[\frac{d\phi}{ds}+(\nu-1)\phi\right]^2$$

$$= -2\Phi\frac{\frac{d\phi}{ds}+(\nu-1)\phi}{\phi^{2(\gamma-1)}\left(\frac{d\phi}{ds}+\nu\phi\right)^{\gamma}} \tag{3.3.28}$$

by choosing

$$\nu = \frac{3(\gamma-2)}{3\gamma-1}. \tag{3.3.29}$$

Now writing

$$\Psi = \frac{d\phi}{ds}+\nu\phi, \tag{3.3.30}$$

which is the same as

$$zF'(z) = e^{\nu s}\Psi(s), \tag{3.3.31}$$

(3.3.28) becomes an algebraic relation in the functions ϕ and Ψ:

$$(\Psi-\phi)^2 + 2\Phi\frac{\Psi-\phi}{\phi^{2(\gamma-1)}\Psi^{\gamma}} + \frac{2}{\gamma-1}\Phi\frac{1}{\phi^{2(\gamma-1)}\Psi^{\gamma-1}} = 0. \tag{3.3.32}$$

If we write

$$D = \frac{\gamma-1}{\gamma+1}, \tag{3.3.33}$$

then (see (3.3.17))

$$\Phi = \frac{2}{\gamma+1}\left(\frac{\gamma-1}{\gamma+1}\right)^{\gamma} = (1-D)D^{\gamma}, \tag{3.3.34}$$

$$\frac{2}{\gamma-1} = \frac{1-D}{D}, \tag{3.3.35}$$

and (3.3.32) can be written as

$$\left(\frac{\phi/\Psi-1}{1/D-1}\right)^2 - \frac{2}{\phi^{2(\gamma-1)}(\Psi/D)^{\gamma+1}}\frac{(\phi/\Psi-1)}{(1/D-1)}$$

$$+ \frac{1}{\phi^{2(\gamma-1)}(\Psi/D)^{\gamma+1}} = 0. \tag{3.3.36}$$

Now if we introduce the variables

$$\xi = \frac{\phi/\Psi - 1}{1/D - 1},$$ (3.3.37)

and

$$\eta = \phi^{2(\gamma-1)}(\Psi/D)^{\gamma+1},$$ (3.3.38)

(3.3.36) simply becomes

$$\xi^2 - 2\frac{\xi}{\eta} + \frac{1}{\eta} = 0$$ (3.3.39)

or

$$\eta = \frac{2\xi - 1}{\xi^2}.$$ (3.3.40)

If we further introduce a parameter θ via

$$\xi = \frac{1+\theta}{2},$$ (3.3.41)

then, from (3.3.40), we have

$$\eta = \frac{4\theta}{(1+\theta)^2}.$$ (3.3.42)

Now, using (3.3.37) and (3.3.38), we can write ϕ and Ψ in terms of the parameter θ:

$$\phi = \theta^{1/(3\gamma-1)}\left(\frac{\theta+1}{2}\right)^{-2/(3\gamma-1)}$$
$$\times \left(\frac{\theta+\gamma}{\gamma+1}\right)^{(\gamma+1)/(3\gamma-1)},$$ (3.3.43)

$$\Psi = \frac{\gamma-1}{\gamma+1}\theta^{1/(3\gamma-1)}\left(\frac{\theta+1}{2}\right)^{-2/(3\gamma-1)}$$
$$\times \left(\frac{\theta+\gamma}{\gamma+1}\right)^{-2(\gamma-1)/(3\gamma-1)}.$$ (3.3.44)

Since ϕ is positive in view of (3.3.3) and (3.3.27) and Ψ is positive since $F'(z)$ and ρ_0 are, it follows from (3.3.38) that η is positive. Equation (3.3.42) then implies that θ is positive.

Thus, we have

$$\theta > 0.$$ (3.3.45)

From (3.3.30), we have

$$\frac{d\phi}{ds} = \Psi - \nu\phi.$$ (3.3.46)

Therefore,

$$s = \int \frac{d\phi}{\Psi - \nu\phi} = \int \frac{d\phi/\phi}{\Psi/\phi - \nu},$$

which, on using (3.3.43) and (3.3.44), becomes

$$s = \int \frac{\frac{1}{3\gamma-1}\frac{d\theta}{\theta} - \frac{2}{3\gamma-1}\frac{d\theta}{\theta+1} + \frac{\gamma+1}{3\gamma-1}\frac{d\theta}{\theta+\gamma}}{\frac{\gamma-1}{\theta+\gamma} + \frac{3(2-\gamma)}{3\gamma-1}} \qquad (3.3.47)$$

or

$$s = C_1 + \frac{\gamma}{2\gamma+1}\ln\theta - \frac{2}{5}\ln(\theta+1)$$
$$+ \frac{13\gamma^2 - 7\gamma + 12}{15(2-\gamma)(2\gamma+1)}\ln[3(2-\gamma)\theta + (2\gamma+1)]. \qquad (3.3.48)$$

To get the constant C_1, we use the boundary conditions $F(z) = 1$, $F'(z) = (\gamma-1)/(\gamma+1)$ at $z = 1$ (see (3.3.12)–(3.3.13)). Since $z = 1$ corresponds to $s = 0$, we have $\phi = 1$ and $\Psi = (\gamma-1)/(\gamma+1)$ at $s = 0$ (see (3.3.27) and (3.3.31)). Using these values at $s = 0$ in (3.3.43) and (3.3.44), we find that $s = 0$ corresponds to $\theta = 1$. This determines C_1 in (3.3.48). Therefore, we have

$$s = \frac{\gamma}{2\gamma+1}\ln\theta - \frac{2}{5}\ln\frac{\theta+1}{2}$$
$$+ \frac{13\gamma^2 - 7\gamma + 12}{15(2-\gamma)(2\gamma+1)}\ln\frac{3(2-\gamma)\theta + (2\gamma+1)}{7-\gamma}. \qquad (3.3.49)$$

The similarity variable z may now be written in terms of θ as

$$z = e^s = \theta^{\gamma/(2\gamma+1)}\left(\frac{\theta+1}{2}\right)^{-2/5}\left[\frac{3(2-\gamma)\theta + (2\gamma+1)}{7-\gamma}\right]^{\frac{13\gamma^2-7\gamma+12}{15(2-\gamma)(2\gamma+1)}}. \qquad (3.3.50)$$

The similarity function $F(z)$, in view of (3.3.27) and (3.3.43), becomes

$$F(z) = e^{\nu s}\phi(s)$$
$$= \theta^{(\gamma-1)/(2\gamma+1)}\left(\frac{\theta+1}{2}\right)^{-2/5}\left(\frac{\theta+\gamma}{\gamma+1}\right)^{(\gamma+1)/(3\gamma-1)}$$
$$\times \left[\frac{3(2-\gamma)\theta + (2\gamma+1)}{7-\gamma}\right]^{-\frac{13\gamma^2-7\gamma+12}{5(2\gamma+1)(3\gamma-1)}}. \qquad (3.3.51)$$

Equation (3.3.51) shows that the condition $F(z) \to 0$ as $z \to 0$, or as $\theta \to 0$, is automatically satisfied. Equations (3.3.50) and (3.3.51) together constitute the parametric form of the solution for the function $F(z)$ in the interval

$$0 < z \le 1 \quad (0 < x \le \Sigma), \qquad (3.3.52)$$

which corresponds to the interval $0 < \theta \le 1$.

To write the final solution for the physical variables in terms of θ, we also need to find $F(z) - zF'(z)$. It is easily checked from (3.3.43) and (3.3.44) that

$$\frac{F(z)}{zF'(z)} - 1 = \frac{\phi}{\Psi} - 1 = \frac{\theta + \gamma}{\gamma - 1} - 1 = \frac{\theta + 1}{\gamma - 1}. \tag{3.3.53}$$

Using (3.3.50)–(3.3.51) in (3.3.53) we have

$$
\begin{aligned}
F(z) - zF'(z) &= e^{\nu s}(\phi - \Psi) \\
&= \frac{2}{\gamma + 1}\theta^{(\gamma-1)/(2\gamma+1)}\left(\frac{\theta + 1}{2}\right)^{3/5} \\
&\quad \times \left(\frac{\theta + \gamma}{\gamma + 1}\right)^{-2(\gamma-1)/(3\gamma-1)} \\
&\quad \times \left[\frac{3(2-\gamma)\theta + (2\gamma+1)}{7-\gamma}\right]^{-\frac{13\gamma^2-7\gamma+12}{5(2\gamma+1)(3\gamma-1)}}.
\end{aligned}
\tag{3.3.54}
$$

We can finally write the solution x, X, ρ, u and p from (3.3.4), (3.3.3), (3.3.5), (3.3.6), and (3.3.15) in terms of t and the parameter θ:

$$
\begin{aligned}
x &= at^{2/5}\theta^{\gamma/(2\gamma+1)}\left(\frac{\theta+1}{2}\right)^{-2/5} \\
&\quad \times \left[\frac{3(2-\gamma)\theta + (2\gamma+1)}{7-\gamma}\right]^{\frac{13\gamma^2-7\gamma+12}{15(2-\gamma)(2\gamma+1)}},
\end{aligned}
\tag{3.3.55}
$$

$$
\begin{aligned}
X &= at^{2/5}\theta^{(\gamma-1)/(2\gamma+1)}\left(\frac{\theta+1}{2}\right)^{-2/5} \\
&\quad \times \left(\frac{\theta+\gamma}{\gamma+1}\right)^{(\gamma+1)/(3\gamma-1)} \\
&\quad \times \left[\frac{3(2-\gamma)\theta + (2\gamma+1)}{7-\gamma}\right]^{-\frac{13\gamma^2-7\gamma+12}{5(2\gamma+1)(3\gamma-1)}},
\end{aligned}
\tag{3.3.56}
$$

$$
\begin{aligned}
\rho &= \frac{\gamma+1}{\gamma-1}\rho_0\theta^{3/(2\gamma+1)}\left(\frac{\theta+\gamma}{\gamma+1}\right)^{-4/(3\gamma-1)} \\
&\quad \times \left[\frac{3(2-\gamma)\theta + 2\gamma+1}{7-\gamma}\right]^{\frac{13\gamma^2-7\gamma+12}{(2-\gamma)(2\gamma+1)(3\gamma-1)}},
\end{aligned}
\tag{3.3.57}
$$

$$
\begin{aligned}
u &= \frac{4}{5(\gamma+1)}at^{-3/5}\theta^{(\gamma-1)/(2\gamma+1)}\left(\frac{\theta+1}{2}\right)^{3/5} \\
&\quad \times \left(\frac{\theta+\gamma}{\gamma+1}\right)^{-2(\gamma-1)/(3\gamma-1)} \\
&\quad \times \left[\frac{3(2-\gamma)\theta + 2\gamma+1}{7-\gamma}\right]^{-\frac{13\gamma^2-7\gamma+12}{5(2\gamma+1)(3\gamma-1)}},
\end{aligned}
\tag{3.3.58}
$$

$$p = \frac{8}{25(\gamma+1)}\rho_0 a^2 t^{-6/5}\left(\frac{\theta+1}{2}\right)^{6/5}$$

$$\times \left(\frac{\theta+\gamma}{\gamma+1}\right)^{-4\gamma/(3\gamma-1)}$$

$$\times \left[\frac{3(2-\gamma)\theta+2\gamma+1}{7-\gamma}\right]^{\frac{13\gamma^2-7\gamma+12}{5(2-\gamma)(3\gamma-1)}}. \tag{3.3.59}$$

Besides,

$$\epsilon_i = \frac{1}{\gamma-1}\frac{p}{\rho}$$

$$= \frac{8}{25(\gamma+1)^2}a^2 t^{-6/5}\theta^{-3/(2\gamma+1)}\left(\frac{\theta+1}{2}\right)^{6/5}$$

$$\times \left(\frac{\theta+\gamma}{\gamma+1}\right)^{-4(\gamma-1)/(3\gamma-1)}$$

$$\times \left[\frac{3(2-\gamma)\theta+(2\gamma+1)}{7-\gamma}\right]^{-\frac{2(13\gamma^2-7\gamma+12)}{5(2\gamma+1)(3\gamma-1)}}, \tag{3.3.60}$$

$$\epsilon_c = \frac{1}{2}u^2$$

$$= \frac{8}{25(\gamma+1)^2}a^2 t^{-6/5}\theta^{2(\gamma-1)/(2\gamma+1)}\left(\frac{\theta+1}{2}\right)^{6/5}$$

$$\times \left(\frac{\theta+\gamma}{\gamma+1}\right)^{-4(\gamma-1)/(3\gamma-1)}$$

$$\times \left[\frac{3(2-\gamma)\theta+2\gamma+1}{7-\gamma}\right]^{-\frac{2}{5}\frac{(13\gamma^2-7\gamma+12)}{5(2\gamma+1)(3\gamma-1)}}, \tag{3.3.61}$$

leading to the excellent physical interpretation for the parameter θ as the ratio of kinetic and internal energies:

$$\frac{\epsilon_c}{\epsilon_i} = \theta. \tag{3.3.62}$$

Using this relation, we write the total explosion energy in terms of ϵ_c' and ϵ_i', the internal and kinetic energies per unit volume:

$$E_0 = \int_0^\Sigma (\epsilon_i' + \epsilon_c')4\pi X^2 dX$$

$$= 4\pi \int_0^\Sigma (\theta+1)\epsilon_i' X^2 dX$$

$$= 8\pi\Sigma^3 \int_0^1 \frac{\theta+1}{2}\epsilon_i'[F(z)]^2 dF(z). \tag{3.3.63}$$

We may note that $\epsilon_c'/\epsilon_i' = \epsilon_c/\epsilon_i = \theta$. If we observe that $\epsilon_i' = \rho\epsilon_i = p/(\gamma-1)$, then, using (3.3.59), we have

$$E_0 = K\rho_0 a^5, \tag{3.3.64}$$

where

$$K = \frac{64\pi}{75(\gamma^2-1)} \int_0^1 \left(\frac{\theta+1}{2}\right)^{11/5} \left(\frac{\theta+\gamma}{\gamma+1}\right)^{-4\gamma/(3\gamma-1)}$$

$$\times \left[\frac{3(2-\gamma)\theta+(2\gamma+1)}{7-\gamma}\right]^{\frac{13\gamma^2-7\gamma+12}{5(2-\gamma)(3\gamma-1)}} d(F^3). \tag{3.3.65}$$

Here, F^3 may be expressed in terms of θ via (3.3.51). The constant a in the shock law (3.3.1) has now been determined. This solution for the blast wave problem in Lagrangian co-ordinates is explicit in terms of the parameter θ, which itself has now been physically interpreted. This is in contrast to the exact implicit solution of the Eulerian equations of motion found by J.L. Taylor (1955) and Sedov (1946) (see section 3.1). It may be noted, however, that this solution does not hold for $\gamma=7$, the exponent corresponding to water. The solution (3.3.55)–(3.3.65) can be approximated and written out more explicitly when $\gamma-1 \neq 0$ is small. Retaining the most dominant terms, this approximate solution is found to be

$$x = at^{2/5}\theta^{1/3+(\gamma-1)/9}, \tag{3.3.66}$$

$$X = at^{2/5}\theta^{(\gamma-1)/3}, \tag{3.3.67}$$

$$\rho = \frac{2}{\gamma-1}\rho_0\theta^{1-2(\gamma-1)/3}\frac{\theta+1}{2}, \tag{3.3.68}$$

$$u = \frac{2}{5}at^{-3/5}\theta^{(\gamma-1)/3}, \tag{3.3.69}$$

$$p = \frac{4}{25}\rho_0 a^2 t^{-6/5}\frac{\theta+1}{2}, \tag{3.3.70}$$

$$\epsilon_i = \frac{2}{25}a^2 t^{-6/5}\theta^{-1+2(\gamma-1)/3}, \tag{3.3.71}$$

$$\epsilon_c = \frac{2}{25}a^2 t^{-6/5}\theta^{2(\gamma-1)/3}. \tag{3.3.72}$$

The form (3.3.55)–(3.3.65) of the solution is also not valid for $\gamma=2$ in view of the exponent $1/(2-\gamma)$ in the last factor in the products for the solution x, ρ, and p, namely,

$$\left[\frac{3(2-\gamma)\theta+(2\gamma+1)}{7-\gamma}\right]^{1/(2-\gamma)}. \tag{3.3.73}$$

This expression may be written as

$$\left[1-(2-\gamma)\frac{3}{7-\gamma}(1-\theta)\right]^{1/(2-\gamma)} \tag{3.3.74}$$

and hence, in the limit $\gamma \to 2$, approximated by $e^{-(3/5)(1-\theta)}$. Therefore, the last factor in each of (3.3.55)–(3.3.61) must be replaced in this limit by $e^{-(2/5)(1-\theta)}, 1, e^{-(6/5)(1-\theta)}, 1, e^{-(6/5)(1-\theta)}, 1$, and 1, respectively. Other factors may be evaluated by setting $\gamma = 2$.

The important features of the solution observed by Taylor (1950) are confirmed here. Most of the material in the blast wave gets accumulated near the shock. This tendency becomes more pronounced as γ tends to 1. This is the basis of an analytic theory of point explosion in an exponential atmosphere, proposed by Laumbach and Probstein (1969), which will be treated in the next section. The density vanishes at the center of explosion, but the pressure tends to $p(0)$ where $0 < p(0) < \infty$. As γ increases from 1 to 2, the pressure ratio $p(0)/p_{shock}$ changes from 1/2 to about 1/4. The pressure changes more rapidly near the shock and becomes almost constant in the region close to the center where density is close to zero. Also, since $\rho \to 0$ and $p \to p(0)$ as the center is approached, the temperture $T(0)$ tends to infinity. Equations (3.3.60) and (3.3.61) show that $\epsilon_i \to \infty$ and $\epsilon_c \to 0$ as the center of the blast is approached.

3.4   Point Explosion in an Exponential Atmosphere

We may continue the Lagrangian co-ordinate approach of section 3.3 to the investigation of a more realistic blast wave model in a cold stratified atmosphere with density varying exponentially with altitude. It is also assumed that the flow field is locally radial (see Figure 3.6), that is, we may neglect

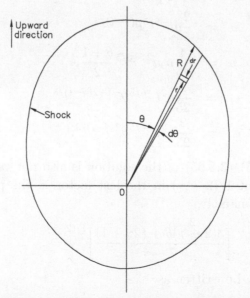

Figure 3.6 The shock envelope at a given time after explosion along with the polar co-ordinate system (Laumbach and Probstein, 1969).

gradients in the θ direction where θ is the polar angle measured from the vertical. This assumption implies that the streamlines from the origin are straight lines; it may fail to hold as the shock ascends to 4–5 scale heights and becomes increasingly asymmetric. It is also assumed that the shock produced by the blast is strong; this assumption also becomes invalid by the time the shock has propagated to 4–6 scale heights, except for $\theta < \pi/4$, even for large energy sources.

Both Taylor (1950) and Von Neumann (1941) pointed out that much of the mass of the blast wave gets concentrated in a small neighbourhood behind the shock. This is the basis of the analytic theory developed by Laumbach and Probstein (1969), which we detail in this section. They attempted to obtain an analytic solution valid for all time; they also deduced scaling laws based on this solution. They compared their analytic results with the numerical solution of Troutman and Davis (1965) and found a good agreement. Laumbach and Probstein (1969) also compared the far field results accruing from their analysis with the asymptotic limits of Raizer (1964) and Hayes (1968a, 1968b) and found favourable agreement.

With the assumptions referred to above, the flow is axisymmetric about the vertical axis through the energy release point, called the origin in Figure 3.6.

We denote the Eulerian co-ordinate of a fluid particle of thickness dr by r. The position of the shock front at a given polar angle θ is denoted by $R(t,\theta)$. The undisturbed density distribution is assumed to be of the form

$$\rho_0 = \rho_B \exp[-(r_0/\Delta)\cos\theta], \qquad (3.4.1)$$

where Δ is the scale height and r_0 the Lagrangian co-ordinate of a particular fluid particle at the burst time $t = 0$; ρ_B is the density at $r_0 = 0$.

Assuming local radiality, the equation of continuity for any polar angle is given by

$$\rho_0 r_0^2 dr_0 = \rho r^2 dr. \qquad (3.4.2)$$

The radial momentum equation in the Lagrangian co-ordinate, on using (3.4.2), becomes

$$\frac{\partial^2 r}{\partial t^2} + \frac{r^2}{\rho_0 r_0^2}\frac{\partial p}{\partial r_0} = 0. \qquad (3.4.3)$$

Since the entropy of a particle after crossing the shock remains constant, we have the energy equation

$$\frac{p(r_0,t)}{p_s(r_0)} = \left[\frac{\rho(r_0,t)}{\rho_s(r_0)}\right]^\gamma, \qquad (3.4.4)$$

where the subscript s denotes conditions immediately behind the shock.

The shock is assumed to be strong. The RH conditions holding across it are

$$\rho_s = \frac{\gamma+1}{\gamma-1}\rho_0,$$ (3.4.5)

$$p_s = \frac{2}{\gamma+1}\rho_0\dot{R}^2,$$ (3.4.6)

with the dot above R denoting differentiation with respect to t.

Since a finite characteristic length Δ appears in the problem, it is not self-similar (see, however, section 3.5 and reference to the work of Hayes (1968) therein).

Integrating (3.4.3) with respect to r_0, we have

$$p(r_0,t;\theta) - p_s(R;\theta) = \int_{r_0}^{R} \frac{1}{r^2}\frac{\partial^2 r}{\partial t^2}\rho_0\bar{r}_0^2 d\bar{r}_0,$$ (3.4.7)

where $p_s(R;\theta)$ is the pressure immediately behind the shock. For a given polar angle, the mass contained within a differential solid angle is constant, so we may write the energy equation for a given solid angle as

$$\frac{E}{4\pi} = \int_0^R \frac{p}{\gamma-1}r^2 dr + \int_0^R \frac{1}{2}\left(\frac{\partial r}{\partial t}\right)^2 \rho_0 r_0^2 dr_0.$$ (3.4.8)

The first integral in (3.4.8) is the internal energy per unit solid angle and, for later use, is written in terms of the Eulerian co-ordinate. The second term in (3.4.8) is the kinetic energy per unit solid angle; E is the total energy of the flow and is assumed to be known and constant; real gas effects and radiative transfer etc. have been ignored.

We expand the Eulerian co-ordinate about the shock,

$$r(r_0,t) = R + \left.\frac{\partial r}{\partial r_0}\right|_R (r_0 - R) + \frac{1}{2}\left.\frac{\partial^2 r}{\partial r_0^2}\right|_R (r_0 - R)^2 + \cdots,$$ (3.4.9)

with the expansion parameterised in t through the Taylor coefficients and R. Implicit in (3.4.9) is the assumption that most of the mass is concentrated near the shock front so that $(r_0 - R) << R$ for $r_0 \sim R$. Only terms up to $(r_0 - R)^2$ are retained in the expansion (3.4.9). Using (3.4.2) and the shock condition (3.4.5), we have

$$\left.\frac{\partial r}{\partial r_0}\right|_R = \left.\frac{\rho_0 r_0^2}{\rho r^2}\right|_R = \frac{\gamma-1}{\gamma+1}.$$ (3.4.10)

From (3.4.9)–(3.4.10), we find by differentiation etc. that

$$r = R + \frac{\gamma-1}{\gamma+1}(r_0 - R) + \frac{1}{2}\left.\frac{\partial^2 r}{\partial r_0^2}\right|_R (r_0 - R)^2,$$ (3.4.11)

$$\frac{\partial r}{\partial t} = \frac{2}{\gamma+1}\dot{R} - \left.\frac{\partial^2 r}{\partial r_0^2}\right|_R (r_0 - R)\dot{R},$$ (3.4.12)

$$\frac{\partial^2 r}{\partial t^2} = \frac{2}{\gamma+1}\ddot{R} + \left.\frac{\partial^2 r}{\partial r_0^2}\right|_R \dot{R}^2 - \left.\frac{\partial^2 r}{\partial r_0^2}\right|_R (r_0 - R)\ddot{R}.$$ (3.4.13)

Therefore, at the shock front, we have

$$r_s = R, \tag{3.4.14}$$

$$\left(\frac{\partial r}{\partial t}\right)_s = \frac{2}{\gamma+1}\dot{R}. \tag{3.4.15}$$

The evaluation of acceleration at the shock requires some calculation. Eliminating $\partial^2 r/\partial t^2$ from the radial momentum equation (3.4.3) and (3.4.13) evaluated at the shock, we get

$$-\frac{\partial^2 r}{\partial r_0^2}\bigg|_R \dot{R}^2 = \frac{2}{\gamma+1}\dot{R}^2\left(\frac{1}{p}\frac{\partial p}{\partial r_0}\right)_R + \frac{2}{\gamma+1}\ddot{R}, \tag{3.4.16}$$

where use has been made of (3.4.6). The pressure gradient term in (3.4.16) must now be found. We differentiate (3.4.4) to obtain

$$\frac{1}{p}\frac{\partial p}{\partial r_0}\bigg|_R = \left(\frac{1}{p_s}\frac{\partial p_s}{\partial r_0} + \frac{\gamma}{\rho}\frac{\partial \rho}{\partial r_0} - \frac{\gamma}{\rho_0}\frac{\partial \rho_0}{\partial r_0}\right)_R. \tag{3.4.17}$$

Using the Rankine-Hugoniot condition (3.4.6), along with (3.4.1) in differentiated form, we obtain

$$\frac{1}{p_s}\frac{\partial p_s}{\partial r_0}\bigg|_R = \frac{1}{\rho_0}\frac{\partial \rho_0}{\partial r_0}\bigg|_R + \frac{2}{\dot{R}}\ddot{R}, \tag{3.4.18}$$

$$\frac{1}{\rho_0}\frac{\partial \rho_0}{\partial r_0}\bigg|_R = -\frac{\cos\theta}{\Delta}. \tag{3.4.19}$$

We still need to find $(\partial\rho/\partial r_0)/\rho$. This is accomplished by taking a logarithmic derivative of (3.4.2) with respect to r_0 and using (3.4.10) at the shock:

$$\frac{1}{\rho}\frac{\partial \rho}{\partial r_0}\bigg|_R = \frac{1}{\rho_0}\frac{\partial \rho}{\partial r_0}\bigg|_R + \frac{4}{\gamma+1}\frac{1}{R} - \frac{\gamma+1}{\gamma-1}\frac{\partial^2 r}{\partial r_0^2}\bigg|_R. \tag{3.4.20}$$

Making use of (3.4.18)–(3.4.20) in (3.4.17) and using the resulting expression in (3.4.16), we determine $(\partial^2 r/\partial r_0^2)_R \dot{R}^2$. Finally, substituting this term in (3.4.13), we determine the acceleration at the shock:

$$\left(\frac{\partial^2 r}{\partial t^2}\right)_s = \frac{4(2\gamma-1)}{(\gamma+1)^2}\ddot{R} - \frac{2(\gamma-1)}{(\gamma+1)^2}\frac{\cos\theta}{\Delta}\dot{R}^2 + \frac{8\gamma(\gamma-1)}{(\gamma+1)^3}\frac{\dot{R}^2}{R}. \tag{3.4.21}$$

If we substitute the expressions for r_s and $(\partial^2 r/\partial t^2)_s$ at the shock from (3.4.14) and (3.4.21) in the integral term in (3.4.7), use (3.4.1) for ρ_0, and carry out the integration with respect to r_0, we obtain

$$p - p_s = \left(\frac{2}{\gamma+1}\right)^2\left(\frac{\Delta}{\cos\theta}\right)^2\frac{\rho_B}{\eta^3}$$

$$\times \left\{ 2(2\gamma - 1)\eta\ddot{\eta} - (\gamma - 1)\eta\dot{\eta}^2 + \frac{4\gamma(\gamma - 1)}{\gamma + 1}\dot{\eta}^2 \right\}$$

$$\times \left\{ \exp(-\frac{r_0}{R}\eta)\left[\left(\frac{r_0}{R}\right)^2\frac{\eta^2}{2} + \left(\frac{r_0}{R}\right)\eta + 1 \right] \right.$$

$$\left. - e^{-\eta}\left[\frac{\eta^2}{2} + \eta + 1 \right] \right\},$$

(3.4.22)

where

$$\eta = (R/\Delta)\cos\theta. \tag{3.4.23}$$

The pressure immediately behind the shock is found in terms of η from (3.4.1) and (3.4.6):

$$p_s = \frac{2\rho_B}{\gamma + 1}\left(\frac{\Delta}{\cos\theta}\right)^2\dot{\eta}^2 e^{-\eta}. \tag{3.4.24}$$

The integrals in (3.4.8) are evaluated by making use again of the basic assumption in the analysis: to the present approximation, r is not a function of r_0 and, therefore, for all r different from R, we may replace r_0 by zero. This implies that all the mass is pulled forward behind the shock, and the only mass that remains inside is that which existed in the vicinity of the origin $r_0 = 0$. Thus, we put $p(r,t) = p(0,t)$ and $r_0 = 0$ in (3.4.22) and hence evaluate the first integral in (3.4.8). The second integral, namely the kinetic energy integral, is evaluated by writing ρ_0 from (3.4.1) and $(\partial r/\partial t)_s$ at the shock from (3.4.15). Carrying out the integration with respect to r, we arrive at the following ODE for η (see (3.4.23)) as a function of time:

$$f(\eta)\ddot{\eta} + g(\eta)\dot{\eta}^2 = \frac{E}{4\pi\rho_B}\left(\frac{\cos\theta}{\Delta}\right)^5, \tag{3.4.25}$$

where

$$f(\eta) = \frac{8}{3}\frac{(2\gamma - 1)\eta}{(\gamma - 1)(\gamma + 1)^2}\left[1 - e^{-\eta}\left(\frac{1}{2}\eta^2 + \eta + 1\right) \right], \tag{3.4.26}$$

$$g(\eta) = \frac{2}{3}\frac{\eta^3 e^{-\eta}}{(\gamma - 1)(\gamma + 1)} + \frac{\gamma - 1}{2(2\gamma - 1)}\left[\frac{7\gamma + 3}{(\gamma + 1)\eta} - 1 \right]f(\eta). \tag{3.4.27}$$

If we change the variable t to

$$t^* = t\left[\frac{E|\cos^5\theta|}{4\pi\rho_B\Delta^5} \right]^{1/2} \tag{3.4.28}$$

in (3.4.25), we get

$$f(\eta)\eta'' + g(\eta){\eta'}^2 = 1 \quad \text{for} \quad 0 \le \theta \le \pi/2, \tag{3.4.29}$$

$$f(\eta)\eta'' + g(\eta)\eta'^2 = -1 \quad \text{for} \quad \pi/2 \le \theta \le \pi, \tag{3.4.30}$$

where now the prime denotes differentiation with respect to t^*. The above scaling has the effect that all motions for $\theta \le \pi/2$ can be obtained from (3.4.29), while those for $\theta > \pi/2$ can be found from (3.4.30). Either solution describes the flow for arbitrary values of E, Δ, and ρ_B for appropriate θ values. The autonomous equations (3.4.29) and (3.4.30) can be changed to linear first order equations in η'^2, namely,

$$\frac{f(\eta)}{2}\frac{d\eta'^2}{d\eta} + g(\eta)\eta'^2 = \pm 1, \tag{3.4.31}$$

and integrated to yield

$$\eta'^2 = \pm 2\exp\left\{-2\int_a^\eta \frac{g(z)}{f(z)}dz\right\}\int_0^\eta \exp\left\{2\int_a^y \frac{g(z)}{f(z)}dz\right\}\frac{dy}{f(y)}, \tag{3.4.32}$$

where a is a zero of the indicated integrals and $\gamma > 1$. The $+$ and $-$ signs in (3.4.32) refer to upward and downward directions, respectively. The constant of integration in (3.4.32) is put equal to zero to satisfy the condition that, as $\Delta \to \infty$ corresponding to $\eta \to 0$, the solution tends to a uniform density solution (see below). The solution (3.4.32) should be supplemented by the equation

$$t^* = \int_0^\eta \frac{d\bar{\eta}}{\bar{\eta}'} \tag{3.4.33}$$

to relate it to the variable t^*. With η and η' thus found, the pressure behind the shock may be found from (3.4.22) and (3.4.24). The density as a function of r_0 and t may now be obtained from the entropy equation (3.4.4). The Eulerian co-ordinate itself is found from (3.4.2) as

$$\frac{r}{R} = \left[3\int_0^{r_0/R}(\rho_0/\rho)\left(\frac{r_0}{R}\right)^2 d\left(\frac{r_0}{R}\right)\right]^{1/3}. \tag{3.4.34}$$

The relation (3.4.34) helps find the pressure and density behind the shock as functions of the Eulerian co-ordinate r. If we let $\eta = R\cos\theta/\Delta \to 0$ (as $\Delta \to \infty$) in (3.4.25), we recover the limiting case of the atmosphere with uniform density:

$$\ddot{R} + \frac{\gamma(5\gamma+1)}{(2\gamma-1)(\gamma+1)}\frac{\dot{R}^2}{R} = \frac{9(\gamma-1)(\gamma+1)^2}{16(2\gamma-1)}\frac{E}{\pi\rho_B}\frac{1}{R^4}. \tag{3.4.35}$$

Equation (3.4.35) is again autonomous. Its first integral may be found to be

$$\dot{R}^2 = \frac{9}{8}\frac{E}{\pi\rho_B}\frac{(\gamma-1)(\gamma+1)^3}{(4\gamma^2-\gamma+3)}\frac{1}{R^3}. \tag{3.4.36}$$

Integrating (3.4.36) we obtain

$$R = \left[\frac{225}{32}\frac{E}{\pi\rho_B}\frac{(\gamma-1)(\gamma+1)^3}{(4\gamma^2-\gamma+3)}\right]^{1/5} t^{2/5}. \tag{3.4.37}$$

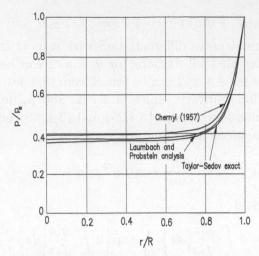

Figure 3.7 Pressure distribution in Eulerian co-ordinates for uniform density atmosphere and $\gamma = 1.4$ (Laumbach and Probstein, 1969).

The $t^{2/5}$ law is thus recovered. Laumbach and Probstein (1969) show that (3.4.37) differs slightly from the exact Taylor-Sedov solution (see Figure 3.7 for the distribution of pressure). The pressure at the center for $\gamma = 1.4$ is 0.37, which compares well with the exact value 0.366 of Taylor (1950). The discrepancy in R is 1.8% for $\gamma = 1.2$ and 2.3% for $\gamma = 1.4$.

As Laumbach and Probstein (1969) point out, due to the severe assumptions made in the analysis, any exact far field or large time results should not necessarily agree with those of their analysis; they nevertheless consider asymptotic forms of their solution both in upward and downward directions and deduce some qualitative features.

In the downward direction, as $-\eta$ becomes large, the asymptotic form of (3.4.25) is found to be

$$\ddot{\eta} - \frac{\gamma}{2\gamma - 1}\dot{\eta}^2 = \frac{3E(\gamma - 1)(\gamma + 1)^2}{16\pi\rho_B(2\gamma - 1)}\left(\frac{\cos\theta}{\Delta}\right)^5 \frac{e^\eta}{\eta^3}. \tag{3.4.38}$$

Writing (3.4.38) as a first order ODE in $\dot{\eta}^2$ and integrating we have

$$\dot{\eta}^2 \sim -\frac{3E(\gamma - 1)(\gamma + 1)^2}{8\pi\rho_B}\left(\frac{\cos\theta}{\Delta}\right)^5 \frac{e^\eta}{\eta^3}\left[1 + O\left(\frac{1}{\eta}\right)\right]$$

$$+ k_1 \exp\left(\frac{2\gamma}{2\gamma - 1}\eta\right), \tag{3.4.39}$$

where k_1 is constant of integration. It is clear that, for $\gamma > 1$ (and η negative), the term involving k_1 is small in comparison with the energy-dependent term so that the aymptotic solution becomes

$$\dot{\eta}^2 \sim -\frac{3E(\gamma - 1)(\gamma + 1)^2}{8\pi\rho_B}\left(\frac{\cos\theta}{\Delta}\right)^5 \frac{e^\eta}{\eta^3}\left[1 + O\left(\frac{1}{\eta}\right)\right]. \tag{3.4.40}$$

Integration of (3.4.40) yields the dominant term for η:

$$\eta = (R/\Delta)\cos\theta \sim -2\ln t. \qquad (3.4.41)$$

Therefore,

$$\dot{R}|\cos\theta| \sim \alpha\Delta/t = 2\Delta/t. \qquad (3.4.42)$$

The form (3.4.42), rather surprisingly, agrees with the (exact) asymptotic self-similar solution obtained by Hayes (1968a) for a plane shock travelling downward, with the assumption that the total energy of the blast is conserved. The coefficient $\alpha = 2$ is found to be independent of γ. The explanation for this agreement is that, under the assumptions made here, the cross-sectional area of the flow increases as R^2 (local radiality), therefore the far field results of the present analysis are appropriately compared with exact asymptotic plane shock solutions. The constant energy constraint is common to both the analyses.

In the upward direction, as η becomes large, the asymptotic form of (3.4.25) is

$$\ddot{\eta} - \frac{\gamma - 1}{2(2\gamma - 1)}\dot{\eta}^2 = \frac{3E(\gamma - 1)(\gamma + 1)^2}{32\pi\rho_B(2\gamma - 1)\eta}\left(\frac{\cos\theta}{\Delta}\right)^5. \qquad (3.4.43)$$

Integrating (3.4.43) we have

$$\dot{\eta}^2 \sim -\frac{3E}{16\pi\rho_B}\frac{(\gamma + 1)^2}{\eta}\left(\frac{\cos\theta}{\Delta}\right)^2\left[1 + O\left(\frac{1}{\eta}\right)\right] + k_2\exp\left(\frac{\gamma - 1}{2\gamma - 1}\eta\right), \qquad (3.4.44)$$

where k_2 is a constant of integration. In the present case the first term containing the energy is asymptotically small compared to the second term; therefore, we have

$$\dot{\eta}^2 \sim k_2\exp\left(\frac{\gamma - 1}{2\gamma - 1}\eta\right). \qquad (3.4.45)$$

An integration of (3.4.45) gives

$$\eta = \frac{R}{\Delta}\cos\theta \sim -\frac{2(2\gamma - 1)}{\gamma - 1}\ln\left[\frac{(\gamma - 1)k_2^{1/2}}{2(2\gamma - 1)}(\tau - t)\right], \qquad (3.4.46)$$

where τ is a constant of integration. Laumbach and Probstein (1969) compared (3.4.46) with the asymptotic results of Hayes (1968a) and Raizer (1964) and found it in good qualitative agreement. Now we compare the results obtained by the approximate theory of Laumbach and Probstein with the numerical solution found by Troutman and Davis (1965) for the vertically ascending parts of the shock wave for $\gamma = 1.4$. The latter investigators obtained their solution for a specific energy yield, scale height and atmospheric density at the burst point, which were appropriately scaled out by Laumbach and Probstein (1969) using (3.4.23) and (3.4.28). The calculations were carried out up to the time the upper part of the shock travelled to three scale heights from the origin of the blast wave. Although Troutman and Davis (1965) had performed their calculations with $\theta = 0$ and $\theta = \pi$, and these were in a sense one-dimensional, comparison of these results with the full two-dimensional calculations showed little difference over the ranges of parameters considered by them.

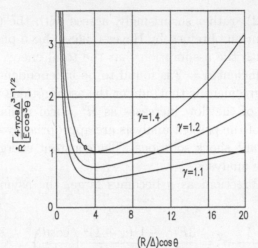

Figure 3.8 Shock velocity of the ascending shock as a function of shock position for $\gamma = 1.1, 1.2, 1.4$ (Laumbach and Probstein, 1969).

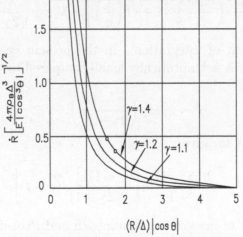

Figure 3.9 Shock velocity of the descending shock as a function of shock position for $\gamma = 1.1, 1.2, 1.4$ (Laumbach and Probstein, 1969).

Figure 3.8 for the shock velocity of the ascending shock shows that it decreases to some minimum value beyond which it accelerates to infinity in a finite time. This happens because the decreasing density ahead of the shock begins to affect the motion of the shock appreciably. On the other hand, the descending shock does not experience any such theoretical limit beyond which it cannot remain strong provided there is a sufficiently large energy source (see Figure 3.9). The behaviour of the ascending part of the shock wave is in conformity with the earlier studies of Kompaneets (1960) and Andriankin et al. (1962).

As pointed out earlier, the analysis of Laumbach and Probstein (1969) works for ascending shock waves for 2 to 3 atmospheric scale heights. At later times, the error involved in the truncation in the series solution discussed here becomes more pronounced, leading eventually to an overestimate for the blow time.

In a later study, Bach, Kuhl and Oppenheim (1975) carried out a perturbation analysis similar to that of Sakurai (1953) (see section 3.2) for the strong blast wave in an exponential medium, which was found to be accurate even when the front had progressed to ten scale heights. They also found a similarity solution of the type found by Raizer (1964) and Hayes (1968a, 1968b), which describes the far field. The near field and far field solutions, it is claimed, matched so well that the extremely difficult analysis of the intermediate regime was deemed unnecessary.

3.5 Asymptotic Behaviour of Blast Waves at a High Altitude

We consider a stratified (plane) atmosphere with the density distribution

$$\rho_0 = \rho_* e^{x/\Delta}, \tag{3.5.1}$$

where Δ is the scale height. The initial pressure is assumed to be zero. An explosion takes place in the high reaches ($x \approx -\infty$) where the density is almost zero, and a (plane) shock propagates downward in the direction of increasing density (the positive x-axis is in the downward direction). The heated gas overtaken by the shock expands into the empty space in the upward direction. The following analysis in Lagrangian co-ordinates is due to Raizer (1964) (see also the book by Zeldovich and Raizer (1967)). We consider the limiting motion of the shock. Here, we have a length scale Δ. The origin of x is arbitrary (see (3.5.1)). This implies that ρ_* is arbitrary by a multiplicative constant. There are no time or density scales. By dimensional arguments, the motion of the shock, $X = X(t)$, is given by

$$D = \dot{X} = \alpha \frac{\Delta}{t}, \tag{3.5.2}$$

where the coefficient α depends on $\gamma = c_p/c_v$ alone. From (3.5.2) we have

$$X = \alpha \Delta \ln t + \text{constant}. \tag{3.5.3}$$

The mass of gas overtaken by the shock is given by

$$M = \int_{-\infty}^{X} \rho_0 dx = \int_{-\infty}^{X} \rho_* e^{x/\Delta} dx = \rho_0(X)\Delta. \qquad (3.5.4)$$

Equations (3.5.1) and (3.5.4) give

$$\dot{M} = \rho_0(X)\dot{X}. \qquad (3.5.5)$$

Using (3.5.2), (3.5.4) and (3.5.5) we obtain, after integration,

$$M = At^\alpha, \qquad (3.5.6)$$

where the constant of integration A characterises the intensity of the shock. Assuming that the shock is strong, we may write the self-similar solution in the form

$$u = \frac{2}{\gamma+1}\alpha\frac{\Delta}{t}v = u_s v, \qquad (3.5.7)$$

$$\rho = \frac{\gamma+1}{\gamma-1}\frac{At^\alpha}{\Delta}q = \rho_s q, \qquad (3.5.8)$$

$$p = \frac{2}{\gamma+1}\alpha^2\frac{\Delta A}{t^{2-\alpha}}f = p_s f, \qquad (3.5.9)$$

where u_s, ρ_s, and p_s are the appropriate quantities just behind the shock front, given by the Rankine-Hugoniot conditions, and the functions v, q and f depend on the similarity variables $\xi = (X(t) - x)/\Delta$, a distance measured from the moving shock, and γ. The Rankine-Hugoniot conditions at the strong shock, $\xi = 0$, give $v = q = f = 1$ there. Raizer (1964) used Lagrangian co-ordinates to study this problem. He introduced the mass Lagrangian co-ordinate

$$m = \int_{-\infty}^{x} \rho(x)dx = \text{const}.M\int_{\xi}^{\infty} q(\xi)d\xi, \qquad (3.5.10)$$

(cf. (3.5.4)) so that ξ, and hence v, q and f, are functions of the new similarity variable

$$\eta = \frac{m}{M} = \frac{m}{At^\alpha}, \qquad (3.5.11)$$

as follows easily from (3.5.10) and (3.5.6).

The planar equations of motion in Lagrangian co-ordinates are

$$u_t + p_m = 0, \qquad (3.5.12)$$

$$(1/\rho)_t - u_m = 0, \qquad (3.5.13)$$

$$p\rho^{-\gamma} = F(m). \qquad (3.5.14)$$

Substituting (3.5.7)–(3.5.9) along with (3.5.11) into (3.5.12)–(3.5.14) we get

$$v + \alpha\eta v' - \alpha f' = 0, \tag{3.5.15}$$

$$\frac{1}{q} + \eta\left(\frac{1}{q}\right)' + \frac{2}{\gamma-1}v' = 0, \tag{3.5.16}$$

$$fq^{-\gamma}\eta^{2/\alpha+\gamma-1} = 1, \tag{3.5.17}$$

where dash denotes differentiation with respect to η. Integrating (3.5.16) and hence eliminating q and v from (3.5.15) and (3.5.17) we get an ODE for $f = f(\eta)$:

$$\frac{df}{d\eta} = \frac{\gamma+1}{2\alpha}\frac{1 - \frac{\gamma-1}{\gamma+1}\left(1 - \frac{2-\alpha}{\gamma}\right)f^{-1/\gamma}\eta^{-(2-\alpha)/\alpha\gamma}}{1 - \frac{\gamma-1}{2\gamma}f^{-\frac{1}{\gamma}-1}\eta^{1-(2-\alpha)/\alpha\gamma}}. \tag{3.5.18}$$

Since the point $x = -\infty$ corresponds to $\eta \to 0$ and since the pressure at $x = -\infty$ is zero, the solution must pass through the point

$$\eta = 0, \quad f = 0. \tag{3.5.19}$$

It must also pass through the shock point

$$\eta = 1, \quad f = 1. \tag{3.5.20}$$

The conditions for the existence of the solution of BVP (3.5.18)–(3.5.20) will determine the value of the exponent α for each γ. This is typical of the self-similar motions of the second kind for which the dimensional arguments alone do not suffice to yield the exponent α and, therefore, the shock motion. This matter is discussed in great detail by Zeldovich and Raizer (1967) who also discuss the present problem in this context. We may observe that the above BVP has explicit solutions for $\gamma = 2, 1$.

For $\gamma = 2$, we may check that $\alpha = 3/2$ so that

$$
\begin{aligned}
M &\sim t^{3/2}, \\
\dot{X} &= \frac{3}{2}\frac{\Delta}{t}, \\
u_s &\sim \frac{1}{t}, \\
\rho_s &\sim t^{3/2}, \\
p_s &\sim \frac{1}{t^{1/2}},
\end{aligned}
\tag{3.5.21}
$$

and the similarity functions are simply

$$f = \eta, \quad q = \eta^{5/3}, \quad v = \frac{3}{2}\left(1 - \frac{1}{3}\eta^{-2/3}\right). \tag{3.5.22}$$

For $\gamma = 1$, it may again be verified that $\alpha = 1$, $f = \eta$, $q = \eta^3$, $v = 1$. Since $1 < \gamma < 2$ for air, these special solutions give a bound for α, $1 < \alpha < \frac{3}{2}$, for this range.

More generally, (3.5.18) must be solved numerically for each γ, starting from the shock point (1,1). α must be found such that the integral curve passes through $(0,0)$.

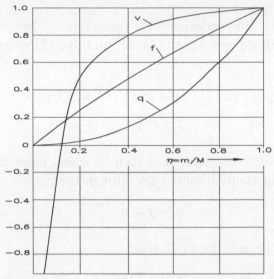

Figure 3.10 Distribution of f, q, v versus m/M behind the shock wave for $\gamma = 1.25$ and $\alpha = 1.345$ (Raizer, 1964).

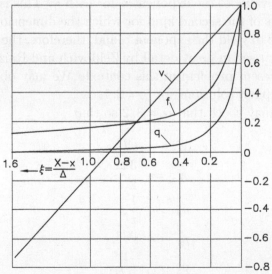

Figure 3.11 Distribution of f, q, v versus $\frac{X-x}{\Delta}$ behind the shock wave for $\gamma = 1.25$ and $\alpha = 1.345$ (Raizer, 1964).

Often a local analysis about the shock point helps to facilitate the numerical solution. Raizer (1964) found the numerical solution for the special value of $\gamma = 1.25$; the corresponding value of α was found to be 1.345. The solution is depicted in Figures 3.10–3.11.

Raizer (1964), referring to earlier studies of Kompaneets (1960), attempted to connect this asymptotic solution to an actual high energy, high altitude explosion and estimated the distance from the point of explosion where this analysis would hold.

However, Wallace Hayes, in the editorial footnote to this analysis in the book of Zeldovich and Raizer (1967), cautioned that such high energy explosions (10^{24} ergs) at high altitudes (100 km) would not be realistically described by hydrodynamical models of a strong explosion because of the large values of the photon mean free path at very low densities.

Hayes (1968) himself studied self-similar strong shocks in an exponential medium using Eulerian co-ordinates; his analysis was close to that of Raizer (1964). Hayes' numerical value of the parameter α (see (3.5.6)) was slightly different from Raizer's. His principal aim was to analyse the approximate approach of Chester (1954), Chisnell (1957) and Whitham (1958) (CCW for short) in the light of his numerical results. He improvised the coefficients in the shock law according to this approximate theory and concluded that, in contrast to the implosion problem, the shock propagation law obtained through the CCW approximation was in error by 15% or more.

3.6 Strong Explosion into a Power Law Density Medium

It is of both physical and mathematical interest to study strong explosions in a medium whose density decreases with distance according to some law, say, $\rho_1 = Ar^{-\omega}$, where A and ω are positive numbers and r is the distance measured from the point of explosion. It turns out that, in the spherically symmetric case that we consider here, the solutions are of two kinds depending on the parameter ω. If $\omega < 5$, we have generalisations of the well-known Taylor-Sedov solution (see section 3.2), which can, in fact, be found in an (implicit) closed form. These constitute the familiar self-similar solutions of the first kind, as defined by Zeldovich and Raizer (1967). Typically, for this class of solutions, no constants with the dimension of length or time occur among the parameters appearing in the boundary conditions. Besides, fluid dynamic equations do not contain any dimensional constants. The parameters with the dimensions of length or time characterising the flow can be constructed from initial/boundary conditions. These parameters describe the length and time scales typical of the early time flow. For large times (and distances) these length and time scales do not characterise the physical processes. The early flows however are fully described by the self-similar solutions of the first kind.

In another class of self-similar solutions, the governing PDEs can still be reduced to ODEs by the similarity transformation, but the parameters appearing in the problem do not fully determine the flow. The unknown parameter (in the shock law in the present case) for the class of problems under study must be found from the solution of an eigenvalue problem (see section 3.5). This solution must satisfy the Rankine-Hugoniot conditions at the shock and pass through an 'appropriate' singular point of the reduced nonlinear ODE in the (nondimensional) sound speed square—particle velocity phase plane. The most famous example of this second kind of self-similar solutions is the so-called Guderley's convergent shock problem (see Zeldovich and Raizer (1967); see also sections 3.5 and 6.1). The curious thing about the present problem—propagation of a strong shock into a (power law) inhomogeneous medium—is that the parameter ω in the density law has different ranges for which either self-similar solutions of the first or those of the second kind exist. There is a transition point $\omega = 3$ where the solution changes its character from the first kind to one of the second kind. This parametric dependence makes this problem particularly interesting. We discuss here the spherically-symmetric case, although the discussion can be easily extended to cylindrical and planar cases. This problem is discussed in the present section and the next.

The equations of motion in spherical symmetry are

$$u_t + uu_r + \frac{1}{\rho}p_r = 0, \tag{3.6.1}$$

$$\rho_t + \rho u_r + u\rho_r + \frac{2\rho u}{r} = 0, \tag{3.6.2}$$

$$\left(\frac{p}{\rho^\gamma}\right)_t + u\left(\frac{p}{\rho^\gamma}\right)_r = 0, \tag{3.6.3}$$

with the usual notation for u, p, ρ, and γ. The strong shock conditions are

$$u_2 = \frac{2}{\gamma + 1}U, \tag{3.6.4}$$

$$\rho_2 = \frac{\gamma + 1}{\gamma - 1}\rho_1, \tag{3.6.5}$$

$$p_2 = \frac{2}{\gamma + 1}\rho_1 U^2, \tag{3.6.6}$$

where the suffixes 2 and 1 denote values immediately behind and ahead of the shock, respectively. U is the shock velocity.

The density distribution ahead of the shock is chosen to be

$$\rho_1 = \frac{A}{r^\omega}, \tag{3.6.7}$$

where A and ω are positive constants. The parameter A has the dimension $[A] = ML^{\omega-3}$. The shock considered here is strong, the undisturbed pressure

p_1 is assumed to be zero. Thus, we have the following 'parameters' appearing in the problem:

$$\omega, \gamma, A, E_0, r, t, \tag{3.6.8}$$

where E_0 is the energy of the explosion, $[E_0] = ML^2T^{-2}$. It is easily checked from a dimensional argument that all the (dimensionless) quantities are functions of the 'parameters',

$$\omega, \gamma, \quad \lambda = \left(\frac{A\alpha}{E_0}\right)^{1/(5-\omega)} \frac{r}{t^{2/(5-\omega)}}, \tag{3.6.9}$$

where α is a constant to be determined. The (nondimensional) solution must depend on the similarity variable λ. The shock locus itself is given by

$$\lambda = \lambda_2 = \text{constant}, \quad r_2 = \lambda_2 \left(\frac{E_0}{A\alpha}\right)^{1/(5-\omega)} t^{2/(5-\omega)}, \tag{3.6.10}$$

$$U = \frac{dr_2}{dt} = \frac{2}{5-\omega}\frac{r_2}{t} = \frac{2\lambda_2^{(5-\omega)/2}}{5-\omega}\left(\frac{E_0}{A\alpha}\right)^{1/2} r_2^{(\omega-3)/2}. \tag{3.6.11}$$

The constant α is chosen such that $\lambda = \lambda_2 = 1$ at the shock. With this choice of α, we have

$$\lambda = \frac{r}{r_2}, \tag{3.6.12}$$

The relation (3.6.11) implies that the shock will decelerate if $\omega < 3$ and accelerate if $\omega > 3$. The case $\omega < 3$, as we shall show in the next section, corresponds to a finite spherical mass containing the center of symmetry; it becomes infinite if $\omega \geq 3$. Equation (3.6.10) shows that the shock starting from $r = 0$ at $t = 0$ propagates with finite velocity for $t > 0$ only if $\omega < 5$. We shall consider the cases $0 < \omega < 3$, $3 \leq \omega < 5$, and $\omega > 5$, separately in the present section and the following. Here we follow the work of Korobeinikov and Riazanov (1959).

Introducing the similarity functions

$$\frac{u}{u_2} = f(\lambda), \tag{3.6.13}$$

$$\frac{\rho}{\rho_2} = g(\lambda), \tag{3.6.14}$$

$$\frac{p}{p_2} = h(\lambda) \tag{3.6.15}$$

into (3.6.1)–(3.6.3), we have the following reduced system of nonlinear ordinary differential equations for $f(\lambda)$, $g(\lambda)$ and $h(\lambda)$:

$$\frac{df}{d\lambda} = \frac{1}{2}\left[4\gamma(\gamma-1)\frac{f}{\lambda} + (3-\omega)(\gamma+1)\right.$$

$$\times \left(f - \frac{\gamma+1}{2}\lambda\right)\frac{fg}{h} - 3(\gamma^2 - 1)\right]$$

$$\times \left[2\left(f - \frac{\gamma+1}{2}\lambda\right)^2\frac{g}{h} - \gamma(\gamma-1)\right]^{-1}, \tag{3.6.16}$$

$$\frac{dg}{d\lambda} = -g\left[\frac{df}{d\lambda} + 2\frac{f}{\lambda} - \frac{\gamma+1}{2}\omega\right]\left(f - \frac{\gamma+1}{2}\lambda\right)^{-1}, \tag{3.6.17}$$

$$\frac{dh}{d\lambda} = h\left[\frac{3(\gamma+1)}{2} - \gamma\left(\frac{df}{d\lambda} + \frac{2f}{\lambda}\right)\right]\left(f - \frac{\gamma+1}{2}\lambda\right)^{-1}. \tag{3.6.18}$$

The boundary conditions (3.6.4)–(3.6.6) at the shock $\lambda = 1$ become

$$f(1) = g(1) = h(1) = 1. \tag{3.6.19}$$

The symmetry condition at the center of explosion $\lambda = 0$ (corresponding to $r = 0$) requires that the particle velocity there is zero so that from (3.6.13) we have

$$f(0) = 0. \tag{3.6.20}$$

The system (3.6.16)–(3.6.18) must be solved subject to the conditions (3.6.19) and (3.6.20). It may be verified that this system admits two intermediate integrals—the energy integral and the integral of adiabacy (see Sedov (1959) for details):

$$\frac{g}{h} = \gamma f^{-2}\left(\frac{f}{\lambda} - \frac{\gamma+1}{2\gamma}\right)\left(\frac{\gamma+1}{2} - \frac{f}{\lambda}\right)^{-1}, \tag{3.6.21}$$

$$g^{\gamma-1} = \left[\frac{2}{\gamma-1}\left(\frac{\gamma+1}{2} - \frac{f}{\lambda}\right)\right]^{1-\frac{\omega\gamma}{3}} h^{1-\frac{\omega}{3}}\lambda^{3-\omega\gamma}. \tag{3.6.22}$$

These integrals satisfy the conditions (3.6.19) and (3.6.20). Using these integrals it is possible to solve the system (3.6.16)–(3.6.20) in an (implicit) closed form in terms of the function

$$F = \frac{f}{\lambda}. \tag{3.6.23}$$

We thus have

$$\lambda(F) = F^{-\delta}\left[\frac{2\gamma}{\gamma-1}\left(F - \frac{\gamma+1}{2\gamma}\right)\right]^{-\alpha_2}$$

$$\times \left[\frac{2(3\gamma-1)}{7-\gamma-(\gamma+1)\omega}\left(\frac{\gamma+1}{2}\frac{5-\omega}{3\gamma-1} - F\right)\right]^{-\alpha_1}, \tag{3.6.24}$$

$$g(F) = F^{\omega\delta}\left[\frac{2\gamma}{\gamma-1}\left(F - \frac{\gamma+1}{2\gamma}\right)\right]^{\alpha_3+\omega\alpha_2}$$

$$\times \left[\frac{2}{\gamma-1}\left(\frac{\gamma+1}{2} - F\right)\right]^{\alpha_5} \times \left[\frac{2(3\gamma-1)}{7-\gamma-(\gamma+1)\omega}\right.$$

$$\times \left(\frac{\gamma + 1}{2} \frac{5 - \omega}{3\gamma - 1} - F \right) \Big]^{\alpha_4 + \omega\alpha_1},$$

$$(3.6.25)$$

$$h(F) = F^{3\delta} \left[\frac{2}{\gamma - 1} \left(\frac{\gamma + 1}{2} - F \right) \right]^{1 + \alpha_5}$$

$$\times \left[\frac{2(3\gamma - 1)}{7 - \gamma - (\gamma + 1)\omega} \right.$$

$$\times \left(\frac{\gamma + 1}{2} \frac{5 - \omega}{3\gamma - 1} - F \right) \Big]^{\alpha_4 + (\omega - 2)\alpha_1},$$

$$(3.6.26)$$

where

$$\delta = \frac{2}{5 - \omega}, \tag{3.6.27}$$

$$\alpha_1 = \frac{\gamma + 1}{3\gamma - 1} - \delta - \alpha_2, \quad \alpha_2 = \frac{1 - \gamma}{2\gamma + 1 - \omega\gamma},$$

$$\alpha_3 = \frac{3 - \omega}{2\gamma + 1 - \gamma\omega}, \quad \alpha_4 = \frac{(3 - \omega)(5 - \omega)}{6 - 3\gamma - \omega}\alpha_1,$$

$$\alpha_5 = \frac{\omega(\gamma + 1) - 6}{6 - 3\gamma - \omega}. \tag{3.6.28}$$

Korobeinikov and Riazanov (1959), who extended the results of Sedov (1959) to cylindrical and plane geometries, used the function F rather than V:

$$V = \frac{4}{(5 - \omega)(\gamma + 1)}F. \tag{3.6.29}$$

It is easily seen from (3.6.23) and (3.6.24) that the solution extends to $\lambda = 0$, the center of explosion, only if

$$1 \geq F \geq \frac{\gamma + 1}{2\gamma}. \tag{3.6.30}$$

First we consider this case. We discuss special (singular) cases of this solution when either the RHS of (3.6.24)–(3.6.26) or α_i ($i = 1, 2, ..., 5$) defined in (3.6.28) tend to infinity and therefore this form of the solution becomes invalid. The coefficients in the former case become infinite when

$$\omega = \omega_1 = \frac{7 - \gamma}{\gamma + 1}. \tag{3.6.31}$$

For this singular case the original system (3.6.16)–(3.6.18) admits a simple exact solution satisfying (3.6.19) and (3.6.20):

$$f = \lambda, \quad g = \lambda, \quad h = \lambda^3. \tag{3.6.32}$$

α_i $(i = 1, 2, 3, 4)$ tend to infinity when

$$\omega \to \omega_2 = \frac{2\gamma + 1}{\gamma}. \tag{3.6.33}$$

In this case, if we use the energy integral (3.6.21) in (3.6.16) and (3.6.17) to eliminate h, we get the system

$$\frac{df}{d\lambda} = -\frac{f}{\lambda} \left[\frac{2\gamma(\gamma - 1)}{\gamma + 1} \left(\frac{f}{\lambda} \right)^2 + \frac{(\omega - 6)\gamma + 3}{2} \frac{f}{\lambda} \right.$$
$$\left. + \frac{(3 - \omega)(\gamma + 1)}{4} \right]$$
$$\times \left[\gamma \left(\frac{f}{\lambda} \right)^2 - (\gamma + 1) \frac{f}{\lambda} + \frac{\gamma + 1}{2} \right]^{-1}, \tag{3.6.34}$$

$$\frac{dg}{d\lambda} = -g \left(\frac{df}{d\lambda} + \frac{2f}{\lambda} - \frac{\gamma + 1}{2} \omega \right) \left(f - \frac{\gamma + 1}{2} \lambda \right)^{-1}, \tag{3.6.35}$$

where $\omega = \omega_2$ is given by (3.6.33). Writing $f = \lambda F$ in (3.6.34) we may integrate it and satisfy the relevant boundary conditions in (3.6.19) and (3.6.20). We thus have

$$\lambda(F) = F^{-2\gamma\delta_2} \left[\frac{2\gamma}{\gamma - 1} \left(F - \frac{\gamma + 1}{2\gamma} \right) \right]^{(\gamma - 1)\delta_2}$$
$$\times \exp \left[-(\gamma + 1)\delta_2 \frac{1 - F}{F - (\gamma + 1)/2\gamma} \right], \tag{3.6.36}$$

where

$$\delta_2 = \frac{1}{3\gamma - 1}. \tag{3.6.37}$$

Using the intermediate integrals (3.6.21)–(3.6.22), we get an explicit form for $g(F)$ and $h(F)$ for this special case:

$$g(F) = F^{2(2\gamma + 1)\delta_2} \left[\frac{2}{\gamma - 1} \left(\frac{\gamma + 1}{2} - F \right) \right]^{[3 - 2(\gamma + 1)]\delta_2}$$
$$\times \left[\frac{2\gamma}{\gamma - 1} \left(F - \frac{\gamma + 1}{2\gamma} \right) \right]^{(1 - 2\gamma)\delta_2}$$
$$\times \exp \left[2(\gamma + 1)\delta_2 \frac{1 - F}{F - (\gamma + 1)/2\gamma} \right], \tag{3.6.38}$$

$$h(F) = F^{6\gamma\delta_2} \left[\frac{2}{\gamma - 1} \left(\frac{\gamma + 1}{2} - F \right) \right]^{\gamma\delta_2}$$
$$\times \left[\frac{2\gamma}{\gamma - 1} \left(F - \frac{\gamma + 1}{2\gamma} \right) \right]^{-3\gamma\delta_2}. \tag{3.6.39}$$

A similar procedure gives the solution for the other singular case $\omega = \omega_3 = 3(2 - \gamma)$ when α_4 and α_5 tend to infinity (see (3.6.28)):

$$\lambda(F) = F^{-2\delta_2} \left[\frac{2}{\gamma - 1} \left(\frac{\gamma + 1}{2} - F \right) \right]^{-\gamma\delta_2}$$

$$\times \left[\frac{2\gamma}{\gamma - 1} \left(F - \frac{\gamma + 1}{2\gamma} \right) \right]^{\delta_2}, \tag{3.6.40}$$

$$g(F) = F^{6(2-\gamma)\delta_2} \left[\frac{2}{\gamma - 1} \left(\frac{\gamma + 1}{2} - F \right) \right]^{(3\gamma-5)\delta_2}$$

$$\times \left[\frac{2\gamma}{\gamma - 1} \left(F - \frac{\gamma + 1}{2\gamma} \right) \right]^{3(\gamma-1)\delta_2}$$

$$\times \exp \left[-3\gamma(\gamma + 1)\delta_2 \frac{1 - F}{\frac{\gamma+1}{2} - F} \right], \tag{3.6.41}$$

$$h(F) = F^{6\delta_2} \left[\frac{2}{\gamma - 1} \left(\frac{\gamma + 1}{2} - F \right) \right]^{2(2\gamma-3)\delta_2}$$

$$\times \exp \left[-3\gamma(\gamma + 1)\delta_2 \frac{1 - F}{\frac{\gamma+1}{2} - F} \right]. \tag{3.6.42}$$

To give an idea of the energy of explosion for these singular cases we observe that

$$E_0 = 4\pi \int_0^{r_2} \left(\frac{\rho u^2}{2} + \frac{p}{\gamma - 1} \right) r^2 dr. \tag{3.6.43}$$

Putting the similarity form of the solution (3.6.13)–(3.6.15) into (3.6.43), we have the form (3.6.10)–(3.6.11) of the shock law provided that

$$\alpha(\gamma, \omega) = \frac{8\pi\delta^2}{(\gamma^2 - 1)} \int_0^1 (h + gf^2)\lambda^2 d\lambda. \tag{3.6.44}$$

For the first singular case (3.6.32) with $\omega = (7 - \gamma)/(\gamma + 1)$, (3.6.44) assumes the explicit form

$$\alpha(\gamma, \omega_1) = \frac{8\pi}{3} \frac{\gamma + 1}{\gamma - 1} \left(\frac{1}{3\gamma - 1} \right)^2. \tag{3.6.45}$$

The expressions for α for other singular values $\omega = \omega_2, \omega_3$ may be found similarly.

Korobeinikov and Riazanov (1959) indicated how these singular solutions may be obtained from the general solution (3.6.24)–(3.6.26) by taking the limit $\omega \to \omega_i$ ($i = 1, 2, 3$). They specifically showed this for the special case $\omega_3 \to 0$, that is, $\gamma \to 2$ (see below (3.6.39)). The main conclusion of their study is that the solution of the problem of a strong explosion into an inhomogeneous (or homogeneous) medium is continuous in γ, as may be expected from the form (3.6.1)–(3.6.3) of the basic equations and the BCs (3.6.4)–(3.6.6), provided $\gamma > 1$.

We continue this study in greater detail in the next section.

3.7 Strong Explosion into Power Law Nonuniform Medium: Self-similar Solutions of the Second Kind

We discussed in the previous section self-similar solutions of the first kind for the parametric regime $0 \leq \omega < 3$ (see (3.6.7)). It turns out that ω is a crucial parameter in deciding the nature of the solutions describing propagation of (point) explosion waves into a medium with density $\rho_0 = K r^{-w}$. Indeed, for $\omega \geq 3$ the so-called self-similar solutions of the second kind make their appearance. Here the shock trajectory is not given by dimensional considerations of the physical parameters that appear in the problem (see section 3.6), but is found from the condition that the self-similar solution must pass through a singular point of the reduced ordinary differential equation in the sound speed–particle velocity plane. The best known example of this kind of solution is Guderley's (1942) solution for converging spherical or cylindrical shock waves. This matter is discussed in great detail in the last chapter of the book by Zeldovich and Raizer (1967). In the sequel we follow the work of Waxman and Shvarts (1993).

Here we seek a self-similar solution in a slightly different form. The flow equations in terms of particle velocity u, sound speed c and particle density ρ may be written for the case of spherical symmetry as

$$u_t + u u_r + \frac{1}{\rho} p_r = 0, \tag{3.7.1}$$

$$\rho_t + u \rho_r + \rho \left(u_r + \frac{2u}{r} \right) = 0, \tag{3.7.2}$$

$$c_t + u c_r + \left(\frac{\gamma - 1}{2} \right) \left(c u_r + \frac{2cu}{r} \right) = 0. \tag{3.7.3}$$

The solution of the system (3.7.1)–(3.7.3) is sought in the form

$$u(r,t) = \dot{R} \xi U(\xi), \tag{3.7.4}$$

$$c(r,t) = \dot{R} \xi C(\xi), \tag{3.7.5}$$

$$\rho(r,t) = B t^{\beta} G(\xi), \tag{3.7.6}$$

where

$$\xi(r,t) = r/R(t). \tag{3.7.7}$$

The shock radius $R(t)$ in the above is assumed to follow the law

$$R(t) = A t^{\alpha}, \tag{3.7.8}$$

where $A > 0$ is a constant, and α is a parameter to be determined.

In the analysis of Waxman and Shvarts (1993) it is assumed that the flow due to the strong explosion takes place over a length scale given by $R(t)$ and that this scale diverges as $t \to \infty$. Moreover, the boundary conditions at the shock front are determined by a single dimensional parameter K, which appears in the undisturbed density distribution, $\rho_0 = Kr^{-\omega}$. In the Taylor-Sedov type solution the second relevant parameter is the energy E of explosion. As explained in section 3.6, the similarity form (3.7.4)–(3.7.7) requires that the nondimensional constants A and B and the similarity exponents α and β are related by

$$
\begin{aligned}
A &= \phi(\gamma, \omega) \left(\frac{E}{K} \right)^{\alpha/2}, \quad \alpha = \frac{2}{5 - \omega}, \\
B &= K A^{-\omega}, \quad \beta = -\alpha\omega.
\end{aligned}
\tag{3.7.9}
$$

ϕ is a dimensionless function of the dimensionless constants γ and ω; it arises from the assumption that the energy of explosion is the energy of flow for all time.

It is clear from (3.7.9) that α becomes infinite when $\omega = 5$. It is positive for $\omega < 5$ and negative for $\omega > 5$.

The main point of the work of Waxman and Shvarts (1993) is to show that, for the present problem, the first kind of solutions in the Taylor-Sedov similarity form are available only when $\omega < 3$. In this case the energy of the explosion is the appropriate second dimensional parameter. For $\omega \geq 3$, it is shown that the self-similar form alone does not give a full description of the flow. In this case the explosion energy becomes infinite and is not the relevant second dimensional parameter. The asymptotic self-similar solution must then be of the second kind. The other relevant parameter in this case is found by using the condition that the self-similar solution passes through an appropriate singular point of the flow equations. It is pointed out that the energy divergence itself does not imply that the Taylor-Sedov type of solution cannot describe the asymptotic flow due to the strong explosion for $\omega \geq 3$. The inconsistency in the present case arises from the fact that, for a Taylor-Sedov solution, the divergence of the energy implies that the flow in some region whose size diverges in proportion to R depends on the initial length and time scales as R diverges; this does not happen for the self-similar solutions of the first kind. The self-similar solution of the second kind holding in this case with $\omega > 3$ possesses infinite initial energy.

We may reduce the system (3.7.1)–(3.7.3) via (3.7.4)–(3.7.8) in the manner of section 3.6 to the (U, C) plane, and also derive the equation connecting ξ and U. For details we refer the reader to the last chapter of Zeldovich and Raizer (1967). We thus have

$$
\frac{dU}{d \log \xi} = \frac{\Delta_1(U, C)}{\Delta(U, C)},
\tag{3.7.10}
$$

$$\frac{dC}{d\log\xi} = \frac{\Delta_2(U,C)}{\Delta(U,C)},\tag{3.7.11}$$

or

$$\frac{dU}{dC} = \frac{\Delta_1(U,C)}{\Delta_2(U,C)}\tag{3.7.12}$$

and

$$\frac{d\log\xi}{dU} = \frac{\Delta(U,C)}{\Delta_1(U,C)},\tag{3.7.13}$$

where the functions Δ, Δ_1, and Δ_2 are defined by

$$\Delta = C^2 - (1-U)^2,\tag{3.7.14}$$

$$\Delta_1 = U(1-U)\left(1 - U - \frac{\alpha-1}{\alpha}\right)$$
$$-C^2\left(3U - \frac{\omega - 2[(\alpha-1)/\alpha]}{\gamma}\right),\tag{3.7.15}$$

$$\Delta_2 = C\left[(1-U)\left(1-U-\frac{\alpha-1}{\alpha}\right)\right.$$
$$-\frac{\gamma-1}{2}U\left(2(1-U)+\frac{\alpha-1}{\alpha}\right) - C^2$$
$$\left.+\frac{(\gamma-1)\omega + 2[(\alpha-1)/\alpha]}{2\gamma}\frac{C^2}{1-U}\right].\tag{3.7.16}$$

The density function G in (3.7.6) is obtained from the first integral

$$C^{-2}(1-U)^\lambda G^{\gamma-1+\lambda}\xi^{3\lambda-2} = \text{constant},\tag{3.7.17}$$

where

$$\lambda = \frac{(\gamma-1)\omega + 2[(\alpha-1)/\alpha]}{3-\omega}.\tag{3.7.18}$$

The strong shock conditions (see section 3.6) in terms of the similarity functions U, C, and G may be written as

$$U(1) = \frac{2}{\gamma+1}, \quad C(1) = \frac{\sqrt{2\gamma(\gamma-1)}}{\gamma+1}, \quad G(1) = \frac{\gamma+1}{\gamma-1}.\tag{3.7.19}$$

Before discussing the case $\omega > 3$, it is instructive to consider the Taylor-Sedov type of solution via energy arguments. The total energy contained in the region $\xi_1 \le \xi \le 1$, corresponding to $\xi_1 R(t) \le r \le R(t)$, is given by

$$E_1 = \int_{\xi_1 R}^{R} dr\, 4\pi r^2 \rho\left(\frac{1}{2}u^2 + \frac{1}{\gamma(\gamma-1)}c^2\right)$$
$$= 4\pi K R^{3-\omega}\dot{R}^2\int_{\xi_1}^{1} d\xi\ \xi^4 G\left(\frac{1}{2}U^2 + \frac{1}{\gamma(\gamma-1)}C^2\right)\tag{3.7.20}$$

or, on using (3.7.8)–(3.7.9),

$$E_1 = \left[4\pi \left(\frac{2}{5-\omega} \right)^2 \phi^{5-\omega} \int_{\xi_1}^1 d\xi \ \xi^4 \right.$$
$$\left. \times G \left(\frac{1}{2} U^2 + \frac{1}{\gamma(\gamma-1)} C^2 \right) \right] \times E. \qquad (3.7.21)$$

Thus, the similarity form of the solution leads to E_1 which is independent of time. This is possible if the work done by a fluid element, which lies at ξ at time t, on the fluid that occupies the domain $\xi > \xi_1$ during the time interval dt, equals the energy that leaves the region $\xi_1 \leq \xi \leq 1$ during the same time through the surface $\xi_1 = $ constant. This argument is expressed mathematically in the following manner. The work done by a fluid element at ξ_1 at time t (that is, at $r_1 = \xi_1 R(t)$) on the fluid that occupies the region $r > \xi_1 R(t)$ in time dt is

$$4\pi r_1^2 u(r_1,t) dt \gamma^{-1} \rho(r_1,t) c^2(r_1,t)$$
$$= 4\pi \gamma^{-1} K R^{2-\omega} \dot{R}^3 \xi_1^5 U(\xi_1) G(\xi_1) C^2(\xi_1) dt. \qquad (3.7.22)$$

The energy that leaves $\xi > \xi_1$ during the same time through the surface $\xi_1 = $ constant is

$$4\pi r_1^2 [\xi_1 \dot{R} - u(r_1,t)] dt \left(\frac{1}{2} \rho(r_1,t) u^2(r_1,t) + \frac{1}{\gamma(\gamma-1)} c^2(r_1,t) \right)$$
$$= 4\pi \gamma^{-1} K R^{2-\omega} \dot{R}^3 \xi_1^5 [1 - U(\xi_1)] G(\xi_1)$$
$$\times \left(\frac{1}{2} U^2(\xi_1) + \frac{1}{\gamma-1} C^2(\xi_1) \right) dt. \qquad (3.7.23)$$

Comparing (3.7.22) and (3.7.23), we have

$$C^2 = \frac{\gamma(\gamma-1)}{2} \frac{U^2(1-U)}{\gamma U - 1}. \qquad (3.7.24)$$

Therefore, the equation for the curve (ξ, U) is obtained from (3.7.13) as

$$\frac{d \log \xi}{dU} = (\gamma+1) \frac{\gamma U^2 - 2U + [2/(\gamma+1)]}{U(\gamma U - 1)[5 - \omega - (3\gamma - 1)U]}. \qquad (3.7.25)$$

Now we show that $U(\log \xi)$ is an increasing function of its argument for $\omega < (7 - \gamma)/(\gamma + 1)$. The numerator in (3.7.25) is always positive for $\gamma > 1$. The term $[5 - \omega - (3\gamma - 1)U]$ in the denominator of (3.7.25) must now be considered. Since ξ decreases from 1 at the shock to 0 at the center, $\log \xi$ varies from 0 to $-\infty$ correspondingly. Thus, $[5 - \omega - (3\gamma - 1)U]$ is positive for $U \leq U(\log \xi = 0) = 2/(\gamma + 1)$. Also, $\gamma U - 1$ is positive for

$1/\gamma < U \leq U(\log \xi = 0)$. Thus, as $\log \xi$ decreases from 0 to $-\infty$ corresponding to ξ decreasing from 1 to 0, $U(\log \xi)$ also decreases and approaches the value $1/\gamma$. We may write (3.7.25) near $U = 1/\gamma$ as

$$\frac{d \log \xi}{dU} = \frac{f(\gamma)}{U - 1/\gamma}, \tag{3.7.26}$$

where $f(\gamma)$ is some positive function of γ. Integrating (3.7.26), we have

$$\xi = \text{constant} \cdot (U - 1/\gamma)^{f(\gamma)}. \tag{3.7.27}$$

Thus ξ tends to zero as U tends to $1/\gamma$. We find therefore that $U(\xi)$ is a strictly increasing function of ξ that approaches $1/\gamma$ as ξ tends to zero.

By differentiating (3.7.24) we find that

$$\frac{dC^2}{dU} = -\frac{\gamma(\gamma - 1)}{2} \frac{U[2\gamma U^2 - (\gamma + 3)U + 2]}{(\gamma U - 1)^2}. \tag{3.7.28}$$

The quadratic in U in the numerator of (3.7.28) has no real roots for $1 < \gamma < 8$ and is always positive. Therefore, C is a decreasing function of U. Also, it follows from (3.7.24) that C tends to infinity as U tends to $1/\gamma$.

The function $[5 - \omega - (3\gamma - 1)U]$ in the denominator of (3.7.25) assumes the value $\left(\frac{7-\gamma}{\gamma+1} - \omega\right)$ at the shock where $U(1) = 2/(\gamma + 1)$. The behaviour of the solution is different depending on the sign of this term (see section 3.6). For $\omega > (7 - \gamma)/(\gamma + 1)$, the self-similar solution has a void in the center. There exists a region $\xi_{in} \leq \xi \leq 1$ which is separated from the vacuum by a tangential or weak discontinuity at $\xi = \xi_{in}$. In this case U tends to 1 and C tends to zero as ξ tends to ξ_{in}.

We now consider specifically the case $\omega \geq 3$ for which $3 > (7-\gamma)/(\gamma+1)$ provided $\gamma > 1$. In this case the solution curve approaches the point $C = 0$, $U = 1$ as ξ tends to ξ_{in}. Considering the local behaviour of the solution in the neighbourhood of this point by analysing (3.7.24), (3.7.25) and (3.7.17), we find that

$$U(\xi) = 1 - \frac{3\gamma + \omega - 6}{\gamma} \log\left(\frac{\xi}{\xi_{in}}\right), \tag{3.7.29}$$

$$C(\xi) = \left[\frac{3\gamma + \omega - 6}{2} \log\left(\frac{\xi}{\xi_{in}}\right)\right]^{1/2}, \tag{3.7.30}$$

$$G(\xi) = \text{const} \times \left[\log\left(\frac{\xi}{\xi_{in}}\right)\right]^{-(\gamma\omega+\omega-6)/(3\gamma+\omega-6)}. \tag{3.7.31}$$

Using this local solution in the energy integral (3.7.21), one may verify that

$$E_1 \to \infty \quad \text{as} \quad \xi_1 \to \xi_{in} \quad \text{for} \quad \omega \geq 3. \tag{3.7.32}$$

Since the energy $E_1(\xi_1)$ in $\xi_1 < \xi < 1$ for the Taylor-Sedov blast wave is independent of time and, since it tends to infinity as $\xi_1 \to \xi_{in}$ for $\omega \geq 3$, there exists a point $\xi_1 = \xi_*$ for which E_1 equals the explosion energy E. We conclude that the explosion from a finite energy can be described by this solution only in the region $\xi_* \leq \xi_0 \leq \xi \leq 1$. There must be a different flow in the range $\xi_{in} \leq \xi \leq \xi_0$. This contradicts the assumptions underlying self-similarity since it implies that there exist initial length and time scales which influence the behaviour of the flow over a scale of order R even as $R \to \infty$. The region $\xi_{in} \leq \xi \leq \xi_0$ corresponds to $\xi_{in}R \leq r \leq \xi_0 R$. Thus the energy of explosion tending to infinity alone does not explain the nonvalidity of the Taylor-Sedov type of solutions for $\omega \geq 3$.

The above discussion also indicates that, for $\omega \geq 3$, the energy which tends to infinity as $\xi \to \xi_{in}$, cannot constitute a relevant dimensional parameter. This is a relevant parameter only for $\omega < 3$.

We observe that the inner boundary of the self-similar solution, $\xi = \xi_{in}$, must be a particle path. For this to be true, $dr_0(t)/dt = u[r = r_0(t), t]$ so that, using the definition (3.7.7) for ξ, we have, for $\xi = \xi_0 = r_0/R$,

$$\frac{d\log\xi_0}{d\log R} = U(\log\xi_0) - 1. \qquad (3.7.33)$$

Since, at the inner boundary, $U(\log\xi_0) \to 1$, $\xi_0 = \xi_{in}$ is a solution of (3.7.33). The curve $r_{in}(t) = \xi_{in}R(t)$ describes the path of a fluid element. Therefore, there is no mass flow through this inner boundary of the self-similar solution. All the mass swept by the shock is contained in the region $\xi_{in} \leq \xi \leq 1$ and the region $\xi < \xi_{in}$ is void. Also, the speed of sound (3.7.30) at the inner boundary is zero, implying that the density is zero there.

For $\omega = (7 - \gamma)/(\gamma + 1)$, the solution in the (U, C) plane degenerates to the point solution $U = 1, C = 0$.

Waxman and Shvarts (1993) argue that even when initial length and time scales are relevant, self-similar solutions defined by (3.7.4)–(3.7.8) can describe an outer region bounded by the surfaces $r = r_1(t)$ and $r = R(t)$. It is assumed that the flow in this region is independent of the flow pattern in the inner region $0 \leq r \leq r_1(t)$. That is, it is assumed that the flow in $r_1(t) \leq r \leq R(t)$ is fully determined by the conditions there at $t = t_0$. It is also assumed that $r_1(t)/R(t) \to 0$ as $R \to \infty$. The integral of the reduced system of ODEs with appropriate shock conditions describes the outer flow in this case. Under these circumstances, the initial length and time scales influence the flow pattern only in the inner (irrelevant) region.

For such a separated flow, the upper boundary of the inner flow $r = r_1(t)$ can be shown to be a C_+ characteristic (Zeldovich and Raizer (1967)); the self-similar solution of the second kind is now constructed following these authors.

The C_+ characteristic

$$\frac{dr_+}{dt} = u + c, \tag{3.7.34}$$

may be written in terms of the similarity variables as

$$\frac{d\xi_+}{dt} = \frac{\dot{R}}{R}\xi_+[U(\xi_+) + C(\xi_+) - 1],$$

or

$$\frac{d\log\xi_+}{d\log R} = U(\xi_+) + C(\xi_+) - 1. \tag{3.7.35}$$

It is easily checked that the value of $U + C$ at the shock $\xi = 1$ is greater than 1 so that the shock point lies above the sonic line $U(\xi_+) + C(\xi_+) = 1$ (see (3.7.19)). The gradients (3.7.10) and (3.7.11) become infinite on the sonic line $U + C = 1$ since $\Delta = 0$ there. For the solution to remain single-valued, the numerators in these equations must also vanish when $U + C = 1$. Thus the parameter α in (3.7.12) is found such that the solution starting from the shock $[U(1), C(1)]$ crosses the sonic line at a singular point. It will now be shown that such a solution exists for all $\omega > 3$.

First, it may be observed that an exact analytic solution of this problem exists for a particular choice of $\omega = \omega_a = 2(4\gamma - 1)/(\gamma + 1)$. In this case the exponent α in the shock law (3.7.8) is found to be $(\gamma + 1)/2$. This exponent is quite different from that obtained from the Taylor-Sedov solution for $\omega = \omega_a$, namely, $2(\gamma + 1)/(7 - 3\gamma)$ (see section 3.6). The analytic form of the solution for $\omega = \omega_a$ satisfying the boundary conditions (3.7.19) at the shock $\xi = 1$ is simply

$$U(\xi) = \frac{2}{\gamma + 1}, \tag{3.7.36}$$

$$C(\xi) = \frac{\sqrt{2\gamma(\gamma - 1)}}{\gamma + 1}\xi^3, \tag{3.7.37}$$

$$G(\xi) = \frac{\gamma + 1}{\gamma - 1}\xi^{-8}. \tag{3.7.38}$$

This solution also passes through the singular point $U = 2/(\gamma + 1)$, $C = 0$ of (3.7.12) as $\xi \to 0$. It crosses the sonic line $U + C = 1$ at $U = 2/(\gamma + 1)$, $C = (\gamma - 1)/(\gamma + 1)$. This singular point also exists generally for all (relevant) values of ω and α and is located at the point P: $U = 1/\alpha$, $C = 0$. All self-similar solutions for $\omega > 3$ cross the sonic line and approach this singular point as $\xi \to 0$. Two examples considered by Waxman and Shvarts (1993) are $\omega = 3.4$ and $\omega = 5.5$. The corresponding values of α obtained by requiring that the solution of (3.7.12) starting from the shock passes through the singular point $(U = 1/\alpha, C = 0)$ are found to be 1.04 and 2.14, respectively. These agree very closely with the asymptotic form of the solution. For the direct numerical simulation of this asymptotic solution, Waxman and Shvarts (1993) chose the initial conditions to be zero velocity everywhere, constant density and pressure for $r < d$, and zero pressure and a density profile proportional to $r^{-\omega}$ for $r > d$. They used the artificial viscosity approach of Richtmyer and Von Neumann (1950). The asymptotic profiles agreed very well with the self-similar solution described here. This is in contrast to the results from Taylor-Sedov type of solutions. The latter exists for $\omega = 3.4$ but not for $\omega = 5.4$, a value greater than 5 (see section 3.6).

We may observe from (3.7.12) that $dU/dC = 0$ when $U = 1$. Also, $dU/dC < 0$ as C decreases from its value $[2\gamma(\gamma-1)]^{1/2}/(\gamma+1)$ at the shock to 0 at the singular point; moreover, $U < 1$ behind the shock. We conclude that the new singular point P $(U = 1/\alpha, C = 0)$ is important only when $\alpha > 1$. Thus the similarity exponent α for this self-similar solution of the second kind is greater than 1 in contrast to that for the Taylor-Sedov type of solutions for $\omega < 3$ when it is less than 1 (see section 3.6).

It is interesting to write the local solution of the present system in the neighbourhood of the singular point P. First, we may approximately write $\Delta, \Delta_1, \Delta_2$ near this point from (3.7.14)–(3.7.16) as

$$\Delta = -\left(1 - \frac{1}{\alpha}\right)^2, \tag{3.7.39}$$

$$\Delta_1 = -\frac{1}{\alpha}\left(1 - \frac{1}{\alpha}\right)\left(U - \frac{1}{\alpha}\right) + \left(\frac{\omega - 2[(\alpha-1)/\alpha]}{\gamma} - \frac{3}{\alpha}\right)C^2, \tag{3.7.40}$$

$$\Delta_2 = -\frac{3}{\alpha}\left(1 - \frac{1}{\alpha}\right)\frac{\gamma-1}{2}C. \tag{3.7.41}$$

With these approximations we may solve (3.7.12) in the form

$$U = \frac{1}{\alpha} + \begin{cases} \text{const.} \times C^{2/3(\gamma-1)}, & \gamma > 4/3, \\ f_1(\gamma,\omega,\alpha)C^2 \log C, & \gamma = 4/3, \\ f_2(\gamma,\omega,\alpha)C^2, & \gamma < 4/3, \end{cases} \tag{3.7.42}$$

where

$$f_1 = \left(\frac{3}{\alpha} - \frac{\omega - 2[(\alpha-1)/\alpha]}{\gamma}\right)\frac{2\alpha}{\alpha-1}, \tag{3.7.43}$$

$$f_2 = \left(\frac{3}{\alpha} - \frac{\omega - 2[(\alpha-1)/\alpha]}{\gamma}\right)\frac{\alpha^2}{(\alpha-1)(3\gamma-4)}. \tag{3.7.44}$$

The solution of (3.7.11) may now be written as

$$C(\xi) = \text{const.} \times \xi^{3(\gamma-1)/2(\alpha-1)}. \tag{3.7.45}$$

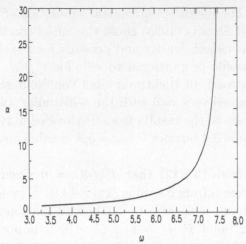

Figure 3.12 The function $\alpha(\omega)$ for $\gamma = 5/3$ for self-similar solutions of second kind (Waxman and Shvarts, 1993).

The function $G(\xi)$ is obtained from (3.7.17) as

$$G(\xi) = \text{const.} \times \xi^{-(\alpha\omega-3)/(\alpha-1)}. \qquad (3.7.46)$$

The second kind of solution presented above exists only for a limited range of ω values. The parameter $\alpha = \alpha(\omega)$ for $\gamma = 5/3$ is shown in Figure 3.12.

It is observed that $\alpha \to 1$ as $\omega \downarrow \omega_g$ for some $\omega_g > 3$ while $\alpha \to \infty$ as $\omega \uparrow \omega_c$ for some ω_c. ω_g and ω_c for $\gamma = 5/3$ are 3.256 and 7.686, respectively. For $3 \le \omega \le \omega_g(\gamma)$, there is no α for which the integral curve in the (U, C) plane crosses the sonic line at the singular point. The intervals $3 \le \omega \le \omega_g(\gamma)$ and $\omega \ge \omega_c$ for the density exponent need further investigation.

Waxman and Shvarts (1993) considered the qualitative nature of the flows in the outer and inner regions, as described above. Their asymptotic nature was confirmed with reference to the numerical solution of the basic system of PDEs with appropriate initial conditions.

3.8   Point Explosion with Heat Conduction

In section 3.1, we discussed the Taylor-Sedov solution for a point explosion in a self-similar form which gave the famous shock law, $r_s \sim t^{2/5}$. However, this solution predicted an anomalous behaviour of infinite temperature and infinite temperature gradient near the center of explosion. What caused this singularity was the omission of the effect of heat transfer by conduction, which is significant at such high temperatures. Indeed, the very early phase $(t \to 0)$ of the phenomenon is essentially described by the heat equation

$$\frac{\partial T}{\partial t} = \frac{1}{r^2} \frac{\partial}{\partial r} \left(\chi r^2 \frac{\partial T}{\partial r} \right), \qquad (3.8.1)$$

where the coefficient of heat conduction is given by

$$\chi = \chi_0 \rho^a T^b. \tag{3.8.2}$$

χ_0, a, and b are constants appropriate to a given material; here

$$a \leq 0, \quad b \geq 1. \tag{3.8.3}$$

Indeed, Barenblatt (1979) considered this equation with $a = 0$, subject to the conditions

$$T(r,0) = 0 \quad (r \neq 0); \quad 4\pi C \int_0^{r_*} T(r,0)r^2 dr = E,$$
$$T(\infty,t) = 0 \quad (t > 0), \tag{3.8.4}$$

and found a self-similar solution of this problem. Here, r_* is a positive constant. The constant C is related to χ_0. This solution describes a spherical heat wave with a sharp front, which moves supersonically, much faster than the hydrodynamic flow. The front of this thermal wave moves according to $r_h(t) \propto t^{1/(3b+2)}$, determined by the relevant form of the similarity variable, namely, $\xi = rt^{-1/(3b+2)}$. Thus the shock wave according to Taylor-Sedov solution and the thermal wave front move according to the same law only if $b = 1/6$. In a more realistic situation, $b \geq 1$, and therefore the trajectories of the shock and the thermal front are different (see Figure 3.13). These fronts intersect at a finite time $t = t_1$ at some $r = r_1$. What actually transpires in an intense explosion is that at the very early times the explosion front is supersonic and is governed essentially by nonlinear heat conduction. This front soon decelerates. There sets in behind it a large hydrodynamic motion including an isothermal shock. As the thermal front decelerates, it is overtaken by the isothermal shock at some time $t = t_1$. At this point the heat front becomes subsonic and the shock front runs ahead of it. The flow

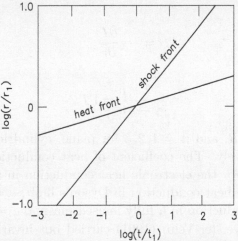

Figure 3.13 Trajectories of shock front and heat front driven by a strong point explosion in a uniform gas (Reinicke and Meyer-ter-Vehn, 1991).

from now onwards is governed by the Taylor-Sedov solution except in a central region where heat conduction still plays an important role and keeps the temperature finite.

In a paper by Reinicke and Meyer-ter-Vehn (1991), an attempt was made to combine these two transport processes—hydrodynamic and thermal conduction—in the framework of self-similarity with both shock front and heat front running against each other. Since the equations of motion with heat conduction do not admit a similarity hypothesis as such, an additional freedom was introduced by regarding the medium ahead of the disturbed region to be stratified according to $\rho_0 \sim \rho^k$ with $k \approx -2$. This enables complete self-similarity with physically reasonable parameters for the coefficient of heat conduction. By varying the energy release E_0 of the explosion, the situations with different ratios r_h/r_s, corresponding to the heat front r_h running behind the shock r_s or vice versa, can be realized. The qualitative features in the self-similar solution—the occurrence of two sharp fronts and the way they shift against each other as a function of E_0—are also characteristics of the non-self-similar solutions. The artifice of varying the density ahead of the disturbed medium proves handy in achieving self-similarity and hence a fairly comprehensive qualitative analysis of the problem. In the sequel we follow the work of Reinicke and Meyer-ter-Vehn (1991).

The gasdynamic equations in different geometries, with the inclusion of the heat conduction effects, are

$$\rho_t + u\rho_r + \frac{\rho}{r^{n-1}}(r^{n-1}u) = 0, \tag{3.8.5}$$

$$u_t + uu_r + \frac{1}{\rho}p_r = 0, \tag{3.8.6}$$

$$e_t + ue_r - \frac{p}{\rho^2}(\rho_t + u\rho_r) = -(\rho r^{n-1})^{-1}(r^{n-1}S)_r, \tag{3.8.7}$$

where

$$S = -\chi\frac{\partial T}{\partial r} \tag{3.8.8}$$

and

$$e = \frac{1}{\gamma - 1}\frac{p}{\rho} = \Gamma T. \tag{3.8.9}$$

Here, $\Gamma = (\gamma - 1)^{-1}R$, and $n = 1, 2, 3$ for plane, cylindrical and spherical symmetries, respectively. The coefficient of heat conduction χ is given by (3.8.2) and (3.8.3). For the electronic heat conduction in a plasma, $a = 0$, $b = 5/2$, for radiative heat conduction in Dyson's limit, $a = -1$, $b = 4$, and for radiative heat conduction in a fully ionised plasma, $a = -2$, $b = 13/2$.

Reinicke and Meyer-ter-Vehn (1991) carried out invariance analysis of the above system of PDEs, considering the boundary conditions at the heat front and the isothermal shock, and arrived at the following form of the similarity solution:

$$u(r,t) \;=\; \left(\frac{\alpha r}{t}\right) U(\xi), \qquad (3.8.10)$$

$$T(r,t) \;=\; \frac{1}{\Gamma}\left(\frac{\alpha r}{t}\right)^2 \Theta(\xi), \qquad (3.8.11)$$

$$\rho(r,t) \;=\; (g_0 r^k) G(\xi), \qquad (3.8.12)$$

where

$$\xi = \frac{r}{(\zeta|t|^\alpha)}. \qquad (3.8.13)$$

ζ and g_0 are constants. The system (3.8.5)–(3.8.9) now reduces to a system of nonlinear ODEs for $U(\xi)$, $\Theta(\xi)$, and $G(\xi)$ provided

$$(a-1)k = (2b-1)(\alpha^{-1}-1)+1. \qquad (3.8.14)$$

The reduced system of ODEs is

$$U' - (1-U)(\ln G)' + (n+k)U = 0, \qquad (3.8.15)$$

$$(1-U)U' + U(\alpha^{-1}-U) = \Theta[\ln(\xi^{2+k}G\Theta)]', \qquad (3.8.16)$$

$$\begin{aligned}
2[U' + nU - \mu(\alpha^{-1}-1)] &- \mu(1-U)[\ln(\xi^2\Theta)]' \\
&= \beta_0\Theta^b G^{a-1}\xi^{(2b-1)/\alpha}((\ln\Theta)'' + [\ln(\xi^2\Theta)]' \\
&\quad \times\{n-2+a[\ln(\xi^k G)]' + (b+1)[\ln(\xi^2\Theta)]'\}),
\end{aligned} \qquad (3.8.17)$$

where $\;' = d/d\ln\xi, \quad \mu = 2/(\gamma-1)$, and

$$\beta_0 = [2\chi_0(\alpha\zeta^{1/\alpha})^{(2b-1)}/\Gamma^{b+1}g_0^{1-a}]\operatorname{sgn}(t).$$

It is convenient to introduce the variables

$$H(\xi) \;=\; \xi^{-\sigma}G(\xi), \qquad \sigma = (2b-1)/\alpha(1-a), \qquad (3.8.18)$$

$$W(\xi) \;=\; -\beta_0 H^{a-1}\Theta^b[\ln(\xi^2\Theta)]'/2. \qquad (3.8.19)$$

It follows from the identity

$$S(r,t)/p(r,t) = \alpha(r/t)W(\xi), \qquad (3.8.20)$$

that W represents a reduced heat velocity. In terms of $W(\xi)$ and $H(\xi)$, equations (3.8.15)–(3.8.17) can be written as

$$(\ln\Theta)' = -2[1 + (W/\beta_0 H^{a-1}\Theta^b)], \qquad (3.8.21)$$

$$U' + (U-1)(\ln H)' = \sigma - (n+k+\sigma)U, \qquad (3.8.22)$$

$$(U-1)U' + \Theta(\ln H)' \; = \; U(\alpha^{-1} - U) + \Theta[(2W/\beta_0 H^{a-1}\Theta^b)$$
$$-k - \sigma],$$

$$(3.8.23)$$

$$W' + U' + W(\ln H)' \; = \; (W/\beta_0 H^{a-1}\Theta^b)$$
$$\times [\mu(U-1) + 2W] + \mu(\alpha^{-1} - 1)$$
$$-nU - (n + k + \sigma)W. \qquad (3.8.24)$$

Solving for the derivatives from (3.8.21)–(3.8.24), we have

$$\frac{d\ln\xi}{N_\xi} = \frac{d\ln H}{N_H} = \frac{dU}{N_U} = \frac{d\ln\Theta}{N_\Theta} = \frac{dW}{N_W}, \qquad (3.8.25)$$

where

$$
\begin{aligned}
N_\xi \;&=\; (1-U)^2 - \Theta, \\
N_H \;&=\; -(n+k+\sigma-1)N_\xi + N_2, \\
N_U \;&=\; (1-U-n-k)N_\xi + (1-U)N_2, \\
N_\Theta \;&=\; AN_\xi, \\
N_W \;&=\; \{(n-1)(1-U) - W + k + \mu(\alpha^{-1}-1) \\
&\quad -(2+A)[W - \mu(1-U)/2]\}N_\xi \\
&\quad -(1-U+W)N_2,
\end{aligned}
\qquad (3.8.26)
$$

and

$$
\begin{aligned}
A \;&=\; (\ln\Theta)' = -2[1 + W/\beta_0 H^{a-1}\Theta^b], \\
N_2 \;&=\; \Theta(A + 3 - n) + (k + n + \alpha^{-1} - 2)(1 - U) \\
&\quad -(\alpha^{-1} - 1).
\end{aligned}
\qquad (3.8.27)
$$

The above form is useful for local singularity analysis, which is crucial for numerical integration of the system (3.8.25). In the five dimensional space $[\ln\xi, U, \Theta^{1/2}, (\ln\Theta)', H]$, the main singularity falls where $N_\xi = 0$, that is, on the 'sonic' hyperplanes

$$U = 1 \pm (\Theta)^{1/2}. \qquad (3.8.28)$$

For the derivative $d(\ln\Theta)/dU$ to be finite, N_U and N_Θ must each vanish along (3.8.28), requiring that N_2 is also zero herewith:

$$(\ln\Theta)'\Big|_{\Theta=(U-1)^2} = n - 3 + \frac{n-2+\alpha^{-1}+k}{U-1} + \frac{\alpha^{-1}-1}{(U-1)^2} \qquad (3.8.29)$$

(see (3.8.27)). An investigation of the second derivative of ξ shows that it reaches its maximum or minimum at the above crossing point and, there-fore, the corresponding solution curves $U(\xi)$, $\Theta(\xi)$, $H(\xi)$, and $W(\xi)$ are double-valued and hence nonphysical. Physical solutions may be obtained by introducing a shock discontinuity connecting points on different sides of the sonic plane.

A local analysis near the center $\xi = 0$ is necessary to identify those (physical) solutions which give a finite pressure there. For the nonplanar cases $n \neq 1$, a local analysis of the system (3.8.21)–(3.8.24) shows that, for $\xi \sim 0$, we have

$$U \sim U_0 \xi^w, \quad G \sim G_0 \xi^{\delta-k}, \quad \Theta \sim \Theta_0 \xi^{-\delta-2}, \tag{3.8.30}$$

where we may identify

$$w = 0, \quad U_0 = (\delta - k)/(\delta + n). \tag{3.8.31}$$

These values lead to one of the following possibilities:

i) $\delta = (n-2)/(b+1-a), \quad n = 3;$ (3.8.32)

ii) $\delta = 0;$ (3.8.33)

iii) $\delta = -2/(b+1-a),$

$$2nU_0 - 2\mu(\alpha^{-1} - 1) + \mu(1 - U_0)\delta = -\delta n \beta_0 \Theta_0^b G_0^{a-1}. \tag{3.8.34}$$

iv) $-2 < \delta < -2/(b+1-a),$

$$2nU_0 - 2\mu(\alpha^{-1} - 1) + \mu(1 - U_0)\delta = 0. \tag{3.8.35}$$

In physical variables, these asymptotic behaviours ($r \to 0$) would have the form

$$
\begin{aligned}
\rho &\sim r^\delta t^{\alpha(k-\delta)}, \\
u &\sim r/t, \\
T &\sim r^{-\delta} t^{\alpha(2+\delta)-2}, \\
S &\sim r^{-1-\delta(b+1-a)} t^{\alpha[\delta(b+1-a)+4+k]-3}.
\end{aligned}
\tag{3.8.36}
$$

The cases (iii) and (iv) place severe constraints on the parameters. Cases (i)–(ii) lead to some interesting solutions. For case(i), the local behaviour near $\xi \sim 0$ can be written out as

$$
\begin{aligned}
U &= U_0 + U_1 \xi + \cdots, \\
G &= G_0 \xi^{\delta-k}(1 + G_1 \xi + \cdots), \\
\Theta &= \Theta_0 \xi^{-\delta-2}(1 + T_1 \xi + \cdots),
\end{aligned}
\tag{3.8.37}
$$

where G_0, Θ_0, and T_1 are free parameters while

$$G_1 = -T_1, \quad U_1 = [(1 - U_0)/(4 + \delta)]T_1 \qquad (3.8.38)$$

(see (3.8.31) for U_0). We may observe from (3.8.37$_3$) that the heat flux near $r = 0$ tends to infinity like r^{-2}.

The case (ii) alone gives finite temperature at $r = 0$. Defining

$$\tilde{\xi} = \xi^2/(z\Theta_0), \quad z = \beta\Theta_0^{b-1}G_0^{a-1}, \text{ a constant}, \qquad (3.8.39)$$

one may write the local solution at $r = 0$ as

$$
\begin{aligned}
U &= -\frac{k}{n} + \sum_{i=1}^{\infty}\left(\sum_{j=0}^{i}\mu_{i,j}z^j\right)\tilde{\xi}^i, \\
G &= G_0\xi^{-k}\left[1 + \sum_{i=1}^{\infty}\left(\sum_{j=0}^{i}\gamma_{i,j}z^j\right)\tilde{\xi}^i\right], \qquad (3.8.40) \\
\Theta &= \Theta_0\xi^{-2}\left[1 + \sum_{i=1}^{\infty}\left(\sum_{j=0}^{i-1}\tau_{i,j}z^j\right)\tilde{\xi}^i\right],
\end{aligned}
$$

where the coefficients $\mu_{i,j}, \gamma_{i,j}$ and $\tau_{i,j}$ are independent of Θ_0, G_0, and β_0. The first coefficients in (3.8.40) are

$$
\begin{aligned}
\mu_{1,j} &= \frac{2 + 2k/n}{n + 2}\gamma_{1,j}, \quad j = 1, 2, \\
\tau_{1,0} &= -\gamma_{1,0} = -\frac{k + \mu(\alpha^{-1} - 1)}{n}, \qquad (3.8.41) \\
\gamma_{1,1} &= -\frac{k}{2n}\left(\frac{1}{\alpha} + \frac{k}{n}\right).
\end{aligned}
$$

For this case, the heat flux at $r = 0$ may be checked to tend to zero like r. Moreover, the temperature at $r = 0$ has a relative maximum or minimum depending on whether $k + 2(\alpha^{-1} - 1)/(\gamma - 1)$ is positive or negative. We may also observe that $\tau_{1,0} + \gamma_{i,0} = 0$ holds for all i. Therefore, reverting to physical variables ρ and T and using (3.8.40) one may check that the pressure becomes independent of r. Thus, the flow is nearly isobaric in the central region. This condition seems to hold even for non-self-similar regimes and is made the basis of an analytic approach by Reinicke and Meyer-ter-Vehn (1991).

It is important to have a local solution in the neighbourhood of the front of the heat wave which runs into a cold medium with the density distribution $\rho_0(r) \propto r^k$. Thus, the boundary conditions at the heat front, which is assumed to be given by $\xi = \xi_f$, are

$$\Theta \to 0, \quad U \to 0, \quad G \to 1, \quad S \to 0 \quad \text{as} \quad \xi \to \xi_f. \qquad (3.8.42)$$

Assuming the local solution satisfying (3.8.42) in the form

$$U = U_0|1-x|^{\beta_1}, \quad G = 1 + G_1|1-x|^{\beta_2},$$
$$\Theta = \Theta_0|1-x|^{\beta_3}, \quad x = \xi/\xi_f, \tag{3.8.43}$$

introducing it into (3.8.15)–(3.8.17), and balancing the terms etc., one may arrive at the following determination of the parameters:

$$\beta_1 = \beta_2 = \beta_3 = 1/b, \quad G_1 = U_0 = \Theta_0,$$
$$\beta_0 U_0^b \xi_f^{(2b-1)/\alpha} \operatorname{sgn}(1-x) = \mu b. \tag{3.8.44}$$

The last equality in (3.8.44) requires that $\beta_0(1-x)$ is positive. Observing the definition of β_0 below (3.8.17), we infer that $t(1-x) > 0$. This implies that, for an exploding heat wave, time must be taken to be positive. The normalisation $G \to 1$ at the front is possible since (3.8.12) contains an arbitrary dimensional constant g_0. It is clear that if the solution joins two singular points this choice can be made use of at one of the singular points only. To get a better approximation, Reinicke and Meyer-ter-Vehn (1991) used the local solution (3.8.43)–(3.8.44) to connect the functions and write the solution in the form

$$G = (1-U)^{-1},$$
$$\Theta = U(1-U), \tag{3.8.45}$$
$$W = [\mu - (\mu+1)U]/2.$$

This leads to the integral

$$\ln x = -\beta_0 \xi_f^{(2b-1)/\alpha} \int_0^U y^{b-1}(1-y)^{b-a}$$
$$\times \frac{1-2y}{\mu - (\mu+1)y} dy. \tag{3.8.46}$$

The approximate solution (3.8.45)–(3.8.46) holds in the neighbourhood of the thermal front $\xi = \xi_f$, that is, $x = 1$. It is obtained analytically from (3.8.15)–(3.8.17) by ignoring smaller terms in this neighbourhood.

Finally, the isothermal shock conditions are (see Marshak (1958))

$$T_2 = T_1,$$
$$\rho_2(u_2 - D) = \rho_1(u_1 - D),$$
$$p_2 + j^2/\rho_2 = p_1 + j^2/\rho_1, \tag{3.8.47}$$
$$E_2 + \frac{p_2}{\rho_2} + \frac{(u_2 - D)^2}{2} + \frac{S_2}{j} = E_1 + \frac{p_1}{\rho_1} + \frac{(u_1 - D)^2}{2} + \frac{S_1}{j},$$

where D is the shock velocity and suffixes 1 and 2 give conditions ahead of and behind the shock (see (3.8.8) for the definition of S).

The system (3.8.47) in terms of the similarity functions Θ, U, G and W may be solved in the form

$$
\begin{aligned}
\Theta_2 &= \Theta_1 =: \Theta, \\
U_2 &= 1 - \Theta/(1 - U_1), \\
G_2 &= [(1 - U_1)^2/\Theta]G_1, \\
W_2 &= \frac{\Theta W_1 - \{[(1 - U_1)^4 - \Theta^2]/2(1 - U_1)\}}{(1 - U_1)^2}.
\end{aligned}
\tag{3.8.48}
$$

We shall now describe the numerical integration of the self-similar system of equations describing a point explosion with heat conduction. We first discuss the parameters involved. The density distribution in the undisturbed medium $\rho_0 = g_0 r^k$ fixes the constants g_0 and k. (The latter appears in the shock conditions too.) The energy of the point explosion E_0 at $r = 0$ is also prescribed. The total energy of the explosion may be written as (see section 3.7)

$$
E_{tot}(t) \propto t^{\alpha(n+2+k)-2}
\tag{3.8.49}
$$

and is therefore constant if

$$
\alpha(n + 2 + k) = 2.
\tag{3.8.50}
$$

For an explosion in a nonheat-conducting uniform medium we have $n = 3$, $k = 0$ and, therefore, $\alpha = 2/5$ from (3.8.50), confirming the famous shock law $R_s = \xi_s \zeta t^{2/5}$. In the present case, the self-similarity condition (3.8.14) and the constancy of total energy (3.8.50) fix the parameters α and k as

$$
\alpha = \frac{2b - 2a + 1}{2b - (n + 2)a + n}, \quad k = -\frac{(2b - 1)n + 2}{2b - 2a + 1},
\tag{3.8.51}
$$

in terms of the exponents a and b in the heat conduction coefficient in (3.8.2) for different geometries $n = 3, 2, 1$. Since $a \leq 0$ and $b \geq 1$ in (3.8.2), (3.8.51) implies that $0 < \alpha < 1$ and $-n < k < 0$. For given values of the parameters a, b, and γ, the solution depends now on the dimensional parameters E_0, g_0, and $2\chi_0/\Gamma^{b+1}$, which combine to give the dimensionless parameter

$$
\lambda = [2\chi_0/\Gamma^{b+1} g_0^{1-a}](E_0/g_0)^{b-1/2}.
\tag{3.8.52}
$$

For sufficiently large explosion energy E_0, λ is large and so the leading front moving into the undisturbed gas ahead is a heat wave and is described by $\xi_f = r_f/(\zeta t^\alpha)$. In this case it turns out that the solution curves in the $(U, \Theta^{1/2})$ plane have to connect points on the two sides of the sonic discontinuity. This transition must therefore be brought about by an isothermal shock. For an appropriate choice of ζ (see (3.8.13)), this shock may be identified as $\xi = \xi_s = 1 < \xi_f$. As the strength of explosion λ decreases, ξ_f decreases until $\xi_f \to \xi_s$; thereafter, the flow is headed by a strong shock wave, the flow resembling the usual hydrodynamic explosion with heat conduction dominating only near the center of explosion.

The integration of (3.8.21)–(3.8.24) may now be performed as follows. Starting at the heat front with $U = 0$, $\Theta = 0$ (cold conditions ahead) and a trial value of β_0, we choose some value $\xi_f > 1$ and integrate in the direction of decreasing ξ (see the approximate solution (3.8.45)). We continue the integration to the point $\xi = 1$ and apply the shock conditions (3.8.48) to cross the sonic line $U = 1 - \sqrt{\Theta}$. We continue the integration towards $\xi = 0$. As we vary the parameter β_0 (see below (3.8.17)), a family of curves is obtained all except one of which run into the upper or lower sonic line and therefore are unphysical. The exceptional integral curve goes through the singular point $(-k/n, \infty)$ in the $(U, \Theta^{1/2})$ plane, which corresponds to $\xi = 0$. From the definition of β_0, λ in (3.8.52), and the energy equation, we find that

$$\lambda(\xi_f) = \beta_0(\xi_f)[I(\xi_f)]^{b-1/2}, \qquad (3.8.53)$$

$$E_0/g_0 = g_0(\alpha\zeta^{1/\alpha})^2 I(\xi_f), \qquad (3.8.54)$$

where

$$I(\xi_f) = 2\pi \int_0^{\xi_f} \xi^{n+k+1} G(\xi)[U^2(\xi) + \mu\Theta(\xi)]d\xi \qquad (3.8.55)$$

(cf. (3.1.25)). Here ζ is obtained in terms of E_0/g_0; E_0 and g_0 are known parameters. The parameter $\lambda(\xi_f)$ fully determines the solution of the problem.

Using the similarity transformation (3.8.10)–(3.8.14), the entropy equation may be written as

$$p/\rho^\gamma = r^{-\epsilon}A(\xi), \qquad (3.8.56)$$

where

$$\epsilon = k(\gamma - 1) + 2(\alpha^{-1} - 1) \qquad (3.8.57)$$

and $A(\xi)$ is the reduced entropy function. If we use (3.8.51) in (3.8.57) to express the latter in terms of a and b only, we may verify, using (3.8.3), that there is a temperature maximum at the center if $\epsilon > 0$. It is a minimum if $\epsilon < 0$. This point has been discussed earlier. ϵ determines the radial distribution of entropy. For the nonheat-conducting case (see section 3.2), the temperature near the center of the explosion is given by

$$T(r,t) \propto r^{-\epsilon/(\gamma-1)}t^{-\alpha(n\gamma-2n-k)/(\gamma-1)}. \qquad (3.8.58)$$

It diverges or vanishes as $r \to 0$, depending on the sign of ϵ. The inclusion of heat conduction makes the temperature at the center finite with a relative maximum or minimum there. $\epsilon > 0$ ($\epsilon < 0$) corresponds to $\gamma < \gamma_0$ ($\gamma > \gamma_0$), where

$$\gamma_0 = 1 + [2n(1 - a) - 2]/[(2b - 1)n + 2]. \qquad (3.8.59)$$

Taylor-Sedov solution (without the effect of heat conduction) exists provided $\gamma < \gamma_1$, where

$$\gamma_1 = (3n - 2 + k)/(n - 2 - k), \tag{3.8.60}$$

(see section 3.2). This condition changes to $\gamma > \gamma_1$ if $\gamma_1 < 0$. For $k = 0$, $n = 3$, $\gamma_1 = 7$, we have the well-known Primakoff solution for water which has constant (reduced) velocity U and temperature Θ and is distinct in character from the Taylor-Sedov solution. Reinicke and Meyer-ter-Vehn (1991) used (3.8.60) for γ_1 with k given by (3.8.51) for the heat-conducting case and found it a 'reasonable' condition here too.

The transition of the heat front to a strong shock is analytically examined by using the approximation (3.8.45)–(3.8.46) in the limit $\xi_f \to \xi_s = 1$. In this limit, β_0 tends to 0 (see below (3.8.17)). It then follows from (3.8.19) that the heat velocity $W = W_2$ behind the shock vanishes since H_2, Θ_2, and $(\Theta')_2$ are finite there. With $W_2 = 0$, the shock conditions (3.8.48) reduce to

$$U_2 = \mu/(1 + \mu), \quad G_2 = \mu + 1, \quad \Theta_2 = \mu/(\mu + 1),$$

corresponding to those with no heat conduction. Here, $\mu = 2/(\gamma - 1)$.

Figures 3.14 and 3.15 show typical self-similar flows in spherical symmetry with a strong heat conduction with $\beta_0 = 7.12 \times 10^7$ and $\xi_f = 2$ and weak heat conduction with $\beta_0 = 2.14 \times 10$ and $\xi_f = 1 + 10^{-7}$. The parameters in

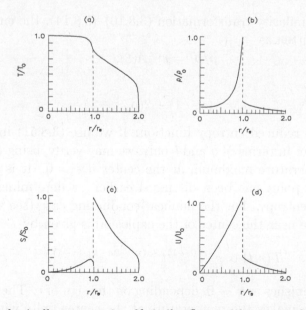

Figure 3.14 Spherically symmetric self-similar flows with a strong heat conduction with $\beta_0 = 7.12 \times 10^7$ and $\xi_f = 2$. The dashed lines mark the shock front (Reinicke and Meyer-ter-Vehn, 1991).

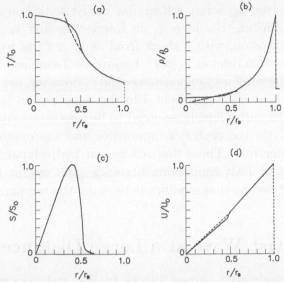

Figure 3.15 Spherically symmetric self-similar flows with a weak heat conduction with $\beta_0 = 2.14 \times 10$ and $\xi_f = 1 + 10^{-7}$. The thin dashed lines mark the shock front; the dashed lines behind the shock represent the solution without any heat conduction (Reinicke and Meyer-ter-Vehn, 1991).

the heat conduction coefficient (3.8.2) are chosen to be $a = -2$, $b = 13/2$, while $\gamma = 5/4$. The cases (a)–(d) in each figure show normalised temperature, density, heat flux and velocity, respectively.

In the conduction-dominated case, the heat front is located at $r/r_s = 2$, far ahead of the shock at $r/r_s = 1$. In this case there is a large heat flux in the region $1 < r/r_s < 2$. The particle velocity and density have large values just behind the shock front at $r = r_s$. Figure 3.15 shows essentially the hydrodynamic case with a strong shock at the head with typical density and velocity discontinuities and vanishing heat flux. The heat wave itself has now retreated to the inner region; there is, however, a makedly sharp front in the temperature and heat flux at $r/r_s \approx 0.5$. It is also observed that the pressure in the heat region is almost constant.

It was found from the computational results that as E_0 is increased by a factor of 14, keeping all other parameters fixed, there is a drastic change from a hydrodynamic (Figure 3.15) to a heat transport mode (Figure 3.14). If E_0 is further multiplied by 2 so that $\xi_f/\xi_s \approx 20$, the total explosive wave assumes the character of Barenblatt's pure heat wave solution. Contrarily, if E_0 is reduced to values below those indicated in Figure 3.15, $\xi_f/\xi_s \to 0$ with the heat wave present near the center only. The work of Reinicke and Meyer-ter-Vehn (1991) concludes with an interesting analysis of the solution in the inner region, based on the observation that the pressure determined by the hydrodynamic motion is almost uniform there and remains unaffected by heat conduction. Assuming the pressure to be a function of time alone, as given by the purely hydrodynamic Taylor-Sedov model, the equations of motion were solved by a non-self-similar analysis; the medium ahead is assumed to be uniform. For large t, an interesting flow regime develops. Here, the total explosion, with a shock front at $r_s \propto t^\alpha$ as well as an inner heat wave region with a front at $r_h \propto t^\delta$, becomes self-similar, and $r_h \to r_s$ as time increases. The similarity exponents α and δ, however, are different with $\delta < \alpha$. The new heat wave exponent δ in the inner region is also different from that when the entire flow is assumed to be heat-dominated. From this approximate analysis the central temperature and the extent of the heat wave region are derived. Thus, the well-known hydrodynamic solution of the problem without heat conduction becomes valid except in the central region where heat conduction modifies it to make the temperature finite.

3.9   The Blast Wave at a Large Distance

We have so far dealt with strong shocks in the early stage of the blast wave propagation. We now discuss the other end—the nearly linear regime when the shock has become rather weak. Whitham (1950) gave a general theory to cover such flows. His method describes the attenuation of spherical shocks at large distances from the origin. It also describes the situation when the disturbances are small from the start. In the latter case, his approach, which involves modification of the linearised form of the solution, gives results which are uniformly valid at all distances from the origin. One basic assumption of the theory is that, since the entropy changes at the shock are of third order in strength, the flow with weak shocks may be considered isentropic; this, however, is the only assumption underlying the theory.

We consider spherically symmetric isentropic flow in air, governed by the equation

$$a^2 \nabla^2 \phi = \phi_{tt} + 2\phi_r \phi_{rt} + (\phi_r)^2 \phi_{rr}. \tag{3.9.1}$$

where ϕ is the velocity potential and a is the velocity of sound, given by Bernoulli's equation

$$a^2 = a_0^2 - (\gamma - 1)\left[\phi_t + \frac{1}{2}\phi_r^2\right]. \tag{3.9.2}$$

a_0 is the velocity of sound in the undisturbed air (see also section 4.1). We may rewrite (3.9.1) as the system

$$v_t - u_r = 0, \tag{3.9.3}$$

$$v_r \left[a_0^2 - (\gamma - 1)u - \frac{1}{2}(\gamma + 1)v^2\right] - u_t - 2vv_t$$

$$+ \frac{2v}{r}\left[a_0^2 - (\gamma - 1)u - \frac{1}{2}(\gamma - 1)v^2\right] = 0, \tag{3.9.4}$$

where $u = \phi_t$ and $v = \phi_r$.

As in the piston problem considered by Taylor (1946) (see section 3.1), the linear theory provides the motivation for the asymptotic results. In this case, the velocity potential is given by $\phi = f_0(a_0 t - r)/r$; therefore, u and v have the form

$$u = \frac{f_1(a_0 t - r)}{r}, \tag{3.9.5}$$

$$v = \frac{f_2(a_0 t - r)}{r} + \frac{f_3(a_0 t - r)}{r^2}. \tag{3.9.6}$$

Equations (3.9.5)–(3.9.6) suggest that u and v may be sought in descending powers of r, with coefficients that are constant on exact characteristics described by $z = $ constant, where z is a function of r and t to be found in the process of solution. Thus, we write

$$u = a_0^2 \left[\frac{f(z)}{r} + \frac{g(z)}{r^2} + \cdots \right], \tag{3.9.7}$$

$$v = -\frac{u}{a_0} + a_0 \left[\frac{b(z)}{r^2} + \frac{c(z)}{r^3} + \cdots \right]. \tag{3.9.8}$$

The characteristics of the system (3.9.3)–(3.9.4) are given by

$$\left(\frac{dt}{dr} \right)^2 \left[a_0^2 - (\gamma - 1)u - \frac{1}{2}(\gamma + 1)v^2 \right] + 2v \frac{dt}{dr} - 1 = 0. \tag{3.9.9}$$

Substitution of (3.9.7)–(3.9.8) into (3.9.9) shows that, if $z = $ constant is an exact characteristic, it must have the form

$$a_0 t = r - z \log r - h(z) - \frac{m(z)}{r} \cdots \cdots . \tag{3.9.10}$$

On the other hand, the $\log r$ term in (3.9.10) would require that the expansions (3.9.7), (3.9.8) and (3.9.10) must be changed with g, b, c, and m replaced by

$$g_1(z) \log r + g_2(z), \quad b_1(z) \log r + b_2(z),$$
$$c_1(z) \log r + c_2(z), \quad m_1(z) \log r + m_2(z), \tag{3.9.11}$$

respectively, if equations (3.9.3) and (3.9.4) are to be satisfied. Substituting (3.9.7), (3.9.8) and (3.9.10) with the changed coefficients (3.9.11) into (3.9.3), (3.9.4) and (3.9.9) and equating coefficients of different products of $1/r$ and $\log r$ to zero lead to the conclusion that the unknown functions of z may be expressed in terms of $h(z)$ and that the function $g_1(z)$ is identically zero. The result is

$$u = a_0^2 \left[-\frac{kz}{r} + \frac{K_1 z^2 + \frac{1}{2} B_1^2}{r^2} + \cdots \right], \tag{3.9.12}$$

$$v = -\frac{u}{a_0} - a_0 \left[\frac{(\frac{1}{2}kz^2 + B_1)\log r + \frac{1}{2}kz^2 + k\int_0^z \zeta h'(\zeta)d\zeta + B_2}{r^2} \right]$$
$$+ \cdots\cdots, \tag{3.9.13}$$

$$a_0 t = r - z\log r - h(z)$$
$$- \frac{(\frac{1}{2}kz^2 + B_1)\log r + K_2 z^2 + \frac{1}{4}(\gamma + 5)B_1 + k\int_0^z \zeta h'(\zeta)d\zeta + B_2}{r}$$
$$+ \cdots\cdots, \tag{3.9.14}$$

where B_1 and B_2 are arbitrary constants, $k = 2/(\gamma + 1)$, $K_1 = k^2 - k/4$, $K_2 = 5/4 + 3k/2$.

It is convenient to impose the shock conditions in the following form: (i) the angle property which states that, to first order in strength of the shock, the angles that the shock makes with the characteristics on each side of it are equal, and (ii) ϕ is continuous along the shock path so that the derivative $\phi_r + U^{-1}\phi_t$ assumes the same value on each side of it; here U is the velocity of the shock (see Figure 3.16). Let the main shock be referred to as S while the secondary shock (if any) behind the main shock may be called S_1. Let the equation of a characteristic C meeting the main shock be written as

$$a_0 t = r - z\log r - h(z) + O(r^{-1}\log r), \quad z = \text{constant}. \tag{3.9.15}$$

Since the characteristic C lies between the two shocks, for a fixed r, t is bounded. It follows that z and $h(z)$ must be bounded in this region. If the equation of the shock S is assumed to be $a_0 t = r - f(r)$, then the angle property via (3.9.14) gives

$$f'(r) = \frac{1}{2}zr^{-1} + O(r^{-2}\log r). \tag{3.9.16}$$

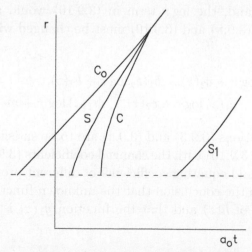

Figure 3.16 The main shock S and the secondary shock S_1 in $(r, a_0 t)$ plane (Whitham, 1950).

From (3.9.15) and the shock locus $a_0 t = r - f(r)$ we get

$$f(r) = z \log r + h(z) + O(r^{-1} \log r). \qquad (3.9.17)$$

For each characteristic meeting the shock we may have a value of the co-ordinate r correspond to z. This relation may be written as $r = r(z)$. Differentiating (3.9.17) with respect to z and substituting for $f'(r)$ from (3.9.16) we have

$$\left[z r^{-1} + O(r^{-2} \log r) \right] \frac{dr}{dz} + 2 \log r = -2 h'(z), \qquad (3.9.18)$$

that is,

$$\frac{d}{dz} \left[z^2 \log r + O(z r^{-1} \log r) \right] = -2 z h'(z). \qquad (3.9.19)$$

Therefore,

$$
\begin{aligned}
z^2 \log r + O(z r^{-1} \log r) &= -2 \int z h'(z) \, dz \\
&= -2 z h(z) + 2 h_1(z) + b^2, \qquad (3.9.20)
\end{aligned}
$$

where

$$h_1(z) = \int_0^z h(z) \, dz \qquad (3.9.21)$$

and b is an arbitrary constant. We thus have

$$\log r = \frac{b^2}{z^2} - \frac{2 h(z)}{z} + 2 \frac{h_1(z)}{z^2} + O(r^{-1} \log^{\frac{3}{2}} r), \qquad (3.9.22)$$

and so, using (3.9.17),

$$f(r) = \frac{b^2}{z} - 2 h(z) + 2 \frac{h_1(z)}{z} + O(r^{-1} \log r). \qquad (3.9.23)$$

The velocity U of the shock in the (r, t) plane is given by $1/U = dt/dr = (1 - f'(r))/a_0$. Applying condition (ii) behind the shock, namely, ϕ is continuous at the shock, we have $a_0 [\phi_r + \phi_t/U] = a_0 v + u - u f'(r) = 0$ immediately behind the shock, since ϕ is zero ahead of the shock. Using (3.9.12), (3.9.13) and (3.9.23) in this equation, we easily check that $B_1 = 0$, $B_2 = -k b^2/2$.

Equation (3.9.22) gives the relation between r and z at any point of the shock. Therefore, as $r \to \infty$, $z \to 0$. This implies that there exists a characteristic C^* behind the shock for which $a_0 t = r - h(0) + O(r^{-1} \log r)$ (see (3.9.14)). This characteristic is thus a straight line in the limit $r \to \infty$. Therefore, the results developed by Whitham (1950) would apply to any flow for which some characteristic behind the front shock is asymptotically straight.

Now we apply these results to analyse an explosion. The distant behaviour of the blast wave may be assumed to be produced by some piston motion $R = R(a_0 t)$. To show that there would exist a straight characteristic between the piston and the shock as $r \to \infty$, we may argue as follows. The function $R(\xi)$ has either a maximum at a finite value ξ_0, or $R'(\xi) > 0$ for all ξ and $R(\xi) \to R_0$ (a constant) as $\xi \to \infty$. The latter is the case for a piston motion simulating an explosion. It follows that the function $R^2 R'(\xi)$ is zero at $\xi = 0$ and is also zero either at some ξ_0 or tends to zero as $\xi \to \infty$. In either case, Rolle's theorem implies that there exists a zero of the function $F(\xi) = -d(R^2 R')/d\xi$ different from $\xi = 0$. We shall presently identify the function $F(\xi)$ by reference to the piston motion. The boundary condition on the piston is that the velocity of the piston is equal to that of the particle on it. If the piston motion is given by $r = R(a_0 t)$, then $\partial \phi / \partial r = v = dR/dt$ on the piston. As stated earlier, the flow here is found from the (linear) solution

$$u = \frac{a_0^2 F(\xi)}{r}, \tag{3.9.24}$$

$$v = -\frac{u}{a_0} - \frac{a_0 \int_0^\xi F(\xi') d\xi'}{r^2}, \tag{3.9.25}$$

where ξ, however, must be identified with the exact characteristics (cf. (3.9.5) and (3.9.6)). Using (3.9.24) for u and the relation $v = dR/dt$ in (3.9.25) and differentiating the resulting expression, we identify $F(\xi)$ as

$$F(\xi) = -\frac{d}{d\xi}[R^2(\xi) R'(\xi)], \tag{3.9.26}$$

if terms of fifth order in R, R', and R'' are ignored. We may, in fact, write the solution (3.9.12)–(3.9.14) for large r in a slightly different form which resembles more closely the system (3.9.24)–(3.9.25). The arbitrary function $h(z)$ in (3.9.10) must contain a term of the form $-z \log d$ where d has the dimension of length. We may write $y = h(z) + z \log d$ and let $z = -k^{-1} F(y)$. Then the expressions (3.9.12)–(3.9.14) may be rewritten as

$$u = a_0^2 \left[\frac{F(y)}{r} + \frac{\frac{1}{2} B_1 + O(F^2)}{r^2} + \cdots \right], \tag{3.9.27}$$

$$v = -\frac{u}{a_0}$$

$$-a_0 \left[\frac{(B_1 + O(F^2)) \log r + \int_{y_0}^{y} F(y') dy' + B_2 + O(F^2)}{r^2} \right] \tag{3.9.28}$$

$$+ \cdots \cdots,$$

$$y = a_0 t - r - k^{-1} F(y) \log(r/d) - \cdots, \tag{3.9.29}$$

where y_0 is the zero of $F(y)$, the value of y for an asymptotically straight characteristic C^*. The coefficient b^2 which appears in the locus of the front shock $a_0 t = t - f(r)$, where (3.9.22)–(3.9.23) hold, is now chosen to be

$$b^2 = -2k^{-1}B_2 = -2k^{-1} \int_0^{y_0} F(y')dy' = 2k^{-1}R^2(y_0)R'(y_0), \qquad (3.9.30)$$

(see (3.9.26) and the discussion below (3.9.23)). The form (3.9.27)–(3.9.29) is now essentially the same as (3.9.24)–(3.9.25) if B_1 and B_2 are taken to be 0 and $\int_0^{y_0} F(y')dy'$, respectively. The nonlinear characteristics are given by (3.9.29).

As an example, we may consider the piston curve $R(a_0 t) = \delta a_0 t/(1+a_0 t)$, $\delta << 1$, so that initial velocity of the piston, δa_0, is small while its radius $R(a_0 t) \to \delta$ as $t \to \infty$. In this case, in view of (3.9.26), we have

$$F(y) = -2\delta^3 y(1-y)/(1+y)^5. \qquad (3.9.31)$$

The (relevant) zero of $F(y)$ is $y_0 = 1$. Using (3.9.30), we have

$$b^2 = 2k^{-1}R^2(1)R'(1) = 0.15\delta^3 \quad \text{for} \quad \gamma = 1.4. \qquad (3.9.32)$$

The excess pressure at a point behind the shock in this case is given by

$$\begin{aligned} \frac{p}{p_0} &= -\frac{\rho_0}{p_0}\frac{\partial \phi}{\partial t} + O\left[\left(\frac{\partial \phi}{\partial r}\right)^2\right], \\ &= \frac{\gamma k z}{r} + O(z^2 r^{-2}), \end{aligned} \qquad (3.9.33)$$

where z is given in terms of r and t by (3.9.15) and p_0 and ρ_0 are undisturbed pressure and density, respectively. Expanding $h(z)$ in (3.9.15) etc., we may write (3.9.33) more explicitly as

$$\frac{p}{p_0} = \frac{2\gamma}{\gamma+1}\frac{r - a_0 t - h(0)}{r \log r} + O(r^{-1}\log^{-3/2} r). \qquad (3.9.34)$$

Therefore, for $\gamma = 1.4$, p/p_0 falls from its higher value at the main shock at a rate $1.16 r^{-1} \log^{-1} r$ atm/sec until the secondary shock is reached where it may presumably return to a smaller value.

Chapter 4

Shock Propagation Theories: Some Initial Studies

4.1 Shock Wave Theory of Kirkwood and Bethe

As we mentioned in Chapter 1, there was much activity in the area of shock waves and explosions during World War II, which continued to grow for several decades. Some of the work in this area was carried out independently by researchers, quite unaware of others' contributions. Thus, occasionally, ideas and assumptions overlapped, as we shall show. In the present chapter we discuss two approximate theories of shock propagation due to Kirkwood and Bethe (1942) and Brinkley and Kirkwood (1947) and the extension of the latter by several other authors. The Kirkwood–Bethe theory has essentially the same underlying ideas as were later proposed and developed by Whitham (1950, 1952, 1956) in his theory of weak shock waves (see section 3.9). We may also refer to the work of Akulichev, Boguslavskii, Ioffe, and Naugol'nykh (1968).

The equations of continuity and motion may be written as

$$\frac{1}{\rho}\frac{d\rho}{dt} = -\nabla \cdot \vec{u}, \qquad (4.1.1)$$

$$\frac{d\vec{u}}{dt} = -\frac{1}{\rho}\nabla p, \qquad (4.1.2)$$

where, as usual, $\vec{u}$, ρ, and p stand for particle velocity, density, and pressure, respectively, at any point r and time t, and $d/dt = \partial/\partial t + \vec{u} \cdot \nabla$. We first derive some thermodynamic relations and then make the assumption that the flow is isentropic. The implications of this assumption have been

137

discussed in detail in section 3.9 with reference to Whitham's work. It is a reasonable assumption when the shocks involved are weak or when the flow being considered is asymptotic in nature so that the shocks, initially strong, have become relatively weak.

Equations (4.1.1) and (4.1.2) should be supplemented by the equation of state

$$p = f(\rho, S) \qquad (4.1.3)$$

and the equation of particle entropy

$$\frac{dS}{dt} = 0. \qquad (4.1.4)$$

We introduce enthalpy or heat content,

$$H = E + p/\rho, \qquad (4.1.5)$$

where E is the internal energy per unit mass of the fluid, and the enthalpy increment

$$w = H - H_0, \qquad (4.1.6)$$

where H_0 is the enthalpy of the undisturbed fluid in front of the shock.

We write the fundamental thermodynamic equation for H or w as

$$dw = T dS + \frac{dp}{\rho}, \qquad (4.1.7)$$

where T is the absolute temperature. Integrating (4.1.7) we have

$$w = \int_{p_0}^{p} \frac{dp'}{\rho(S_0, p')} + \int_{S_0}^{S} T(p_0, S') dS', \qquad (4.1.8)$$

where p_0 and S_0 are (undisturbed) pressure and entropy ahead of the shock, respectively, and the line integrals in (4.1.8) are on paths of constant entropy and constant pressure, respectively. Using (4.1.8), we may write (4.1.1), (4.1.2) and (4.1.4) as

$$\nabla \cdot \vec{u} = -\frac{1}{c^2} \frac{dw}{dt}, \qquad (4.1.9)$$

$$\frac{\partial \vec{u}}{\partial t} - \vec{u} \times (\nabla \times \vec{u}) = -\nabla \Omega + T \nabla S, \qquad (4.1.10)$$

$$\frac{dS}{dt} = 0, \qquad (4.1.11)$$

where

$$\Omega = w + u^2/2, \quad c^2 = \left(\frac{\partial p}{\partial \rho}\right)_S. \qquad (4.1.12)$$

Ω may be referred to as kinetic enthalpy; c is the speed of sound. Kirkwood and Bethe (1942) observed that if the (nondimensional) pressure across the shock in water is up to 50,000, the Rankine-Hugoniot conditions show that the dissipated enthalpy $\int_{S_0}^{S} T(p_0, S')dS'$ is only a few percent of the total. Therefore, they replaced (4.1.8) by

$$w = \int_{p_0}^{p} \frac{dp'}{\rho(S_0, p')}, \tag{4.1.13}$$

thus ignoring the entropy changes across the shock.

The governing equations (4.1.9)–(4.1.12) then simplify and become

$$\nabla \cdot \vec{u} = -\frac{1}{c^2} \frac{dw}{dt}, \tag{4.1.14}$$

$$\frac{\partial \vec{u}}{\partial t} - \vec{u} \times \nabla \times \vec{u} = -\nabla \Omega, \tag{4.1.15}$$

$$\Omega = w + u^2/2, \quad c^2 = \left(\frac{\partial p}{\partial \rho}\right)_{S_0}. \tag{4.1.16}$$

Assuming the initial flow to be irrotational, equation (4.1.15) would imply that $\nabla \times \vec{u} = 0$ for all time. (This is always true for the flows which we consider in what follows). This permits the introduction of a velocity potential ψ such that

$$\vec{u} = -\nabla \psi, \tag{4.1.17}$$

so that (4.1.14)–(4.1.15) become

$$\nabla^2 \psi - \frac{1}{c^2} \frac{dw}{dt} = 0, \tag{4.1.18}$$

$$\Omega = \frac{\partial \psi}{\partial t}. \tag{4.1.19}$$

Eliminating w from (4.1.18) with the help of (4.1.16), (4.1.17) and (4.1.19), we arrive at the equation

$$\nabla^2 \psi - \frac{1}{c^2} \frac{\partial^2 \psi}{\partial t^2} = \frac{1}{c^2} \left[\frac{\vec{u}}{2} \cdot \nabla u^2 - \frac{du^2}{dt} \right], \tag{4.1.20}$$

where

$$\nabla^2 \psi = \frac{\partial^2 \psi}{\partial r^2} + \frac{2}{r} \frac{\partial \psi}{\partial r}. \tag{4.1.21}$$

Equation (4.1.20) and the variables ψ, Ω, G and u may be expressed in terms of ϕ as follows:

$$\psi = \frac{\phi}{r}, \quad \Omega = \frac{G}{r}, \quad G = \frac{\partial \phi}{\partial t}, \tag{4.1.22}$$

$$u = \frac{\phi}{r^2} - \frac{1}{r} \frac{\partial \phi}{\partial r}, \tag{4.1.23}$$

$$\frac{\partial^2 \phi}{\partial r^2} - \frac{1}{c^2} \frac{\partial^2 \phi}{\partial t^2} = \frac{1}{c^2} \left[\frac{u}{2} \frac{\partial u^2}{\partial r} - \frac{du^2}{dt} \right]. \tag{4.1.24}$$

Two approximations of (4.1.24) may be noted. If the medium is incompressible so that $c \to \infty$, (4.1.24) reduces to $\partial^2\phi/\partial r^2 = 0$ which, with the boundary condition $\Omega \to 0$ as $r \to \infty$ (see (4.1.22)) leads to the solution $\phi = \phi(t)$, implying that

$$\Omega = \frac{G(t)}{r}, \quad G(t) = \phi'(t),$$

$$u = \frac{\phi(t)}{r^2}. \tag{4.1.25}$$

We observe that in this case the function $r\Omega$ propagates outward with infinite velocity. In the acoustic limit, the nonlinear terms in (4.1.24) may be ignored while c may be replaced by its undisturbed value c_0. Thus, (4.1.24) becomes

$$\phi_{rr} - \frac{1}{c_0^2}\phi_{tt} = 0, \tag{4.1.26}$$

with the relevant solution $\phi = \phi(t - r/c_0)$. The limiting solution in this case of small amplitude waves becomes

$$\Omega = \frac{G(t - r/c_0)}{r},$$

$$u = \frac{\phi(t - r/c_0)}{r^2} + \frac{\Omega}{c_0}. \tag{4.1.27}$$

In the present limit the function $r\Omega$ propagates with the constant speed c_0 of sound in the undisturbed medium. Since it is not possible to write an exact form for the speed with which $r\Omega$ propagates, a plausible choice is the local sound speed $c + u$ in the moving medium (cf. Whitham's approach in section 3.9). Now we give Sachdev's (1976) modified form of Kirkwood–Bethe theory.

Let $G = r\Omega = r\partial\psi/\partial t$ propagate with the exact speed of the positive characteristics so that

$$\frac{\partial G}{\partial t} + (u + c)\frac{\partial G}{\partial r} = 0. \tag{4.1.28}$$

We observe from (4.1.16) and (4.1.19) that

$$G(r,t) = r\frac{\partial\psi}{\partial t} = r\left(w + \frac{u^2}{2}\right), \tag{4.1.29}$$

where

$$w = \int_{p_0}^{p} \frac{dp}{\rho}. \tag{4.1.30}$$

(see (4.1.13)). We consider specifically the motion produced by a spherical piston with the equation $R = R(t)$ so that $u = dR/dt$ and $d^2R/dt^2 = u_t + uu_r$ there.

We may write the equations of continuity and motion explicitly in the present case, assuming that the flow is isentropic:

$$u_t + uu_r + \frac{1}{\rho}p_r = 0, \tag{4.1.31}$$

$$\rho_t + u\rho_r + \rho u_r + \frac{2\rho u}{r} = 0. \tag{4.1.32}$$

We substitute (4.1.29) into (4.1.28) and get an equation in w. We may also express p and ρ in terms of w with the help of (4.1.30) and the equation of isentropy, $p = k\rho^\gamma$. The three equations (4.1.28), (4.1.31), and (4.1.32) can then be combined to write the rate of change of the function w along the piston path $dr/dt = u$:

$$\frac{r}{c}\left(1 - \frac{u}{c}\right)(w_t + uw_r) + \left(\frac{u}{c} + 1\right)w$$

$$= r\left(1 - \frac{u}{c}\right)(u_t + uu_r) + \frac{3}{2}\left(1 - \frac{u}{3c}\right)u^2. \tag{4.1.33}$$

Since, in the present case, the function w is related to c^2 by

$$w = \frac{c_0^2}{(\gamma - 1)}\left(\frac{c^2}{c_0^2} - 1\right), \tag{4.1.34}$$

(4.1.33) may be transformed to express the rate of change of sound speed c along the piston path:

$$\frac{dc}{dt} = \frac{\partial c}{\partial t} + u\frac{\partial c}{\partial r}$$

$$= \frac{\gamma - 1}{2R\left(1 - \frac{u}{c}\right)}\left\{R\frac{d^2R}{dt^2}\left(1 - \frac{u}{c}\right) + \frac{3}{2}\left(\frac{dR}{dt}\right)^2\left(1 - \frac{u}{3c}\right)\right.$$

$$\left. - \frac{1}{(\gamma - 1)}\left(1 + \frac{u}{c}\right)(c^2 - c_0^2)\right\}, \tag{4.1.35}$$

where, as observed earlier, $R = R(t)$, $u = dR/dt$, $u_t + uu_r = d^2R/dt^2$ along the piston path.

The next step is to connect the piston motion with that of the shock via the Rankine-Hugoniot conditions and the basic assumption (4.1.28) underlying the present theory. According to the latter,

$$G(r, t) = G(R, \tau) = G_R(\tau), \tag{4.1.36}$$

where

$$t = \tau + \int_R^r \frac{dr}{u + c}. \tag{4.1.37}$$

τ is the characteristic variable which labels individual characteristics. This variable is specified (there is some freedom in this choice, cf. Whitham

(1956)) such that it is equal to the time t when a given positive characteristic or wavelet originates from the piston. Since we are dealing here with weak shocks, the relations holding along them give (see Whitham (1974) p. 176))

$$G = c_0 r u (1 + \beta u), \tag{4.1.38}$$

$$u + c = c_0(1 + 2\beta u), \quad \beta = \frac{\gamma + 1}{4c_0}. \tag{4.1.39}$$

To draw the individual characteristics from a given piston path, we take recourse to the following procedure. For a given piston motion $R = R(t)$, equations (4.1.34) and (4.1.35) give c, w, and $G = R(w + u^2/2)$ as functions of τ, which is equal to t at the piston. Then, (4.1.37) gives $t = t(r)$ as the locus of the characteristics. Substituting (4.1.38)–(4.1.39) in (4.1.37), remembering that G is constant along a positive characteristic, and integrating from the piston we get

$$t = \tau + \frac{\beta G}{c_0^2}\left[\frac{1 + 2\beta u}{\beta u(1 + \beta u)} - \frac{1 + 2\beta U}{\beta U(1 + \beta U)} - 2\ln\frac{(1 + \beta u)\beta U}{(1 + \beta U)\beta u}\right], \tag{4.1.40}$$

where, in accordance with (4.1.38)–(4.1.39), we have

$$\beta u = \frac{1}{2}\left[\left(1 + \frac{\gamma + 1}{r c_0^2}G\right)^{1/2} - 1\right], \tag{4.1.41}$$

$$\beta U = \frac{1}{2}\left[\left(1 + \frac{\gamma + 1}{R c_0^2}G\right)^{1/2} - 1\right]. \tag{4.1.42}$$

Here, $U = \frac{dR}{dt}$ at the piston. Thus, (4.1.40) represents a relation between t, τ, G, r, and $R = R(t)$.

It is interesting to observe that if we assume the perturbation to be small so that $\beta u \to 0$ in (4.1.40), we simply get

$$t - \tau = \frac{r - R}{c_0}. \tag{4.1.43}$$

Equation (4.1.43) shows that the time $t - \tau$ for the perturbation to propagate from the expanding sphere R to the point r is independent of G. If we retain the next term in the expansion of (4.1.40) in βu, we obtain

$$t - \tau = \frac{r - R}{c_0} - \frac{2\beta G}{c_0^2}\ln\frac{r}{R}. \tag{4.1.44}$$

This is exactly the result obtained by Whitham (1950, 1956) in his theory of weak shocks (see section 3.9). Here the time of propagation of the wavelet from the expanding sphere to the point r depends on G. This result clearly exhibits the nonlinear effect which leads to the distortion of the wave.

To derive the shock locus we first observe that, to second order in u, the shock velocity is given by

$$U_s = c_0 \left(1 + \beta u + \frac{1}{2}\beta^2 u^2 \right), \tag{4.1.45}$$

where u is the particle velocity behind the shock. Let the shock locus be described by $\tau = T(r)$. If this functional relation is found, (4.1.40) gives the shock locus in the (r, t) plane. Now, we put $\tau = T(r)$ in (4.1.40), differentiate the resulting equation with respect to r and write dt/dr equal to U_s^{-1}, where the shock velocity U_s is given by the Rankine–Hugoniot relation (4.1.45). We thus obtain

$$\frac{dT}{dr} = \frac{\frac{\gamma+1}{r^2}\hat{G}L_r + \left(1 + \beta u + \frac{1}{2}\beta^2 u^2\right)^{-1}}{D}, \tag{4.1.46}$$

where

$$
\begin{aligned}
D &= 1 + \left[\frac{1+2\beta u}{u(1+\beta u)} - \frac{1+2\beta U}{U(1+\beta U)} - 2\beta\ln\frac{(1+\beta u)U}{(1+\beta U)u} \right] \frac{d\hat{G}}{dT} \\
&\quad + \frac{(\gamma+1)}{r}L_r\frac{d\hat{G}}{dT} - \left(\frac{\gamma+1}{R}\frac{d\hat{G}}{dT} - \frac{(\gamma+1)}{R^2}\hat{G}\frac{dR}{dT} \right)L_R,
\end{aligned}
$$

$$
\begin{aligned}
L_r &= \frac{\hat{G}}{4\beta\left(1 + \frac{\gamma+1}{r}\hat{G}\right)^{1/2}} \left[\frac{2\beta}{(1+\beta u)u} - \frac{1+2\beta u}{u^2(1+\beta u)} \right. \\
&\quad \left. - \frac{\beta(1+2\beta u)}{u(1+\beta u)^2} - \frac{2\beta^2}{(1+\beta u)} + \frac{2\beta}{u} \right], \tag{4.1.47}
\end{aligned}
$$

$$\hat{G} = G/c_0^2, \tag{4.1.48}$$

and L_R is obtained from L_r in (4.1.47) by replacing r by R and u by U. βu and βU are defined by (4.1.41) and (4.1.42), respectively. R, G and H (see (4.1.6)) are functions of τ as obtained from the piston motion $R = R(t)$, (4.1.36), and (4.1.34). In (4.1.46), these are evaluated at $\tau = T(r)$. Equation (4.1.46) is solved for $T = T(r)$, subject to some initial condition at the shock. Equation (4.1.40), along with $\tau = T(r)$ thus found, gives the locus of the shock in the (r, t) plane.

Cole (1948), in his review of Kirkwood–Bethe theory, remarked that the speed with which the function G propagates may better be taken to be $c + \sigma$ rather than $c + u$, where $\sigma = \int_0^\rho (c/\rho)\,d\rho$. The discussion of this matter in the appendix to his book, however, does not confirm this claim. $c + u$, we believe, is a good choice in the light of Whitham's more recent work (Whitham (1974)).

4.2   The Brinkley-Kirkwood Theory

The work discussed in section 4.1 due to Kirkwood and Bethe (1942) (as also the related theory of Whitham (1950)) has implicit in it the assumption that shocks are of small strength. Both these theories are of second order in shock strength and ignore the finite entropy jump across the shock. In another war time work due to Brinkley and Kirkwood (1947)—BK for short—this simplifying assumption is not made. Shocks of all strengths are permitted and the Rankine-Hugoniot relations are exactly satisfied. This theory, however, also suffers from two limitations: (i) it is local and gives the shock trajectory and the flow immediately behind it; it does not give the flow in the entire domain behind the shock, (ii) it uses a physical argument regarding nonacoustic decay of waves of finite amplitude, associated with the finite entropy increment experienced by the fluid passing through the shock front and the accompanying dissipation of energy. As a shock passes through a fluid element, it leaves in its path a residual energy increment determined by the entropy jump due to the passage of the shock. As a consequence, the energy carried by the shock decreases with distance as it travels away from the source. This physical description is written as a partial differential equation which, together with other governing equations, is used to find the decay of the shock in terms of pressure ratio across it (cf. Von Neumann's approach in section 3.3). This physical argument however is not unique.

The equations of motion here are written in a hybrid Euler-Lagrangian form as

$$\frac{\rho r^\alpha}{\rho_0 R^\alpha} u_R + \frac{\alpha u}{R} = -\frac{1}{\rho c^2} p_t, \qquad (4.2.1)$$

$$\frac{R^\alpha}{r^\alpha} u_t = -\frac{1}{\rho_0} p_R, \qquad (4.2.2)$$

$$u = \left(\frac{\partial r}{\partial t}\right)_R, \qquad (4.2.3)$$

where u is particle velocity, p is the pressure in excess of the undisturbed pressure p_0, ρ is the density and ρ_0 is the undisturbed density. r is the Eulerian co-ordinate at time t of an element of fluid which has the Lagrangian co-ordinate R. c is the speed of sound, equal to $[(\partial p/\partial \rho)_s]^{1/2}$. The coefficient α is equal to 0, 1, 2 for plane, cylindrical, and spherical geometry, respectively. Equations (4.2.1)–(4.2.3) can easily be written from the Eulerian system by making use of the conservation of mass relation, $\rho r^\alpha dr = \rho_0 R^\alpha dR$. The partial derivative $\partial/\partial R$ denotes derivative following the particle. The basic PDEs must be solved subject to the conditions on the piston curve in the (R,t) plane and to the Rankine-Hugoniot conditions across the shock, namely,

$$p = \rho_0 u U, \qquad (4.2.4)$$

$$\rho(U - u) = \rho_0 U, \tag{4.2.5}$$

$$\Delta H = (p/2)\left(\frac{1}{\rho_0} + \frac{1}{\rho}\right), \tag{4.2.6}$$

where ΔH is the specific enthalpy increment in the fluid element as it is overtaken by the shock and U is the shock velocity. The enthalpy H in (4.2.6) is equal to $E + (p + p_0)/\rho = \gamma(p + p_0)/(\gamma - 1)\rho$, where E is the internal energy. The shock locus, $R = R(t)$, must be found as part of the solution. The main idea of the BK theory is to write four partial differential equations from the basic system and the physical argument referred to above and hence determine the partial derivatives p_t, p_R, u_t, u_R, which may be combined to give the total derivative

$$\frac{d}{dR} = \frac{\partial}{\partial R} + \frac{1}{U}\frac{\partial}{\partial t}, \tag{4.2.7}$$

of the excess pressure along the shock path. Two of these PDEs are obtained by localising (4.2.1)–(4.2.2) at the shock $r = R(t)$, the third is obtained by differentiating the relation (4.2.4) along the shock path, and the fourth is found from the energy argument stated above. Thus, from (4.2.1) and (4.2.2) localised at the shock, and (4.2.4) differentiated in the direction (4.2.7), we obtain

$$\frac{\rho}{\rho_0} u_R + \frac{1}{\rho c^2} p_t + \frac{\alpha u}{R} = 0, \tag{4.2.8}$$

$$u_t + \frac{1}{\rho_0} p_R = 0, \tag{4.2.9}$$

$$u_t + U u_R - \frac{g}{\rho_0} p_R - \frac{g}{\rho_0 U} p_t = 0, \tag{4.2.10}$$

where

$$g = \rho_0 U \frac{du}{dp} = 1 - \frac{p}{U}\frac{dU}{dp}. \tag{4.2.11}$$

We observe that, with the help of the Rankine-Hugoniot conditions (4.2.4)–(4.2.6), all the coefficients in (4.2.8)–(4.2.11) may be expressed as functions of excess pressure p across the shock.

Now we translate the energy argument into a PDE. We denote by w_0 the adiabatic work done per unit area by the initial generating surface on the fluid exterior to itself. Thus, we have

$$w_0 a_0^\alpha = \int_{a_0}^R \rho_0 r_0^\alpha E[p(r_0)] dr_0 + \int_{t_0(R)}^\infty r^\alpha u'(p' + p_0) dt$$

$$= I_1 + I_2, \tag{4.2.12}$$

where u' and p' denote particle velocity and excess pressure behind the shock front, respectively; the corresponding values immediately behind the shock will be denoted by unprimed letters. $t_0(R)$ is the time of arrival of the shock front at the point R. $E(p)$ is the jump in the specific energy of a fluid element as it crosses the shock; this is related to the entropy jump across the shock at the pressure p. a_0 is the Lagrangian co-ordinate of the initial generating surface. The two terms in (4.2.12) represent, respectively, the increased internal energy of fluid at pressure p_0 within a radius R and the work done on the generating spherical surface (for all time).

We may write

$$\int_{t_0(R)}^{\infty} p_0 r^\alpha u' dt = p_0 \Delta V + p_0 \int_{a_0}^{R} \left(\frac{\rho_0}{\rho} - 1 \right) r_0^\alpha dr_0, \qquad (4.2.13)$$

where ρ is density of the fluid and ΔV is the volume swept out by the generating surface per unit area of the initial generating surface. In writing (4.2.13) we have used the relation $\rho r^2 dr = \rho_0 r_0^2 dr_0$. Putting (4.2.13) into the second term I_2 of (4.2.12), we have

$$w_0 a_0^\alpha = p_0 \Delta V + \int_{a_0}^{R} \rho_0 r_0^\alpha h[p(r_0)] dr_0 + \int_{t_0(R)}^{\infty} r^\alpha u' p' dt, \qquad (4.2.14)$$

where $h(p) = E + p_0 \Delta(1/\rho)$ is the specific enthalpy of an element of fluid traversed by a shock wave which has excess pressure p and which returns to pressure p_0 along its new adiabatic. As $t \to \infty$ and $R \to \infty$, the second integral in (4.2.14) is assumed to tend to zero. Now subtracting (4.2.14) from its limiting form as $t \to \infty$, we get

$$D(R) = \int_{t_0(R)}^{\infty} r^\alpha u' p' dt, \qquad (4.2.15)$$

where

$$D(R) = \int_{R}^{\infty} \rho_0 r_0^\alpha h[p(r_0)] dr_0. \qquad (4.2.16)$$

It is clear from (4.2.15) that $a_0^{-\alpha} D(R)$ is the shock wave energy at R per unit area of the initial generating surface. This is also the work done (per unit area) on the fluid exterior to R. (The BK theory assumes that no secondary shock is formed in the flow behind.)

The energy time integral (4.2.15) can be normalised by writing it in the form

$$D(R) = R^\alpha p u \mu \nu, \qquad (4.2.17)$$

where

$$\frac{1}{\mu} = -\left(\frac{\partial}{\partial t} \log p' u' r^\alpha \right)_{t=t_0(R)}$$
$$= -\frac{1}{p} p_t - \frac{1}{u} u_t - \frac{\alpha u}{R},$$

$$\nu = \int_0^\infty f(R,\tau)d\tau, \quad \tau = \frac{t - t_0(R)}{\mu}, \tag{4.2.18}$$

$$f(R,\tau) = \frac{r^\alpha p' u'}{R^\alpha pu}.$$

Brinkley and Kirkwood (1947) succinctly describe the integrand in (4.2.18): the function $f(R,\tau)$ is the energy time integrand, normalised by its peak value $R^\alpha pu$ at the shock front and expressed as a function of R and a reduced time τ which normalises its initial slope to -1 if μ does not vanish.

The function $f(R,\tau)$ is not known. It is found from the data on explosions that this function is a slowly varying function of R. It is also known that initial pressure-time and energy-time curves of an explosion wave rapidly decrease with time. It is therefore assumed that

$$f(R,\tau) = f(\tau) = e^{-\tau} \tag{4.2.19}$$

so that $\nu = 1$. More realistic estimates for ν have been suggested by other investigators (Sachdev (1971, 1972)). Choosing $\nu = 1$ and eliminating μ from (4.2.17) with the help of (4.2.18), we get the fourth equation involving the derivatives of u and p:

$$\frac{1}{u}u_t + \frac{1}{p}p_t + \frac{\alpha u}{R} = -\frac{R^\alpha pu}{D(R)}. \tag{4.2.20}$$

Solving for $\partial p/\partial R$ and $\partial p/\partial t$ from the four equations (4.2.8)–(4.2.10) and (4.2.20) and combining them appropriately, we get the total derivative $dp/dR = \partial p/\partial R + (\partial p/\partial t)/U$ along the shock path; the corresponding derivative for dD/dR is obtained from (4.2.16). We thus arrive at the following coupled system of ODEs for D and p along the shock curve:

$$\frac{dD}{dR} = -R^\alpha L(p), \tag{4.2.21}$$

$$\frac{dp}{dR} = -\nu \frac{R^\alpha p^3}{D} M(p) - \frac{\alpha p}{2R} N(p), \tag{4.2.22}$$

where

$$\begin{aligned}
L(p) &= \rho_0 h(p), \\
M(p) &= \frac{1}{\rho_0 U^2} \frac{G}{2(1+g) - G}, \\
N(p) &= \frac{4(\rho_0/\rho) + 2(1 - \rho_0/\rho)G}{2(1+g) - G}, \\
G &= 1 - (\rho_0 U/\rho c)^2, \quad g = 1 - \frac{p}{U}\frac{dU}{dp}.
\end{aligned} \tag{4.2.23}$$

The functions $L(p), M(p)$, and $N(p)$ can be expressed entirely as functions of pressure with the help of the equation of state and the Rankine-Hugoniot conditions. The system (4.2.21)–(4.2.23) applies both to air and water, if the corresponding equations of state are used, and takes full account of the finite entropy jump at the shock. This system is clearly not integrable in a closed form and must be solved numerically with appropriate initial conditions. However, we may first check its asymptotic form as $p \to 0$, that is, as the shock degenerates to a sound wave. In this limit (assuming the undisturbed medium to be air with the ideal adiabatic equation of state $p = p_0[(\rho/\rho_0)^\gamma - 1]$) and $\gamma = c_p/c_v$, we check that

$$
\begin{aligned}
\lim_{p \to 0} L(p) &= \frac{\gamma + 1}{12\gamma^2} \frac{p^3}{p_0^2}, \\
\lim_{p \to 0} M(p) &= \frac{\gamma + 1}{8\gamma^2} \frac{p}{p_0^2}, \\
\lim_{p \to 0} N(p) &= 1, \\
\lim_{p \to 0} \nu &= \frac{2}{3}.
\end{aligned}
\tag{4.2.24}
$$

The last relation follows if it is assumed that

$$
\begin{aligned}
f(\tau) &= (1 - \tau/2)^2 & \tau \le 2, \\
&= 0 & \tau > 2.
\end{aligned}
\tag{4.2.25}
$$

Equations (4.2.21)–(4.2.22) now become

$$
\frac{dD}{dR} = -\frac{\gamma + 1}{12\gamma^2} \frac{R^\alpha p^3}{p_0^2},
\tag{4.2.26}
$$

$$
\frac{dp}{dR} + \frac{\alpha p}{2R} = -\frac{(\gamma + 1)}{12\gamma^2} \frac{R^\alpha p^4}{Dp_0^2}.
\tag{4.2.27}
$$

The system (4.2.26)–(4.2.27) may be integrated for different geometries:

Spherical ($\alpha = 2$):

$$
\begin{aligned}
Rp &= P_1[\log(R/R_1)]^{-1/2}, \\
D &= [(\gamma + 1)/6\gamma^2 p_0^2] P_1^2 Rp.
\end{aligned}
\tag{4.2.28}
$$

Cylindrical ($\alpha = 1$):

$$
\begin{aligned}
\sqrt{R}p &= P_1[2(R^{1/2} - R_1^{1/2})]^{-1/2}, \\
D &= [(\gamma + 1)/6\gamma^2 p_0^2] P_1^2 \sqrt{R}p.
\end{aligned}
\tag{4.2.29}
$$

Plane ($\alpha = 0$):

$$
\begin{aligned}
p &= P_1[(R - R_1)]^{-1/2}, \\
D &= [(\gamma + 1)/6\gamma^2 p_0^2] P_1^2 p.
\end{aligned}
\tag{4.2.30}
$$

Here P_1 and R_1 are constants.

The above results for air may be extended to water if p_0 is replaced by the characteristic pressure B in the Tait equation of state for water, namely, $p = B[(\rho/\rho_0)^\gamma - 1]$ where, however, $\gamma \neq c_p/c_\gamma$ but is the exponent of (ρ/ρ_0) in the Tait equation. We observe that the asymptotic result (4.2.28) for $\alpha = 2$ is in agreement with the results of Kirkwood–Bethe theory and Whitham's weak shock theory (see sections 4.1 and 3.9).

The theory of Brinkley and Kirkwood (1947) attracted considerable attention, particularly in the astrophysical context, since, analogous to the approximate approach of Chester (1954), Chisnell (1957) and Whitham (1958), it offered a convenient way of finding shock decay and shock locus for explosions. It was generalised to apply to nonuniform media by several authors (Nadezhin and Frank-Kamenetskii (1965) and Kogure and Osaki (1962)) and hence used to study shock wave phenomena in the stars.

Sachdev (1971, 1972) showed how to extend the BK theory so that it gave correct asymptotic limits both when the shock becomes weak (see (4.2.26)–(4.2.30)) and when it is infinitely strong and has the Taylor-Sedov limiting solution in its early stages of propagation. For the latter purpose, the following thermodynamic argument was used to find the decay of the shock. As a particle crosses the shock, its entropy and internal energy increase. This particle, with the new value of entropy, expands adiabatically until it comes to its original pressure (but higher temperature) and then it radiates energy at constant pressure, finally assuming its ambient value of pressure and specific volume. This path in the thermodynamic variables was suggested by Schatzman (1949) and is in conformity with pressure–specific volume relation used by Taylor (1950) to determine the fraction of explosion energy which is degraded as heat and is thus not available for doing work as the shock propagates. This process is expressed as an equation which describes the dissipation of the energy of explosion as heat. Sachdev (1971, 1972) exploited the arbitrariness of the parameter ν in (4.2.17) to obtain the correct limiting behaviour in the strong shock limit. It was shown that if, in (4.2.21), the shock strength is allowed to tend to infinity, the rate of decay of shock energy, dD/dR, tends to zero. Equation (4.2.22) with $D = $ constant then leads to a solution which has exactly the same form as the Taylor-Sedov solution. The similarity parameter ν is now made to depend on γ and is so chosen that the form of the solution in the strong shock limit exactly coincides with the Taylor-Sedov solution. The value of ν in the weak shock limit is $2/3$ and does not change significantly with γ. Comparison of the analytic results so obtained with the numerical solution of Lutzky and Lehto (1968) showed that the two agreed very well in the strong shock limit if ν is appropriately chosen as a function of γ. The above argument was shown to hold even when the medium ahead of the shock is nonuniform.

4.3   Pressure Behind the Shock: A Practical Formula

An approach closely related to that of Brinkley and Kirkwood (1947), simpler, less rigorous, but useful nevertheless is due to Theilheimer (1950). The time history of the pressure behind the shock may be written down in an empirical form as

$$p(t) = p_0 + p_s e^{-t/\theta}, \tag{4.3.1}$$

where θ is a (dimensional) constant. Here, p_s and p_0 are pressure behind the shock at $t = 0$ and the undisturbed pressure, respectively. It is clear from (4.3.1) that

$$\theta = -\frac{(p - p_0)}{(\partial p/\partial t)}. \tag{4.3.2}$$

Theilheimer (1950) exploited the definition (4.3.2) to derive the 'constant' θ as a function of γ. To that end, he made use of basic equations of motion and the Rankine-Hugoniot conditions. From (4.3.1)–(4.3.2) we immediately find that

$$\theta = -\frac{p_s}{(\partial p/\partial t)_{t=0^+}}. \tag{4.3.3}$$

That is, θ defines the initial pressure decay behind the shock front. Introducing the speed of sound $a = \sqrt{(\partial p/\partial \rho)_S}$, equations of motion for spherical symmetry may be written in the form

$$\rho(u_t + u u_r) + p_r = 0, \tag{4.3.4}$$

$$p_t + u p_r + a^2 \rho u_r + \frac{2a^2 \rho u}{r} = 0. \tag{4.3.5}$$

We may write the derivatives of p and u along the shock as

$$\frac{dp}{dR} = p_r + \frac{1}{U} p_t, \tag{4.3.6}$$

$$\frac{du}{dR} = u_r + \frac{1}{U} u_t. \tag{4.3.7}$$

Equations (4.3.4)–(4.3.7) can be solved for $\partial p/\partial t$:

$$\frac{\partial p}{\partial t} = \frac{U \left\{ \frac{2\rho u a^2}{R}(U - u) + \frac{dp}{dR}[a^2 + u(U - u)] + \frac{du}{dR} a^2 \rho U \right\}}{a^2 - (U - u)^2}. \tag{4.3.8}$$

If we know the quantities behind the shock, we may use (4.3.8) to find the initial decay in the pressure-time history. By introducing the nondimensional pressure behind the shock,

$$\overline{p}_s = \frac{p_s}{p_0} = \frac{p - p_0}{p_0}, \tag{4.3.9}$$

equation (4.3.8) may be rewritten in terms of θ (see (4.3.2)) as

$$\frac{1}{\theta} = \frac{-U\left\{\frac{2\rho u a^2}{R p_0}(U - u) + \frac{d\overline{p}_s}{dR}[a^2 + u(U - u)] + \frac{du}{dR}\frac{a^2 \rho U}{p_0}\right\}}{\overline{p}_s[a^2 - (U - u)^2]}. \qquad (4.3.10)$$

Using the Rankine-Hugoniot conditions and explicit equation of state for air, (4.3.10) may be simplified further. For example, for $\gamma = 1.4$, Theilheimer wrote (4.3.10) as

$$\frac{1}{\theta} = -a_0 \left(\frac{6\overline{p}_s + 7}{7}\right)^{1/2} \left[\frac{7(\overline{p}_s + 1)}{3\overline{p}_s R} + \frac{d\overline{p}_s}{dR}\left(\frac{7}{3\overline{p}_s^2} + \frac{7}{2\overline{p}_s} + \frac{2}{\overline{p}_s + 7}\right)\right]. \qquad (4.3.11)$$

Equation (4.3.11) gives θ explicitly in terms of dimensionless shock over-pressure $\overline{p}_s$ and its derivative with respect to shock radius R. If the shock 'line', $\overline{p}_s$ versus R, is known from experiment or theory, then θ may be found. Theilheimer (1950) computed θ from (4.3.11) by using an empirical fit to the 'shock line' for Pentolite spheres obtained earlier by Stoner and Bleakney (1948).

With θ thus obtained, (4.3.1) gives an 'empirical' time history of pressure behind the shock. This may be used for comparison with overpressure gauge records and may thus provide an independent check of the initial decay rate of these records.

Chapter 5

Some Exact Analytic Solutions of Gasdynamic Equations Involving Shocks

5.1   Exact Solutions of Spherically Symmetric Flows in Eulerian Co-ordinates

It is illuminating to study attempts at the exact solution of one-dimensional gasdynamic equations, which would throw light on the structure of the solutions, the blast waves being one class of solutions of these equations. In the present section we consider the work in this context by McVittie (1953) and in the following section that by Keller (1956). The former is in Eulerian co-ordinates while the latter uses Lagrangian co-ordinates (cf. Taylor (1950) and Von Neumann (1941) for the blast wave problem).

McVittie (1953) gave explicit form of the solution for the equations of motion and continuity in three dimensions but then restricted his analysis to the spherically symmetric case. We discuss here this special case. We write the equations of motion and continuity in spherical symmetry:

$$q_t + qq_r + \frac{1}{\rho}p_r = 0, \qquad (5.1.1)$$

$$\rho_t + \frac{1}{r^2}(r^2\rho q)_r = 0. \qquad (5.1.2)$$

Here $q, p,$ and ρ are particle velocity, pressure and density at the point r at

153

time t. Using (5.1.1), we may write (5.1.2) in the alternative form

$$(\rho q)_t + (\rho q^2 + p)_r + \frac{2}{r}\rho q^2 = 0. \tag{5.1.3}$$

Motivated by "the Einstein's gravitational potential function," McVittie (1953) showed that the expressions

$$q = -\phi_{rt}/\nabla^2\phi, \tag{5.1.4}$$
$$\rho = -\nabla^2\phi, \tag{5.1.5}$$
$$p = P(t) - \phi_{tt} + 2\int \frac{I}{r}dr + I, \tag{5.1.6}$$

in terms of an arbitrary function $\phi(r,t)$, satisfy (5.1.1) and (5.1.3) identically. Here,

$$I = (\phi_{rt})^2/\nabla^2\phi, \tag{5.1.7}$$

and

$$\nabla^2\phi = \left(\frac{\partial^2}{\partial r^2} + \frac{2}{r}\frac{\partial}{\partial r}\right)\phi. \tag{5.1.8}$$

$P(t)$ is an arbitrary function of t.

The above statement may be verified by substituting (5.1.4)–(5.1.8) directly into (5.1.1) and (5.1.3). We emphasise that the solution (5.1.4)–(5.1.8) of (5.1.1) and (5.1.3) holds for an arbitrary 'appropriately' smooth function ϕ. This 'solution' is expressed in terms of an integral which itself involves the function I, defined by (5.1.7). It must satisfy the third equation, namely, the particle isentropy, as well as appropriate boundary conditions, say, at the shock or vacuum front, to yield a physically meaningful solution. Here we follow McVittie's analysis and shall detail its generalisation in section 5.3.

McVittie (1953) restricted the form of ϕ to describe 'progressive waves',

$$\phi = -f(t)w(x), \tag{5.1.9}$$

where f is an arbitrary function of t,

$$x = rt^{-\alpha}, \tag{5.1.10}$$

and α is a constant. $w(x)$ is an arbitrary function of x. Using the above definitions, we may write the derivatives of ϕ etc. as

$$\phi_{tt} = -\alpha^2 ft^{-2}\left\{\frac{\alpha g^2 - g + tg_t}{\alpha}w\right.$$
$$\left. - \left(2g - \frac{1}{\alpha}\right)xw_x + x(xw_x)_x\right\}, \tag{5.1.11}$$
$$\phi_{rt} = -\alpha ft^{-(\alpha+1)}\{(g-1)w_x - xw_{xx}\},$$
$$\nabla^2\phi = -ft^{-2\alpha}x^{-2}(x^2w_x)_x,$$

where

$$g = \frac{tf_t}{\alpha f}.$$ (5.1.12)

For convenience, we introduce the function

$$v = x^{-(g-1)}w_x.$$ (5.1.13)

Therefore, (5.1.11) becomes

$$
\begin{aligned}
\phi_{tt} &= -\alpha^2 f t^{-2}\Big\{ \frac{\alpha g^2 - g + tg_t}{\alpha} w \\
&\quad - \left(2g - \frac{1}{\alpha}\right) x^g v + x(x^g v)_x \Big\}, \\
\phi_{rt} &= \alpha f t^{-(\alpha+1)} x^g v_x, \\
\nabla^2 \phi &= -f t^{-2\alpha} x^{-2}(x^{g+1}v)_x.
\end{aligned}
$$ (5.1.14)

Now the function I in (5.1.7) and the integral $2\int (I/r)dr$ become

$$I = -\alpha^2 f t^{-2} x^{g+2} u,$$ (5.1.15)

$$2\int \frac{I}{r}dr = -2\alpha^2 f t^{-2}\int x^{g+1}u\,dx,$$ (5.1.16)

where

$$u = \frac{x^g(v_x)^2}{(x^{g+1}v)_x} = \frac{v_x^2}{xv_x + (g+1)v}.$$ (5.1.17)

The 'solution' (5.1.4)–(5.1.6) now has the form

$$q = \alpha t^{\alpha-1} x^{g+2}\frac{v_x}{(x^{g+1}v)_x} = \alpha t^{\alpha-1} x^{g+2}\frac{(x^{-(g-1)}w_x)_x}{(x^2 w_x)_x},$$ (5.1.18)

$$\rho = f t^{-2\alpha} x^{-2}(x^{g+1}v)_x = f t^{-2\alpha} x^{-2}(x^2 w_x)_x,$$ (5.1.19)

$$
\begin{aligned}
p - P &= \alpha^2 f t^{-2}\Big\{ \frac{\alpha g^2 - g + tg_t}{\alpha} w - \left(2g - \frac{1}{\alpha}\right) x^g v + x(x^g v)_x \\
&\quad - x^{g+2}u - 2\int x^{g+1}u\,dx \Big\} \\
&= \alpha^2 f t^{-2}\Big\{ \frac{\alpha g^2 - g + tg_t}{\alpha} w - \left(2g - \frac{1}{\alpha}\right) xw_x + x(xw_x)_x \\
&\quad - x^{g+2}u - 2\int x^{g+1}u\,dx \Big\},
\end{aligned}
$$ (5.1.20)

where v and u are defined in terms of w and f by (5.1.12), (5.1.13) and (5.1.17).

Now one may find different subclasses of solutions which have special properties. For example, one may assume that the velocity q is proportional to r^μ, requiring g to be a constant. Here we shall give a few simple examples due to McVittie (1953) and give more general results in section 5.3.

First we discuss the solutions for which the particle isentropy equation, $\partial S/\partial t + q\partial S/\partial x = 0$, has not been imposed. In the special case for which $q \propto r$, the particle velocity (5.1.18) becomes

$$q = \alpha t^{-1}\frac{(-g+1)w_x + xw_{xx}}{2w_x + xw_{xx}}r. \tag{5.1.21}$$

Two simple cases may be identified.
Case A.

$$xw_{xx} = (\lambda - 1)w_x \tag{5.1.22}$$

so that

$$w = \frac{x^\lambda}{\lambda} \tag{5.1.23}$$

and (5.1.21) reduces to

$$q = \frac{a(t)}{t}r, \tag{5.1.24}$$

where

$$a(t) = \alpha\frac{\lambda - g}{\lambda + 1} = \frac{\alpha\lambda - tf_t/f}{\lambda + 1}. \tag{5.1.25}$$

Here λ is a constant. In the present case, $w(x)$ is determined while $f(t)$ remains arbitrary.
Case B. If we choose $g = -1$, the numerator in (5.1.21) becomes equal to the denominator. We thus have

$$q = \alpha rt^{-1}. \tag{5.1.26}$$

Here, the function $w(x)$ remains arbitrary.

In both these cases the solution (5.1.18)–(5.1.20) can be written out explicitly.
Case A. Here,

$$w = \frac{x^\lambda}{\lambda}, \quad v = x^{\lambda-g}, \quad u = \frac{(\lambda - g)^2}{\lambda + 1}x^{\lambda-g-2},$$
$$\int ux^{g+1}dx = \frac{(\lambda - g)^2}{\lambda(\lambda + 1)}x^\lambda. \tag{5.1.27}$$

The solution becomes

$$\begin{aligned}
q &= \frac{a}{t}r, \\
\rho &= (\lambda + 1)ft^{-2\alpha}x^{\lambda-2}, \\
p &= \alpha^2 ft^{-2}\{b(t)x^\lambda + Q(t)\},
\end{aligned} \tag{5.1.28}$$

where α and λ are arbitrary constants, $a(t)$ is related to $f(t)$ by (5.1.25), $Q(t)$ is a new (arbitrary) function of t replacing $P(t)$, and

$$b(t) = \frac{\lambda + 1}{\lambda \alpha^2} \{ -a^2 + a - t a_t \}. \qquad (5.1.29)$$

Case B. Since $g = -1$ in this case, (5.1.12) gives $f = t^{-\alpha}$. We, therefore, have

$$
\begin{aligned}
q &= \alpha \frac{r}{t}, \\
\rho &= t^{-3\alpha} F(x), \qquad\qquad (5.1.30)\\
p &= \alpha^2 t^{-\alpha-2} \{ G(x) + Q(t) \},
\end{aligned}
$$

where $Q(t)$ is again an arbitrary function of t and

$$
\begin{aligned}
F(x) &= x^{-2}(x^2 w_x)_x, \\
G(x) &= \frac{\alpha+1}{\alpha} w + \left(2 + \frac{1}{\alpha} \right) x w_x \qquad (5.1.31)\\
&\quad + x(x w_x)_x - xu - 2 \int u\, dx.
\end{aligned}
$$

For case A we observe from (5.1.28) that α and λ are arbitrary, but if the density is to remain bounded at $r = 0$ for $t > 0$ we must have $\lambda \geq 2$. Further, if the pressure at $r = 0$ is zero, we also require that $Q = 0$. For case B, $w(x)$ must be chosen such that $F(0)$ is either zero or constant so that there is no singularity of density at $r = 0$. To keep the pressure at $r = 0$ nonsingular, we must choose $G(0) = 0$. We again require that $Q \equiv 0$.

A case common to both A and B is obtained when pressure and density are both functions of time alone. In this case we must have

$$\lambda = 2, \quad b(t) = 0, \qquad (5.1.32)$$

while $Q(t) \neq 0$. We then have

$$a = t(\tau + t)^{-1}, \quad f = \sigma t^{2\alpha}(1 + t/\tau)^{-3} \qquad (5.1.33)$$

from (5.1.25) and (5.1.29); τ and σ are constants of integration.

The solution in this case assumes the simple form

$$
\begin{aligned}
q &= \frac{r}{\tau + t}, \\
\rho &= 3\sigma(1 + t/\tau)^{-3}, \qquad (5.1.34)\\
p &= P(t).
\end{aligned}
$$

We may observe that, in the present case, pressure and density are independent of each other. This is a 'cosmological' solution which has similar behaviour even when self-gravitation of the mass of gas is allowed for. This solution, however, is not of much practical interest.

Now we impose the particle adiabacy condition on the nonsingular solutions from cases A and B. For case A, we require that $Q(t) = 0$ and $\lambda > 2$ so that

$$
\begin{aligned}
q &= \frac{a(t)}{t}r, \\
\rho &= (\lambda + 1)ft^{-2\alpha}x^{\lambda-2}, \\
p &= \alpha^2 ft^{-2}b(t)x^{\lambda}.
\end{aligned}
\tag{5.1.35}
$$

Substituting (5.1.35) into the particle adiabacy condition

$$
\frac{dS}{dt} = \left(\frac{\partial}{\partial t} + q\frac{\partial}{\partial r}\right)(\ln p - \gamma\ln\rho) = 0,
\tag{5.1.36}
$$

we have

$$
\frac{1}{t}\left\{-2 + (3\gamma - 1)a + t\frac{b_t}{b}\right\} = 0
$$

or

$$
t\frac{b_t}{b} = 2 - (3\gamma - 1)a.
\tag{5.1.37}
$$

Here, $\gamma = c_p/c_v$.

Using the definition (5.1.29) of $b(t)$ in (5.1.37) and introducing the transformation $z = \ln t$, we get

$$
a_{zz} + \{-3 + (3\gamma + 1)a\}a_z + \{2 - (3\gamma - 1)a\}(1 - a)a = 0.
\tag{5.1.38}
$$

If we use (5.1.29) in (5.1.37) and write

$$
a = (tj_t/j),
\tag{5.1.39}
$$

we obtain the third order equation

$$
jj_{ttt} + (3\gamma - 2)j_t j_{tt} = 0.
\tag{5.1.40}
$$

This equation can be integrated to yield

$$
j_t = \left\{\frac{2C}{3(1 - \gamma)}j^{3(1-\gamma)} + 2D\right\}^{1/2}
\tag{5.1.41}
$$

or

$$
t = \int\left[\frac{2C}{3(1 - \gamma)}j^{3(1-\gamma)} + 2D\right]^{-1/2}dj + E,
\tag{5.1.42}
$$

where C, D, and E are arbitrary constants.

Thus, the adiabatic cases of the solutions (5.1.35) hold if $a(t)$ and $b(t)$ are given by equations (5.1.29), (5.1.39) and (5.1.42).

For case B, if we substitute (5.1.30) into (5.1.36), we get

$$\frac{1}{t(G+Q)}\{[\alpha(3\gamma-1)-2]G + [\alpha(3\gamma-1)-2]Q + tQ_t\} = 0. \qquad (5.1.43)$$

Two solutions of (5.1.43) are easily obtained:

(i) If $G(x)$ is not a constant, then

$$\alpha = 2/(3\gamma-1),$$
$$Q = \text{constant.} \qquad (5.1.44)$$

(ii) If $G = \epsilon$, a constant, then

$$Q = -\epsilon + \delta t^{-\alpha(3\gamma-1)+2}, \qquad (5.1.45)$$

where δ is a constant of integration. With this choice of Q, (5.1.30) shows that $p = \alpha^2 \delta t^{-3\alpha\gamma}$, a function of t alone. We ignore this case and look at the alternative (i) with $Q = 0$. The adiabatic motions in this case are governed by

$$q = \frac{2}{3\gamma-1}\frac{r}{t},$$
$$\rho = t^{-6/(3\gamma-1)}F(x), \qquad (5.1.46)$$
$$p = \alpha^2 t^{-2/(3\gamma-1)-2}G(x),$$

where $F(x)$ and $G(x)$ are to be computed from (5.1.31), (5.1.13) and (5.1.17) with $\alpha = 2/(3\gamma-1)$ and $w(x)$ arbitrary.

$a = 2/(3\gamma-1)$ is a constant solution of (5.1.38), which is common to both cases A and B; this is in addition to the cosmological solution referred to earlier without the condition of particle isentropy. With this choice of a, equation (5.1.25) and (5.1.29) give

$$f = \sigma(\lambda+1)^{-1}t^{\alpha\lambda-\{2(\lambda+1)\}/(3\gamma-1)}, \qquad (5.1.47)$$
$$b = 6(\lambda+1)\lambda^{-1}\alpha^{-2}(\gamma-1)(3\gamma-1)^{-2}, \qquad (5.1.48)$$

where σ is a constant of integration different from that in (5.1.34). Introducing these expressions in (5.1.35), we have

$$q = \frac{2}{3\gamma-1}\frac{r}{t},$$
$$\rho = \sigma t^{-\{2(\lambda+1)\}/(3\gamma-1)}r^{\lambda-2} \quad (\lambda > 2), \qquad (5.1.49)$$
$$p = 6\sigma\lambda^{-1}(\gamma-1)(3\gamma-1)^{-2}t^{-2-\{2(\lambda+1)\}/(3\gamma-1)}r^{\lambda}.$$

This solution may also be obtained from case B by simply choosing

$$\alpha = 2/(3\gamma-1), \quad w(x) = \sigma\lambda^{-1}(\lambda+1)^{-1}x^{\lambda}, \quad Q = 0. \qquad (5.1.50)$$

The cosmological solution (5.1.34) also satisfies the adiabatic condition (5.1.36) provided $P(t) = P_0(1 + t/\tau)^{-3\gamma}$, where P_0 is constant. With this definition of $P(t)$, the pressure and density in (5.1.34) become related, in contrast to the more general 'cosmological' case.

McVittie (1953) showed how each of the shock conditions may be fitted to the solution (5.1.49). We shall discuss only the case for which all of these conditions can be satisfied. Assuming the medium ahead of the shock to be quiescent and uniform with pressure p_0 and density ρ_0, the Rankine-Hugoniot conditions across the shock $h = h(t)$ may be written as

$$\rho_1(q_1 - h_t) = -\rho_0 h_t, \qquad (5.1.51)$$

$$p_1 + \rho_1(q_1 - h_t)^2 = p_0 + \rho_0 h_t^2, \qquad (5.1.52)$$

$$\frac{\gamma}{\gamma - 1}\frac{p_1}{\rho_1} + \frac{1}{2}(q_1 - h_t)^2 = \frac{\gamma}{\gamma - 1}\frac{p_0}{\rho_0} + \frac{1}{2}h_t^2, \qquad (5.1.53)$$

where h_t is the velocity of the shock. The solution (5.1.49) involving two arbitrary constants σ and λ must be subjected to the conditions (5.1.51)–(5.1.53). For brevity, the following notation is introduced:

$$\mu = \frac{2}{3\gamma - 1}, \quad \beta = \mu(\lambda + 1),$$

$$\delta = \lambda - 2, \quad \eta = \frac{1}{3}\rho_0(\lambda + 1),$$

$$\mu_1 = \rho_0\mu^2\sigma^{-2/\delta},$$

$$\mu_2 = \frac{\mu}{\lambda}(1 - \mu)\sigma^{-2/\delta}$$

$$= \frac{6(\gamma - 1)\sigma^{-2/\delta}}{\lambda(3\gamma - 1)^2}, \qquad (5.1.54)$$

$$\mu_3 = \frac{2}{\mu}\left\{\left(\frac{\gamma}{\gamma - 1}\right)\mu_2\sigma^{2/\delta} + \frac{1}{2}\mu^2\right\}$$

$$= \frac{2\gamma}{\gamma - 1}\frac{1 - \mu}{\lambda} + \mu,$$

$$\mu_4 = \frac{2\gamma}{\gamma - 1}\frac{1}{\mu}\frac{p_0}{\rho_0}.$$

The motion inside the shock $r = h(t)$ may now be written from (5.1.49) as

$$q = \mu\frac{r}{t},$$

$$\rho = \sigma t^{-\beta}r^\delta \quad (\delta > 0), \qquad (5.1.55)$$

$$p = \sigma^{1+2/\delta}\mu_2 t^{-2-\beta}r^{2+\delta}.$$

It is also convenient to introduce the density behind the shock in the form $H(t) = \rho_1 = \sigma t^{-\beta}h^\delta$ or

$$h = \sigma^{-1/\delta}t^{\beta/\delta}H^{1/\delta}. \qquad (5.1.56)$$

Also from (5.1.51), we have

$$h_t = \rho_1 q_1 (\rho_1 - \rho_0)^{-1}. \tag{5.1.57}$$

Using $(5.1.55)_2$, (5.1.56) and (5.1.57), an equation for $H(t)$ is obtained:

$$t \frac{\rho_0 - H}{H} H_t = (\beta - \gamma\mu)H - \rho_0\beta. \tag{5.1.58}$$

The solution of (5.1.58) is

$$\beta \ln t - \frac{\delta}{3}\ln(\nu\mu) = -\ln H + \frac{\mu\delta}{\mu\delta - \beta}\ln\{(\delta\mu - \beta)H + \rho_0\beta\}, \tag{5.1.59}$$

where ν is a constant of integration. Since $(\delta\mu - \beta)/\delta\mu = -3/\delta$, we may rewrite (5.1.59) as

$$3(\eta - H) = \nu t^{-(3\beta)/\delta} H^{-3/\delta}. \tag{5.1.60}$$

H, given by (5.1.60), satisfies (5.1.51). Now we turn to the second condition (5.1.52). Eliminating h_t from (5.1.51) and (5.1.52) we have

$$\rho_0 \rho_1 q_1^2 = (p_1 - p_0)(\rho_1 - \rho_0). \tag{5.1.61}$$

Using the solution (5.1.55) at $r = h(t)$, (5.1.56) and (5.1.54) in (5.1.61), we have

$$\mu_1 = (H - \rho_0)\{\mu_2 - p_0 t^{2(1-\beta/\delta)} H^{-(1+2/\delta)}\}. \tag{5.1.62}$$

This must be an identity for any solution of (5.1.60). McVittie (1953) chose this solution to be

$$\nu = 0, \quad H = \text{constant} = \eta. \tag{5.1.63}$$

The conditions (5.1.62) and (5.1.63) then lead to two possibilities:

(i) $\beta = \delta, \quad p_0 \neq 0,$

$$\mu_1 = (\eta - \rho_0)\{\mu_2 - p_0\eta^{-(1+2/\delta)}\}. \tag{5.1.64}$$

(ii) $\beta \neq \delta, \quad p_0 \approx 0,$

$$\mu_1 = (\eta - \rho_0)\mu_2. \tag{5.1.65}$$

McVittie (1953) showed that case (i) is incompatible with the third shock condition (5.1.53). We, therefore, consider only case (ii) of a strong shock for which $p_0 \approx 0$.

In this case, (5.1.54) and (5.1.65) with $\mu = \frac{2}{3\gamma-1}$ give

$$\lambda = \frac{2(\gamma - 1)}{\gamma - 3}. \tag{5.1.66}$$

Now, $\delta = \lambda - 2 = 4/(\gamma - 3)$ and, since $\lambda > 2$ and, therefore, $\delta > 0$, we have $\gamma \geq 3$. From (5.1.54) we have

$$\eta = \frac{1}{3}\frac{3\gamma - 5}{\gamma - 3}\rho_0, \tag{5.1.67}$$

even as σ remains arbitrary. Thus, the first two conditions across the shock $r = h(t)$ are satisfied provided (5.1.66)–(5.1.67) hold and $\gamma \geq 3$. The third shock condition (5.1.53), after the use of (5.1.55) along with (5.1.54) and some simplification, reduces to

$$\mu_3 h^2 t^{-2} - t^{-1}\frac{dh^2}{dt} = \mu_4. \tag{5.1.68}$$

This has the integral

$$h = \left(\kappa t^{\mu_3} - \frac{\mu_4}{2 - \mu_3}t^2\right)^{1/2} \tag{5.1.69}$$

which, on using (5.1.56), becomes

$$H^{2/\delta} = \sigma^{2/\delta}\left(\kappa t^{\mu_3 - (2\beta/\delta)} - \frac{\mu_4}{2 - \mu_3}t^{2(1-\beta/\delta)}\right). \tag{5.1.70}$$

This gives the function $H(t)$ such that the third condition at the shock is also satisfied. Since $H = \eta$ and $\lambda = 2(\gamma - 1)/(\gamma - 3)$, we must have

$$\mu_3 = \frac{3\gamma^2 - 7\gamma - 2}{(3\gamma - 1)(\gamma - 1)}, \quad \mu_3 - \frac{2\beta}{\delta} = \frac{\gamma - 7}{(3\gamma - 1)(\gamma - 1)},$$

$$\mu_4 = \frac{\gamma(3\gamma - 1)}{\gamma - 1}\frac{p_0}{\rho_0}. \tag{5.1.71}$$

Furthermore, since $p_0 \approx 0$, and, therefore, $\mu_4 \approx 0$, (5.1.70) becomes

$$\left(\frac{\sigma}{\rho_0}\right)^{2/\delta} = \kappa^{-1}t^{-(\gamma-7)/\{(3\gamma-1)(\gamma-1)\}}\left(\frac{\lambda + 1}{3}\right)^{2/\delta}. \tag{5.1.72}$$

The LHS of (5.1.72) is independent of t; we must, therefore, have

$$\gamma = 7, \quad \kappa = \left(\frac{\sigma}{\rho_0}\right)^{-2/\delta}\left(\frac{\lambda + 1}{3}\right)^{2/\delta}. \tag{5.1.73}$$

With this value of γ, we have $\lambda = 3$ (see (5.1.66)) and the special solution (5.1.55) reduces to the well-known Primakoff solution for explosion in water:

$$q = \frac{1}{10}\frac{r}{t}, \quad \rho = \sigma t^{-2/5}r, \quad p = \frac{3\sigma}{100}t^{-\frac{12}{5}}r^3, \tag{5.1.74}$$

where

$$0 \leq r \leq \frac{4}{3}\frac{\rho_0}{\sigma}t^{2/5}; \tag{5.1.75}$$

see (5.1.56) and the definition of the constants in (5.1.71) and (5.1.54).

It is clear from the above that while the basic idea of expressing the solution of the equations of motion and continuity in terms of a 'potential function' is very ingenious, there is a need to generalise the work of McVittie (1953). All the solutions found by him have a linear velocity profile. The particle adiabacy condition further restricts the class of explicit solutions. This class of solutions must be enlarged so that they can be applied to other physically realistic situations. This is what we attempt to do in section 5.3.

5.2   Exact Solutions of Gasdynamic Equations in Lagrangian Co-ordinates

A study related to that of McVittie (1953) is due to Keller (1956) who apparently was not aware of the former work. The approach here, however, is quite distinct and applies to all geometries—planar, cylindrical, and spherical. The basic idea is to use the single second order nonlinear partial differential equation governing the Eulerian co-ordinate with the Lagrangian co-ordinate h and time t as independent variables (see Courant and Friedrichs (1948)). First, product solutions were sought without reference to boundary conditions. These solutions depend upon an arbitrary function which is related to the entropy distribution in the gas. Applications of isentropic and nonisentropic solutions include flows with shocks of finite and infinite strength and vacuum fronts. We shall bring out the relationship of these solutions with those of McVittie (1953).

Let us introduce the Lagrangian co-ordinate of a particle, namely,

$$h = \int_{y(0,t)}^{y(h,t)} r^{n-1} \rho(r,t) dr, \quad n = 1, 2, 3, \tag{5.2.1}$$

where $y(h,t)$ is the radius of the particle with Lagrangian co-ordinate h at time t, and $n = 1, 2, 3$ for planar, cylindrical and spherical symmetry, respectively. In the latter two cases y represents the distance from the axis and center of symmetry, respectively. From the definition of y, the velocity u of a particle is given by

$$u = y_t. \tag{5.2.2}$$

Differentiating (5.2.1) with respect to h, we have the density ρ and specific volume τ given by

$$\tau = \rho^{-1} = y^{n-1} y_h. \tag{5.2.3}$$

We assume that the flow is inviscid and nonconducting so that the entropy s of a particle is independent of time and, therefore,

$$s = s(h), \tag{5.2.4}$$

The function $s(h)$ is given either by the initial data or is determined by the shock motion imparting different entropy values to different particles.

The thermodynamic relation for a polytropic gas or liquid has the form

$$p = p(\rho, s) = g(\tau, s)$$
$$= g_0 + A(s)\tau^{-\gamma}, \qquad (5.2.5)$$

where the function $A(s)$, the adiabatic exponent $\gamma = c_p/c_v$, and the internal pressure g_0 are assumed to be known. The advantage of the Lagrangian co-ordinate system is that the equation of particle isentropy does not need to be imposed explicitly.

In terms of the quantities defined above, the equation of motion is simply

$$y_{tt} = -y^{n-1}[g_\tau(y^{n-1}y_h)_h + g_s s_h] \qquad (5.2.6)$$

or, for a polytropic gas obeying (5.2.5), we have

$$y_{tt} = \gamma A(s)(y^{n-1}y_h)^{-\gamma-1}(y^{n-1}y_h)_h y^{n-1} - A_h(y^{n-1}y_h)^{-\gamma}y^{n-1}. \qquad (5.2.7)$$

One disadvantage of (5.2.7) is that the exponents of y and its derivative are in general nonintegral. Equation (5.2.7) is a second order nonlinear PDE for $y(h,t)$, where the function A is assumed to be a known function of s and hence of h (see Courant and Friedrichs (1948)). Once solutions of (5.2.7) are known, the physical quantities may be found from (5.2.2)–(5.2.5).

Looking for product solutions of (5.2.7) we write

$$y(h,t) = f(h)j(t). \qquad (5.2.8)$$

Substituting (5.2.8) into (5.2.7) and separating the variables, we have

$$j'' - \lambda j^{n(1-\gamma)-1} = 0, \qquad (5.2.9)$$
$$-A[(f^{n-1}f')^{-\gamma}]' f^{n-2} - A'(f^{n-1}f')^{-\gamma}f^{n-2} = \lambda, \qquad (5.2.10)$$

where λ is the separation parameter and prime denotes derivative with respect to t in (5.2.9) and with respect to h in (5.2.10). Equation (5.2.9) is easily integrated to give

$$(j')^2 = \frac{2\lambda}{n(1-\gamma)}j^{n(1-\gamma)} + a \quad (\gamma \neq 1), \qquad (5.2.11)$$

$$(j')^2 = 2\lambda \log j + a, \quad (\gamma = 1), \qquad (5.2.12)$$

where a is the constant of integration. Excluding the case $j = $ constant which is possible only if $\lambda = a = 0$ or if $j = 0$, we have an implicit solution for j given by

$$\int_{j_0}^{j} \left[\frac{2\lambda}{n(2-\gamma)}j^{n(1-\gamma)} + a \right]^{-1/2} dj = t \quad (\gamma \neq 1), \qquad (5.2.13)$$

$$\int_{j_0}^{j} [2\lambda \log j + a]^{-1/2} dj = t \quad (\gamma = 1). \qquad (5.2.14)$$

Writing (5.2.10) more explicitly, we have

$$\gamma A[f'' f^{n-1} + (n-1)f^{n-2}(f')^2](f^{n-1}f')^{-\gamma-1}f^{n-2}$$
$$-f^{n-2}(f^{n-1}f')^{-\gamma}A' = \lambda. \tag{5.2.15}$$

Introducing the inverse function $h = h(f)$ and writing $h'(f) = q(f)$, (5.2.15) may be written as

$$\gamma A[-q'q^{-3}f^{n-1} + (n-1)f^{n-2}q^{-2}][f^{n-1}q^{-1}]^{-\gamma-1}f^{n-2}$$
$$-f^{n-2}(f^{n-1}q^{-1})^{-\gamma}A'(h) = \lambda. \tag{5.2.16}$$

Assuming that $\gamma \neq 1$ and introducing further the functions

$$z = q^{\gamma-1}, \quad B(f) = A[h(f)], \tag{5.2.17}$$

in (5.2.16), we have

$$z' + z[-(n-1)(\gamma-1)f^{-1} + \frac{\gamma-1}{\gamma}(\log B)']$$
$$+\frac{\lambda(\gamma-1)}{\gamma B}f^{(n-1)(\gamma-1)+1} = 0, \tag{5.2.18}$$

where prime denotes differentiation with respect to f. The solution of (5.2.18) may be written as

$$z = f^{(n-1)(\gamma-1)}B^{(1-\gamma)/\gamma}\left[G - \frac{\lambda(\gamma-1)}{\gamma}\int^f fB^{-1/\gamma}df\right], \tag{5.2.19}$$

where G is a constant. We have q from (5.2.17) and (5.2.19). Since $h'(f) = q$, we also have

$$h = \int_{f_0}^f f^{n-1}B^{-1/\gamma}\left[G - \frac{\lambda(\gamma-1)}{\gamma}\int^f fB^{-1/\gamma}df\right]^{\gamma/(\gamma-1)}df. \tag{5.2.20}$$

Equation (5.2.20) gives f as a function of h implicitly. If we define

$$F(f) = \left[G - \frac{\lambda(\gamma-1)}{\gamma}\int^f fB^{-1/\gamma}df\right]^{\gamma/(\gamma-1)}, \tag{5.2.21}$$

we may solve for $B(f)$ provided $\lambda \neq 0$:

$$B(f) = (-\lambda f)^\gamma (F')^{-\gamma}F. \tag{5.2.22}$$

The flow variables may now formally be written from (5.2.2)–(5.2.5), (5.2.8), (5.2.20) and (5.2.21), where we assume that $\gamma \neq 1$ and $\lambda \neq 0$ and where $f = y/j(t)$. Thus, we have

$$u(y,t) = y\frac{j'}{j}, \tag{5.2.23}$$
$$\tau(y,t) = -\lambda y j^{n-1}/F'(yj^{-1}), \tag{5.2.24}$$
$$p(y,t) = g_0 + j^{-n\gamma}F(yj^{-1}). \tag{5.2.25}$$

We may observe that, provided $j(t)$ is governed by (5.2.13) and (5.2.14), equations (5.2.23)–(5.2.25) give solutions of gasdynamic equations in Eulerian co-ordinates for $n = 1, 2, 3$ for arbitrary choice of the function F. This is the same form as obtained by McVittie (1953). Indeed, the expressions (5.2.13) and (5.2.14) for $j(t)$ may be verified to be essentially the solution of the Abel equation derived by McVittie (1953) (see section 5.1). In the solution (5.2.23)–(5.2.25), the function F must be chosen such that τ (and hence density) given by (5.2.24) is positive, that is, F must be monotonic in the region where y is of one sign. For the excluded case $\gamma \neq 1$, $\lambda = 0$, the solution is given by

$$u(y,t) = y\frac{j'}{j} = \frac{y}{t}, \tag{5.2.26}$$

$$\tau(y,t) = j^n B^{1/\gamma}(yj^{-1})G^{1/(1-\gamma)}$$
$$= t^n b(yt^{-1}), \tag{5.2.27}$$

$$p(y,t) = g_0 + j^{-n\gamma}G^{\gamma/(1-\gamma)}$$
$$= g_0 + lt^{-n\gamma}, \tag{5.2.28}$$

where an arbitrary function b and a constant l have been introduced. The solution in the present case becomes simpler since (5.2.13) now integrates to give

$$j(t) = \pm a^{1/2}t, \tag{5.2.29}$$

where we have assumed that $j(0) = 0$.

Now we consider the case $\gamma = 1$. With $B(f)$ from (5.2.17), (5.2.16) becomes

$$q'q^{-1} - (n-1)f^{-1} + \frac{B'}{B} + \frac{\lambda f}{B} = 0 \tag{5.2.30}$$

and integrates to give

$$q(f) = f^{n-1}B^{-1}(f)\exp \int^f -\lambda f B^{-1}(f)df, \tag{5.2.31}$$

and, since $h'(f) = q(f)$, we have

$$h(f) = \int_{f_0}^f [f^{n-1}B^{-1}(f)\exp \int^f -\lambda f B^{-1}(f)df]df. \tag{5.2.32}$$

Equation (5.2.32) gives $f = f(h)$ implicitly. Again, if we define

$$F(f) = \exp \int^f -\lambda f B^{-1}(f)df, \tag{5.2.33}$$

then, for $\lambda \neq 0$, we have

$$B(f) = -\lambda f F(F')^{-1}. \tag{5.2.34}$$

The special cases $\gamma = 1, \lambda \neq 0$ and $\gamma = 1, \lambda = 0$ may be obtained from the above appropriately.

The forms (5.2.23)–(5.2.25) and (5.2.26)–(5.2.28) each represent non-isentropic solutions of the basic gasdynamic equation and involve an arbitrary function. The pressure in the latter solution is a function of time alone. These solutions are explicit if the integrals in (5.2.13) and (5.2.14) can be found explicitly.

Now we consider isentropic solutions as a special case. Here the function $A(= B)$ is constant. Then, from (5.2.21) and (5.2.33), we have

$$F(f) = \left[G - \frac{\lambda(\gamma - 1)}{2\gamma B^{1/\gamma}} f^2 \right]^{\gamma/(\gamma-1)} \qquad \gamma \neq 1, \qquad (5.2.35)$$

$$F(f) = \exp - \frac{\lambda f^2}{2B} \qquad \gamma = 1. \qquad (5.2.36)$$

The solution (5.2.23)–(5.2.25) with $\gamma \neq 1$, $\lambda \neq 0$, and $f = yj^{-1}$ becomes

$$u(y, t) = y j' j^{-1}, \qquad (5.2.37)$$

$$\tau(y, t) = j^n B^{1/\gamma} \left[G - \frac{\lambda(\gamma - 1)}{2\gamma B^{1/\gamma}} y^2 j^{-2} \right]^{-1/(\gamma-1)}, \qquad (5.2.38)$$

$$p(y, t) = g_0 + j^{-n\gamma} \left[G - \frac{\lambda(\gamma - 1)}{2\gamma B^{1/\gamma}} y^2 j^{-2} \right]^{\gamma/(\gamma-1)}. \qquad (5.2.39)$$

Here, B and G are constants while $j(t)$ is given by (5.2.13). For $\lambda = 0$ and all $\gamma \neq 1$, the solution (5.2.26)–(5.2.28) holds with b a constant.

For $\gamma = 1$ and all λ, equation (5.2.37) still holds but (5.2.38)–(5.2.39) become

$$\tau(y, t) = j^n B \exp \frac{-\lambda j^{-2} y^2}{2B}, \qquad (5.2.40)$$

$$p(y, t) = g_0 + j^{-n} \exp \frac{-\lambda j^{-2} y^2}{2B}, \qquad (5.2.41)$$

where B is a constant. $j(t)$ is now given by (5.2.14).

For an example of isentropic flows we choose $a = 0$, $\gamma \neq 1$ in (5.2.13) and integrate to get

$$j(t) = \left\{ \left[\frac{2\lambda}{n(1 - \gamma)} \right]^{1/2} \left(\frac{n(\gamma - 1) + 2}{2} \right) t \right\}^{2/\{n(\gamma-1)+2\}}. \qquad (5.2.42)$$

With $G = 0$ in (5.2.38)–(5.2.39) we obtain a simple explicit solution for isentropic flows:

$$u(y, t) = \frac{2}{n(\gamma - 1) + 2} y t^{-1}, \qquad (5.2.43)$$

$$\tau(y,t) = \left[\frac{\gamma B}{n}\left(\frac{n(\gamma-1)+2}{\gamma-1}\right)^2\right]^{1/(\gamma-1)}$$
$$\times (yt^{-1})^{-2/(\gamma-1)}, \tag{5.2.44}$$

$$p(y,t) = g_0 + B\left[\frac{\gamma B}{n}\left(\frac{n(\gamma-1)+2}{\gamma-1}\right)^2\right]^{\gamma/(1-\gamma)}$$
$$\times (yt^{-1})^{2\gamma/(\gamma-1)}. \tag{5.2.45}$$

As an application we again consider the propagation of a strong shock with the equation $y = R(t)$ into a quiet medium with variable density $\rho_0(y)$ and pressure zero. The shock conditions in this case are

$$\frac{\rho}{\rho_0} = \frac{\gamma+1}{\gamma-1}, \tag{5.2.46}$$

$$u = \left[\frac{2p}{(\gamma+1)\rho_0}\right]^{1/2}, \tag{5.2.47}$$

$$\dot{R} = \left[\frac{(\gamma+1)p}{2\rho_0}\right]^{1/2}, \tag{5.2.48}$$

where p is the pressure behind the shock. We wish to find the functions $F(yj^{-1})$, $R(t)$ and $\rho_0(y)$ such that the solution, given by (5.2.23)–(5.2.25), satisfies the Rankine-Hugoniot conditions (5.2.46)–(5.2.48). To that end, we insert the former into the latter and obtain

$$\frac{F'(Rj^{-1})}{-\lambda Rj^{n-1}\rho_0(R)} = \frac{\gamma+1}{\gamma-1}, \tag{5.2.49}$$

$$\frac{Rj'}{j} = \left[\frac{2g_0 + 2j^{-n\gamma}F(Rj^{-1})}{(\gamma+1)\rho_0(R)}\right]^{1/2}, \tag{5.2.50}$$

$$\dot{R} = \left[\frac{(\gamma+1)\{g_0 + j^{-n\gamma}F(Rj^{-1})\}}{2\rho_0(R)}\right]^{1/2}. \tag{5.2.51}$$

From (5.2.50)–(5.2.51) we have

$$\frac{\dot{R}}{R} = \frac{\gamma+1}{2}\frac{j'}{j}. \tag{5.2.52}$$

Integrating (5.2.52) we get

$$R(t) = R_0[j(t)]^{(\gamma+1)/2}, \tag{5.2.53}$$

where R_0 is a constant of integration. Putting (5.2.53) into (5.2.50) and using (5.2.11) satisfied by $j(t)$, we have

$$F(x) = \frac{\gamma+1}{2}\rho_0(x^{(\gamma+1)/(\gamma-1)}R_0^{2/(1-\gamma)})$$

$$\times \left[\frac{2\lambda}{n(1-\gamma)} (xR_0^{-1})^{-2n} + D \right]$$
$$\times R^{(-2n\gamma)/(\gamma-1)} x^{[(2n+2)\gamma-2]/(\gamma-1)}$$
$$-g_0(xR_0^{-1})^{(2n\gamma)/(\gamma-1)}. \tag{5.2.54}$$

Thus, the function $R(t)$ is given by (5.2.53) while $F(x)$ and $\rho_0(x)$ are related by (5.2.54). The solution so obtained has R_0, D, and λ as arbitrary constants. The conditions (5.2.47)–(5.2.48) across the shock are already satisfied. We substitute (5.2.53) and (5.2.54) into the third shock condition (5.2.46) to find F. We let $g_0 = 0$ for simplicity. We thus obtain

$$F(x) = F_0 x^n \left[\frac{2\lambda R_0^{2n}}{n(1-\gamma)} + Dx^{2n} \right]^{-1/2}. \tag{5.2.55}$$

From (5.2.55) and (5.2.54) we get the undisturbed density as

$$\rho_0(R) = \frac{2F_0 R_0^{(2n\gamma)/(\gamma-1)}}{\gamma+1} \left[\frac{2\lambda R_0^{2n}}{n(1-\gamma)} \right.$$
$$\left. + DR_0^{(4n)/(\gamma+1)} R^{[2n(\gamma-1)/(\gamma+1)]} \right]^{-3/2}$$
$$\times R^{[(n-2)\gamma-3n+2]/(\gamma+1)}$$
$$\times R^{[2(n-2)\gamma-6n+4]/(\gamma+1)(\gamma-1)}. \tag{5.2.56}$$

F_0 in (5.2.55) and (5.2.56) is an arbitrary constant. The corresponding solution with $g_0 \neq 0$ is somewhat more complicated. The above solution involves four arbitrary constants—F_0, λ, R_0, and D—and describes a shock moving according to (5.2.53), where $j(t)$ is given by (5.2.13). The flow behind the shock is given by (5.2.23)–(5.2.25) with $F(x)$ defined by (5.2.55). This flow may be viewed as produced by a piston moving along one of the particle paths, $Y(t) = Y_0 j(t)$, say. The above solution was derived subject to the constraints that $\gamma \neq 1$ and $\lambda \neq 0$.

As in the paper of McVittie (1953), the well-known Primakoff solution may be obtained as a special case if it is assumed that $\rho_0(R) = $ constant. It follows from (5.2.56) that we must now have

$$D = 0, \quad \gamma = \frac{3n-2}{n-2}. \tag{5.2.57}$$

For the spherically symmetric case, $n = 3$, (5.2.57) gives $\gamma = 7$. In this case, we get from (5.2.13) and (5.2.55),

$$j = \left[\frac{5(-\lambda)^{1/2}}{3} t \right]^{1/10}, \quad F(x) = x^3 \frac{3F_0 R_0^{-3}}{(-\lambda)^{1/2}}. \tag{5.2.58}$$

The solution (5.2.23)–(5.2.25) with the functions $j(t)$ and $F(x)$ given by (5.2.58) coincides exactly with the solution of the point explosion problem in water found first by Primakoff (Courant and Friedrichs (1948)). Here we assume that $\lambda < 0$. Now we may compare the approaches and results of McVittie (1953) and Keller (1956). McVittie (1953) wrote a very general form of the solution of equations of continuity and motion in Eulerian co-ordinates in spherical symmetry in terms of a potential function but then restricted its form to a product of a function of time and a function of a similarity variable. He obtained solutions which have a linear distribution of particle velocity. In the two forms he wrote, one involves a function of time governed by an Abel equation while the other has an arbitrary function of the similarity variable. Keller (1956) adopted an entirely different approach. He attempted to solve the single second order nonlinear PDE for the Eulerian co-ordinate with the Lagrangian co-ordinates as independent variables. His product solutions again end up with a particle velocity which is linear in distance. Both these approaches deserve further investigation and generalisation. We take up this matter in the next section.

5.3   Exact Solutions of Gasdynamic Equations with Nonlinear Particle Velocity

As we pointed out earlier in section 5.2, the investigations of McVittie (1953) and Keller (1956) are both restricted to flow velocities which are linear functions of the spatial co-ordinate. There is considerable scope to extend their results such that the velocity distribution is given by a more realistic nonlinear function. We recall that Taylor (1950) chose particle velocity behind the shock heading a blast wave as a sum of a linear term and a nonlinear term in the similarity variable (see section 3.2). He then found an approximate solution analytically which was in excellent agreement with the numerical solution. This work was later extended by Sakurai (1953, 1954) to other geometries. We shall generalise the work of McVittie (1953) in two ways. First, we shall write (5.1.18) or (5.1.21) as $q = \alpha(r/t)a(x)$, where α is a constant, and find an equation for the function $a(x)$ such that the adiabatic condition (5.1.36) is satisfied; the solutions (5.1.18)–(5.1.20) already satisfy equations of momentum and continuity. We shall discuss in some detail a particular form of $a(x)$ which is similar to that of Taylor (1950); our solution, however, would be exact and satisfy all the equations of motion. The second generalisation is motivated by the simple form (5.1.19) for the density distribution. Several results of McVittie (1953) and Keller (1956) may be identified as special cases of these more general solutions.

We begin by writing the particle velocity in the formal solution of equations of motion and continuity (5.1.18)–(5.1.20) as

$$q = \alpha r t^{-1} a(x), \qquad (5.3.1)$$

where

$$a(x) = \frac{(1-g)w_x + xw_{xx}}{2w_x + xw_{xx}}. \tag{5.3.2}$$

For this case, the function $f(t)$ has already been chosen to be $f = t^{\alpha g}$ where g is a constant. We may write (5.3.2) as

$$w'' = A(x)w', \tag{5.3.3}$$

where $' = d/dx$ and

$$A(x) = \frac{1 - g - 2a}{(a-1)x}. \tag{5.3.4}$$

The function $a = a(x)$ is arbitrary so far. It will be determined to satisfy the equation of particle isentropy. It follows from (5.3.3) that

$$w(x) = c_1 + c_0 \int^x e^{\int^y A(z)dz} \, dy, \tag{5.3.5}$$

where c_0 and c_1 are constants of integration. With this form for $w(x)$ we may write the solution (5.1.18)–(5.1.20) as

$$q(x,t) = \alpha t^{\alpha-1}Q(x), \tag{5.3.6}$$
$$\rho(x,t) = -c_0(1+g)t^{\alpha(g-2)}R(x), \tag{5.3.7}$$
$$p(x,t) = p_0 + \alpha^2 t^{\alpha g-2}P(x), \tag{5.3.8}$$

where

$$Q(x) = xa(x),$$
$$R(x) = \frac{1}{x(a-1)}e^{\int^x A(z)dz}, \tag{5.3.9}$$
$$P(x) = g\left(g - \frac{1}{\alpha}\right)c_1 + c_0\left[\left(\frac{1}{\alpha} - g\right) + (1+g)a(x)\right]xe^{\int^x A(z)dz}$$
$$+ c_0 \int^x \left[2(1+g)\frac{a^2(y)}{a(y)-1} + \left(g^2 - \frac{g}{\alpha}\right)\right]e^{\int^y A(z)dz} \, dy.$$

Here, p_0, c_0 amd c_1 are constants.

We may observe that for any function $F = F(x)$, $x = rt^{-\alpha}$, and $q(x,t)$ given by (5.3.6) and (5.3.9)$_1$, we have

$$\left(\frac{\partial}{\partial t} + q\frac{\partial}{\partial r}\right)F = \alpha(a-1)\frac{x}{t}F'(x). \tag{5.3.10}$$

Therefore, substituting the expressions for q, p and ρ from (5.3.6)–(5.3.9) into the equation of particle isentropy,

$$\left(\frac{\partial}{\partial t} + q\frac{\partial}{\partial r}\right)(\log p - \gamma \log \rho) = 0, \tag{5.3.11}$$

and making use of (5.3.10), we obtain

$$\beta + \alpha x(a-1)\left(\frac{P'}{P} - \gamma\frac{R'}{R}\right) = 0, \tag{5.3.12}$$

where

$$\beta = \alpha g(1-\gamma) + 2(\alpha\gamma - 1). \tag{5.3.13}$$

We let $c_0 = 1$, $c_1 = 0$ in $(5.3.9)_2$. From $(5.3.9)_2$, we have

$$\frac{R'}{R} = \frac{2 - g - 3a - xa'}{x(a-1)}. \tag{5.3.14}$$

Here we assume that $a(x) \neq 1$. This singular case would be treated separately.

From (5.3.12) and (5.3.14), we have

$$P' = \frac{1}{C(x)}P(x), \tag{5.3.15}$$

where

$$C(x) = \frac{\alpha x(a-1)}{2 - \alpha g - 3\alpha\gamma a - \alpha\gamma x a'}. \tag{5.3.16}$$

Differentiating (5.3.15) we get

$$(1 - C'(x))P' = C(x)P''. \tag{5.3.17}$$

From $(5.3.9)_3$ with $c_0 = 1$, $c_1 = 0$ it follows that

$$P'(x) = (1+g)\left[(1 - 1/\alpha) + \frac{(1 - 1/\alpha)}{a - 1} + a + xa'\right]e^{\int^x A(z)dz}, \tag{5.3.18}$$

$$P''(x) = \frac{(1+g)}{x(a-1)^2}e^{\int^x A(z)dz} \times \left[x^2 a^2 a'' - 2x^2 aa'' + x^2 a'' - (1+g)xaa'\right.$$
$$\left. -2a^3 + \left(1 + \frac{2}{\alpha} - g\right)a^2 + \left(\frac{1}{\alpha} + g\right)xa' - \frac{1}{\alpha}(1-g)a\right]. \tag{5.3.19}$$

Substituting (5.3.18) and (5.3.19) into (5.3.17), we obtain

$$(1 - C')[(1 - (1/\alpha))(a/(a-1)) + a + xa'] = \frac{C}{x(a-1)^2}$$
$$\times [x^2 a^2 a'' - 2x^2 aa'' + x^2 a'' - (1+g)xaa' - 2a^3$$
$$+(1 + (2/\alpha) - g)a^2 + (g + (1/\alpha))xa' - (1/\alpha)(1-g)a]. \tag{5.3.20}$$

Substituting for $C(x)$ and $C'(x)$ from (5.3.16) into (5.3.20), we get

$$[(2/\alpha - g - \gamma(3a + xa'))^2 - (g - 2/\alpha) - (2/\alpha - g + 3\gamma)a$$
$$-(2/\alpha - g - 3\gamma)xa' + \gamma x^2 a'^2 - \gamma x^2 aa'' + 3\gamma a^2 + \gamma x^2 a'']$$
$$\times [a^2 - (a/\alpha) + xaa' - xa'] - [(2/\alpha - g) - \gamma(3a + xa')]$$
$$\times [x^2 a^2 a'' - 2x^2 aa'' + x^2 a'' - (1+g)xaa' - 2a^3 + (1 + 2/\alpha - g)a^2$$
$$+(g + 1/\alpha)xa' - (1/\alpha)(1-g)a] = 0. \tag{5.3.21}$$

Thus, the solution (5.3.6)–(5.3.8) of the equations of motion also satisfies particle isentropy (5.3.11) provided the function $a(x) \neq 1$ satisfies (5.3.21). This function is related to $w(x)$ via (5.3.2). Indeed we may derive a third order ODE for w' which however is not much simpler to handle.

Motivated by the approximate form of the similarity solution of Taylor (1950), we consider the special case

$$a(x) = \lambda_0 x^\mu + \lambda_1, \tag{5.3.22}$$

where μ, λ_0 and λ_1 are constants. With this choice of $a(x)$, the function $A(x)$ in (5.3.4) becomes

$$
\begin{aligned}
A(x) &= \frac{1 - g - 2\lambda_0 x^\mu - 2\lambda_1}{x(\lambda_0 x^\mu + \lambda_1 - 1)} \\
&= -\frac{(1+g)}{\lambda_0} x^{-(1+\mu)} - 2x^{-1}, \quad \lambda_1 = 1 \\
&= \frac{(1 - g - 2\lambda_1)}{\lambda_1 - 1} \frac{1}{x} - \frac{(1+g)}{\lambda_1 - 1} \frac{\lambda_0 x^{\mu-1}}{\lambda_0 x^\mu + \lambda_1 - 1}, \quad \lambda_1 \neq 1.
\end{aligned}
\tag{5.3.23}
$$

Therefore, we have

$$\int^x A(z)dz = \frac{(1+g)}{\lambda_0 \mu} x^{-\mu} - 2\log x + A_0, \quad \lambda_1 = 1$$

and

$$\int^x A(z)dz = \log A_0 + \log \frac{(\lambda_0 x^\mu + \lambda_1 - 1)^{\frac{1+g}{\mu(1-\lambda_1)}}}{x^{\frac{1-g-2\lambda_1}{1-\lambda_1}}}, \lambda_1 \neq 1. \tag{5.3.24}$$

The solution (5.3.6)–(5.3.8) now assumes the following forms for $\lambda_1 = 1$ and $\lambda_1 \neq 1$, respectively.

$\lambda_1 = 1.$

$$
\begin{aligned}
q(x,t) &= \alpha t^{\alpha-1} x(\lambda_0 x^\mu + 1), \\
\rho(x,t) &= -c_0 A_0 \lambda_0^{-1}(1+g) t^{\alpha(g-2)} x^{-(\mu+3)} e^{\frac{(1+g)}{\lambda_0 \mu} x^{-\mu}}, \\
p(x,t) &= p_0 + \alpha^2 t^{\alpha g-2} P(x),
\end{aligned}
\tag{5.3.25}
$$

where

$$
\begin{aligned}
P(x) &= g\left(g - \frac{1}{\alpha}\right)c_1 + c_0 A_0 \left(\frac{1}{\alpha} - g + (1+g)(\lambda_0 x^\mu + 1)\right) \\
&\quad \times x^{-1} e^{\frac{(1+g)}{\lambda_0 \mu} x^{-\mu}} + c_0 A_0 \int^x \left(\frac{2(1+g)(\lambda_0 s^\mu + \lambda_1)^2}{\lambda_0 s^\mu + \lambda_1 - 1}\right. \\
&\quad \left. + g^2 - \frac{g}{\alpha}\right) s^{-2} e^{\frac{1+g}{\lambda_0 \mu} s^{-\mu}} ds
\end{aligned}
\tag{5.3.26}
$$

and p_0 is a constant.

$\lambda_1 \neq 1.$

$$q(x,t) = \alpha t^{\alpha-1} x(\lambda_0 x^\mu + \lambda_1),$$

$$\rho(x,t) = -c_0 A_0(1+g)t^{\alpha(g-2)}$$

$$\times \frac{(\lambda_0 x^\mu + \lambda_1 - 1)^{\frac{1+g}{\mu(1-\lambda_1)}-1}}{x^{1+\frac{1-g-2\lambda_1}{1-\lambda_1}}}, \tag{5.3.27}$$

$$p(x,t) = p_0 + \alpha^2 t^{\alpha g - 2} P(x),$$

where

$$P(x) = g\left(g - \frac{1}{\alpha}\right)c_1 + c_0 A_0\left[\left(\frac{1}{\alpha} - g\right)\right.$$

$$+ (1+g)(\lambda_0 x^\mu + \lambda_1)\frac{(\lambda_0 x^\mu + \lambda_1 - 1)^{\frac{1+g}{\mu(1-\lambda_1)}}}{x^{-1+\frac{(1-g-2\lambda_1)}{1-\lambda_1}}}\right]$$

$$+ c_0 A_0 \int^x \left[2(1+g)\frac{(\lambda_0 s^\mu + \lambda_1)^2}{(\lambda_0 s^\mu + \lambda_1 - 1)}\right.$$

$$\left. + \left(g^2 - \frac{g}{\alpha}\right)\right]\frac{(\lambda_0 s^\mu + \lambda_1 - 1)^{\frac{1+g}{\mu(1-\lambda_1)}}}{s^{\frac{1-g-2\lambda_1}{1-\lambda_1}}}ds. \tag{5.3.28}$$

We consider the special case $g = 2/\alpha$, $\lambda_1 = 0$, which leads to a simple solution. To satisfy the adiabatic condition (5.3.11) with $a(x) = \lambda_0 x^\mu$, we must find $C(x)$ from (5.3.16):

$$C(x) = -\frac{x}{\gamma(3+\mu)} + \frac{x^{1-\mu}}{\lambda_0 \gamma(3+\mu)}. \tag{5.3.29}$$

Equation (5.3.21) in the present case becomes

$$-\lambda_0^3(1+\mu)(1+\gamma(3+\mu))x^{3\mu} + \lambda_0^2[(1/\alpha+\mu)$$
$$\times(1+\gamma(3+\mu) + (1-\mu)(1+\mu)x^{2\mu}] - \lambda_0(1-\mu)(1/\alpha+\mu)x^\mu$$
$$= \lambda_0^3(\mu(\mu-1)-2)x^{3\mu} + \lambda_0^2[1-2\mu(\mu-1)-(1+2/\alpha)\mu]x^{2\mu}$$
$$+ \lambda_0[\mu(\mu-1) + (3\mu/\alpha) - 1/\alpha(1-2/\alpha)]x^\mu. \tag{5.3.30}$$

This equation is identically satisfied if

$$\mu = -1/\alpha,$$

and

either $\mu + 1 = 0$ or $1 + \gamma(3+\mu) = 2 - \mu.$ \tag{5.3.31}

These relations yield

$$\mu = -1, \quad \alpha = 1, \tag{5.3.32}$$

both when $\gamma \neq 1$ and when $\gamma = 1$.

It is clear from (5.3.14) that the case $a(x) = 1$ is singular and must be treated separately. Using (5.1.13) and (5.1.17), we find that the solution (5.1.18)–(5.1.20) of the equations of continuity and momentum for this case becomes

$$
\begin{aligned}
q &= \alpha t^{\alpha-1} x = \alpha \frac{r}{t}, \\
\rho &= t^{-3\alpha} \frac{2w_x + x w_{xx}}{x}, \\
p &= p_0 + \alpha(1-\alpha) t^{-2-\alpha}(w + x w_x + Q(t)),
\end{aligned}
\tag{5.3.33}
$$

where p_0 is a constant. When the particle adiabacy condition (5.3.11) is imposed on the solution (5.3.33), we get

$$
\frac{\alpha(3\gamma - 1) - 2}{t} + \frac{Q'(t)}{w + x w_x + Q(t)} = 0.
\tag{5.3.34}
$$

This equation may be satisfied if either of the following conditions hold:

(i) $Q'(t) = 0$, i.e., $Q = $ constant, and $\alpha = 2/(3\gamma - 1)$.

(ii) $(xw)_x = c_0$, a constant, and so $w = c_0 + c_1/x$, where c_1 is a constant of integration. Moreover, $Q(t) = -c_0 + c_2 t^{2-\alpha(3\gamma-1)}$, where c_2 is another constant.

Correspondingly, we have the following representations of the physical quantities. For case (i), we have

$$
\begin{aligned}
q &= \frac{2}{3\gamma - 1} t^{\frac{2}{3\gamma-1}-1} x, \\
\rho &= t^{\frac{-6}{3\gamma-1}} \frac{2w_x + x w_{xx}}{x}, \\
p &= p_0 + \frac{6(\gamma - 1)}{(3\gamma - 1)^2} t^{-2-\frac{2}{3\gamma-1}}((xw)_x + c_0),
\end{aligned}
\tag{5.3.35}
$$

where p_0 and c_0 are constants. In (5.3.35), w is an arbitrary function of $x = rt^{-\alpha}$.

For case (ii), we have the particle adiabatic solution

$$
\begin{aligned}
q &= \alpha t^{\alpha-1} x, \\
\rho &= 0, \\
p &= p_0 + \alpha c_2 (1 - \alpha) t^{-3\alpha\gamma}.
\end{aligned}
\tag{5.3.36}
$$

McVittie (1953) studied the subcase (i) of this singular solution with $a = 1$. Thus, his analysis for the particle adiabatic motions relates only to this special case.

Now we indicate how to generalise the solutions of McVittie (1953) from a different point of view. The form (5.1.19) of the solution for the density function suggests that, provided this function is analytic in the neighbourhood of $x = 0$, we may write

$$\left[t^{2\alpha}/f(t)\right]\rho = x^{-2}(x^2 w_x)_x = \sum_{n=0}^{\infty} a_n x^n. \qquad (5.3.37)$$

where a_n are arbitrary constants. Integrating (5.3.37) twice we have

$$w(x) = \sum_{n=0}^{\infty} b_n x^{n+2}, \qquad (5.3.38)$$

where

$$b_n = \frac{a_n}{(n+2)(n+3)}. \qquad (5.3.39)$$

The particle velocity and pressure may then be found from (5.1.18) and (5.1.20). It may easily be verified that case B of McVittie (1953), given by (5.1.30) and (5.1.31), is obtained if we choose $g = -1$ and $f = t^{-\alpha}$. However, the function $F(x)$, which is arbitrary therein, is now specified as an analytic function $\sum_{n=0}^{\infty} a_n x^n$ about $x = 0$; the velocity profile, however, is now rendered more general. We may impose the particle isentropy condition (5.3.11) on the present form of the solution to determine the coefficients a_n. Since the details are rather complicated, we shall publish these results subsequently.

Chapter 6

Converging Shock Waves

6.1 Converging Shock Waves: The Implosion Problem

The generation and propagation of converging shock waves is a fascinating phenomenon which has a long history and continues to arouse interest. It was first treated by Guderley (1942). This may be visualised in two ways—either arising from a converging spherical or cylindrical piston, or by the instantaneous release of energy on a rigid spherical or cylindrical wall. In the latter case we may imagine a spherical or cylindrical chamber of radius R_0 containing a test gas at initial pressure p_0 and initial density ρ_0. At time $t = t_i < 0$, a finite amount of energy E_0 or, for the cylindrical case E_0 per unit length, is released instantaneously at a radius R_0, generating a strong shock wave. At subsequent times, the shock wave collapses towards the center or axis of symmetry.

We first consider the simplest situation when the resulting flow is self-similar, as considered first by Guderley (1942). Here, however, we follow the more recent work of Chisnell (1998) who has studied this problem for a long time (see Chisnell (1957)). This problem is also an excellent example of self-similar solutions of the second kind when the similarity exponent is not obtained from the dimensional considerations alone but must be determined by solving an eigenvalue problem for a nonlinear ordinary differential equation. There are other attractive features of this problem—a rich analytic structure and asymptotic character of the solution relating to its stability. We may mention that not all questions regarding the nature of the possible solutions of this problem have been answered.

Chisnell (1998), unlike previous investigators, was able to get accurate analytic and numerical results—the former, though approximate, give an 'extremely good' value of the similarity exponent and yield a simple analytic description of flow variables at all points behind the converging shock.

We write the equations of symmetric (spherical and cylindrical) adiabatic equations of motions as

$$\rho_t + u\rho_r + \rho r^{1-s}(ur^{s-1})_r = 0, \tag{6.1.1}$$

$$u_t + uu_r + \frac{1}{\rho}p_r = 0, \tag{6.1.2}$$

$$\left(\frac{\partial}{\partial t} + u\frac{\partial}{\partial r}\right)(\ln(p/\rho^\gamma)) = 0, \tag{6.1.3}$$

where the last equation describes constancy of entropy along a particle line. $s = 2, 3$ for cylindrical and spherical symmetry, respectively. We may replace p in (6.1.2) and (6.1.3) by the speed of sound via $c^2 = \gamma p/\rho$. Introducing now the variables

$$\rho = \rho_0 G, \quad u = \frac{r}{t}V, \quad c^2 = \frac{r^2}{t^2}Z, \tag{6.1.4}$$

the partial differential equations (6.1.1)–(6.1.3) assume the form

$$tG_t + VrG_r + GrV_r = -sVG, \tag{6.1.5}$$

$$tV_t + VrV_r + \frac{1}{\gamma}\frac{Z}{G}rG_r + \frac{1}{\gamma}rZ_r = V - V^2 - \frac{2Z}{\gamma}, \tag{6.1.6}$$

$$t\frac{Z_t}{Z} + rV\frac{Z_r}{Z} - \frac{\gamma-1}{G}(tG_t + VrG_r) = 2 - 2V. \tag{6.1.7}$$

Introducing the similarity variable

$$\xi = \frac{r}{R}, \tag{6.1.8}$$

where $R(t)$ is the distance of the shock from the origin or axis of symmetry at time $t(< 0)$, the derivatives in (6.1.5)–(6.1.7) change according to

$$\frac{\partial}{\partial r} = \frac{1}{R}\frac{\partial}{\partial \xi}, \quad \frac{\partial}{\partial t} = \frac{\partial}{\partial t} - \xi\frac{\dot{R}}{R}\frac{\partial}{\partial \xi}, \tag{6.1.9}$$

where $\dot{R} = dR/dt$. The explicit time dependence in the resulting equations will occur only in the form $t\dot{R}/R$. If we let

$$t\frac{\dot{R}}{R} = \alpha \quad \text{or} \quad R = A(-t)^\alpha, \tag{6.1.10}$$

where α and A are constants, and let G, V and Z depend on ξ alone, we get the system of ODEs

$$\xi V' + (V - \alpha)\xi\frac{G'}{G} = -sV, \tag{6.1.11}$$

$$(V - \alpha)\xi V' + \frac{Z}{\gamma}\xi\frac{G'}{G} + \frac{1}{\gamma}\xi Z' = V - V^2 - \frac{2Z}{\gamma}, \tag{6.1.12}$$

$$(\gamma - 1)Z\xi\frac{G'}{G} - \xi Z' = -\frac{2Z(1 - V)}{V - \alpha}. \tag{6.1.13}$$

Solving for V', G' and Z', we have

$$\xi V' = \frac{\Delta_1}{\Delta}, \tag{6.1.14}$$

$$\xi \frac{G'}{G} = \frac{\Delta_2}{\Delta}, \tag{6.1.15}$$

$$\xi Z' = \frac{\Delta_3}{\Delta}, \tag{6.1.16}$$

where

$$\Delta = -Z + (V - \alpha)^2, \tag{6.1.17}$$

$$\Delta_1 = -\Delta \left\{ sV - \frac{2(1 - \alpha)}{\gamma} \right\} - (\alpha - V)Q(V), \tag{6.1.18}$$

$$\Delta_2 = \frac{2(1 - \alpha)}{\gamma(\alpha - V)} \Delta - Q(V), \tag{6.1.19}$$

$$\Delta_3 = \frac{Z}{V - \alpha} \left[2\Delta \left\{ \alpha - V + \frac{1 - \alpha}{\gamma} \right\} \right.$$
$$\left. + (\gamma - 1)(\alpha - V)Q(V) \right], \tag{6.1.20}$$

and

$$Q(V) = sV(V - \alpha) + \frac{2(1 - \alpha)}{\gamma}(\alpha - V) - V(V - 1). \tag{6.1.21}$$

We may reduce the discussion of the above system to the (Z, V) plane since

$$\frac{dZ}{dV} = \frac{\Delta_3}{\Delta_1}, \tag{6.1.22}$$

and relate other variables via

$$\frac{1}{G} \frac{dG}{dV} = \frac{\Delta_2}{\Delta_1}, \tag{6.1.23}$$

$$\frac{1}{\xi} \frac{d\xi}{dV} = \frac{\Delta}{\Delta_1}. \tag{6.1.24}$$

Once the solution of (6.1.22) is known, (6.1.23) and (6.1.24) give G as a function of Z or V and relate (V, Z, G) to the variable ξ.

Assuming that the shock produced is strong, the Rankine-Hugoniot conditions across it are

$$\frac{\rho}{\rho_0} = \frac{\gamma + 1}{\gamma - 1}, \tag{6.1.25}$$

$$u = \frac{2}{\gamma + 1} \dot{R}, \tag{6.1.26}$$

$$c^2 = \frac{2\gamma(\gamma - 1)}{(\gamma + 1)^2} \dot{R}^2. \tag{6.1.27}$$

On using (6.1.4) and (6.1.10), these conditions at the shock $\xi = 1$ become

$$G_s = \frac{\gamma + 1}{\gamma - 1}, \quad V_s = \frac{2\alpha}{\gamma + 1}, \quad Z_s = \frac{2\gamma(\gamma - 1)\alpha^2}{(\gamma + 1)^2}, \tag{6.1.28}$$

where the suffix s denotes conditions immediately behind the shock. Since $\xi = \frac{r}{R} = (r/A(-t)^\alpha)$, the point far behind the shock where r is large corresponds to $\xi = \infty$ so that

$$V(\infty) = 0, \quad Z(\infty) = 0, \tag{6.1.29}$$

stating that the particle velocity and sound speed both are zero there.

Now we pose the BVP problem in the (V, Z) plane. According to (6.1.17), $\Delta = 0$ is a parabola touching the V-axis at $V = \alpha$. Also, in view of (6.1.28), Δ has a negative value $-\alpha^2(\gamma - 1)/(\gamma + 1)$ at the shock $\xi = 1$. It is equal to $\alpha^2 > 0$ at $\xi = \infty$. The solution curve must cross the parabola to reach the point $(0, 0)$ which represents the point far behind the shock. To avoid infinite slope at the point of crossing, the denominators Δ_i $(i = 1, 2, 3)$ in (6.1.14)–(6.1.16) must also vanish there. According to (6.1.18), $\Delta_1 = 0$ when $\Delta = 0$ provided $Q(V) = 0$. $Q(V)$ (see (6.1.21)) is a quadratic in V. For a given value of γ, there is a value of the parameter α for which a solution of (6.1.22), starting at a singular point on $\Delta = 0$, passes through the shock point (6.1.28). Only one of the zeros of $Q(V)$ would permit that. Besides, one would have to resort to iteration to solve this eigenvalue problem with α as the eigenvalue. One could alternatively start from the shock and find α such that the solution passes through the 'appropriate' singular point (see Zeldovich and Raizer (1967)). After solving the problem in the (Z, V) plane one could numerically integrate (6.1.23)–(6.1.24) to complete the solution. Since the approximate solution given by Chisnell (1998) still requires an iteration to find the exponent α, we content ourselves here with a summary of his analytic approach. Equation (6.1.22) is written as

$$\frac{1}{Z}\frac{dZ}{dV} = \frac{2\Delta(\alpha - V + (1 - \alpha)/\gamma) + (\gamma - 1)(\alpha - V)Q}{\Delta(sV - 2(1 - \alpha)/\gamma)(\alpha - V) + (\alpha - V)^2 Q}. \tag{6.1.30}$$

Q is given by (6.1.21). The basic idea is to examine the local behaviour of the function Z at the two singular points, guess a trial function Z_T which has the right behaviour at these points and hence substitute it into Δ on the right hand side of (6.1.30). This enables an integration of (6.1.30) in a closed form.

Using the auxilary equations (6.1.23)–(6.1.24) along with the conditions at the shock, the (approximate) analytic form of the solution behind the shock is determined. Unfortunately it introduces another unknown constant into the solution, namely V_0, the value of V at the other singular point (a zero of $Q(V) = 0$) which also must be found as part of the solution. However, a relation between α and V_0 is found such that ultimately one has to find

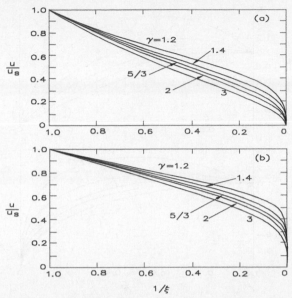

Figure 6.1 The particle velocity u/u_S versus $1/\xi = R(t)/r$ for (a) spherical and (b) cylindrical symmetries for different values of γ (Chisnell, 1998).

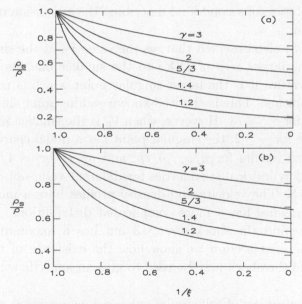

Figure 6.2 The density ρ_s/ρ versus $1/\xi = R(t)/r$ for (a) spherical and (b) cylindrical symmetries for different values of γ (Chisnell, 1998).

only the parameter V_0/α using iteration. It is observed that, for $\gamma = 1.4$, a six figure accuracy for α/V_0 is obtained, with just one iteration. For $\gamma = 3$, however, one has to iterate five times to obtain an accurate value of α/V_0 to five decimal places. A local analysis in the neighbourhood of the appropriate singular point was also performed earlier by Sakurai (1959) in the context

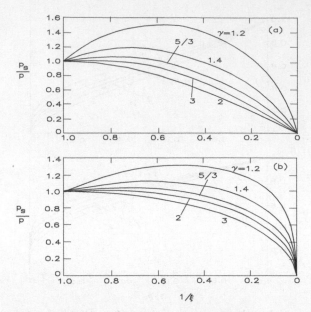

Figure 6.3 The pressure p_S/p versus $1/\xi = R(t)/r$ for (a) spherical and (b) cylindrical symmetries for different values of γ (Chisnell, 1998).

of self-similar solution of second kind describing the expansion of a shock at the edge of a star.

Chisnell (1998) also observed that, as the value V_0 at the singular point changes from the larger zero of $Q(V)$ to the smaller one, its nature also changes. In literature it is the larger singular point which is referred to in this class of problems. This is the well-known saddle point singular point. This is the case for $\gamma \leq 5/3$. However, when V_0 is the smaller zero of $Q(V)$, which is the case for $\gamma \geq 2$, the singular point has a nodal character.

The numerical results for u/u_S, ρ_S/ρ, and p_S/p versus $1/\xi = R(t)/r$ for spherical and cylindrical symmetries for different values of γ are shown in Figures 6.1–6.3. The velocity and density profiles have a monotonic behaviour. The pressure has a more complicated distribution. It decreases monotonically behind the shock if $\gamma \geq 3$ but has a maximum behind it if $\gamma \leq 2$. In the next section we show how the existence of the pressure maximum behind the shock may be used to give an analytic solution of the problem.

An interesting review of self-similar spherical compression waves in gas dynamics with applications to inertial confinement fusion (ICF) was given by Meyer-ter-Vehn and Schalk (1982). The analysis here follows closely the original work of Guderley (1942) on imploding shock waves. The relation between different isentropic and nonisentropic self-similar waves describing imploding and exploding flows is brought out by placing them all on Guderley's original chart of solutions. This includes the cumulative isentropic solutions of Kidder (1974) where all matter finally collapses into a point and the noncumulative isentropic solutions discussed by Ferro Fontan et al.

(1975, 1977) which contain the reflected shock after the imploding shock has reached the center, and finally the imploding shock solution and its extension to nonisentropic imploding shells. The latter is closest to the situation in ICF target implosions. It was shown that the solution in the center behind the reflected shock after shell collapse is of the same origin as the famous blast wave solution of Taylor (1950) and Sedov (1946).

An important comment on the intermediate asymptotic nature of self-similar solutions of the second kind in the context of the converging shock solution was made by Meyer-ter-Vehn and Schalk (1982). It may be observed that a large class of non-self-similar spherically imploding waves with rather general boundary conditions outside and a shock front propagating into the undisturbed gas approaches the self-similar solution asymptotically for radii r and times t close enough to the collapse point $r = 0$ and $t = 0$. It is also known that, in Guderley's solution, the shock velocity and strength as well as other parameters behind the shock, such as temperature, tend to infinity upon convergence. This is clearly unphysical and comes about because of the neglect of such effects as heat conduction and radiation. Therefore, the real shock implosions will deviate from the self-similar solution in the neighbourhood of the center of implosion. This is typically the intermediate region where the self-similar solution is approached by more realistic, non-self-similar solutions, leading to the term 'intermediate asymptotics'. Only numerical attempts have been made to study this aspect of self-similar solutions of the second kind and more work needs to be done to establish the intermediate asymptotic character of these solutions.

Meyer-ter-Vehn and Schalk (1982) carefully analysed the singular points (six in number) in the reduced particle velocity–sound speed plane and interpreted various solutions arising from the joining of these singularities by separatrices or otherwise. The medium ahead of the converging fronts is assumed to be variable, $\rho_0 \sim r^K$, so that the solution, in general, depends upon four parameters, namely, K, n, the dimensionality parameter equal to 3, 2, 1 for spherical, cylindrical and plane geometries, respectively, $\gamma = c_p/c_v$, and the similarity exponent α. A variety of solutions were discussed with regard to their dependence on these parameters.

6.2 Spherical Converging Shock Waves: Shock Exponent via the Pressure Maximum

In an interesting paper, Fujimoto and Mishkin (1978) made an effective use of the existence of a pressure maximum behind the shock to find analytically the exponent in the converging shock law; this is in contrast to the analysis of the last section. We consider the spherically symmetric converging shock waves, governed by the system

$$\rho_t + u\rho_r + +\rho u_r + \frac{2\rho u}{r} = 0, \tag{6.2.1}$$

$$u_t + u u_r + \frac{1}{\rho} p_r = 0, \tag{6.2.2}$$

$$\frac{\partial}{\partial t}(p\rho^{-\gamma}) + u \frac{\partial}{\partial r}(p\rho^{-\gamma}) = 0, \tag{6.2.3}$$

and write the self-similar solution in the form

$$p = \rho_0 \dot{R}^2 P(\xi), \quad \rho(r,t) = \rho_0 \Re(\xi), \quad u(r,t) = \dot{R} U_1(\xi), \tag{6.2.4}$$

where

$$\xi = \frac{r}{R}, \tag{6.2.5}$$

and $R = R(t)$ is the distance of the converging shock from the center of implosion. Introducing (6.2.4)–(6.2.5) and the auxiliary function $U(\xi)$ via

$$U_1(\xi) = U(\xi) + \xi, \tag{6.2.6}$$

into (6.2.1)–(6.2.3) we obtain the system of ODEs

$$-\Re^{-1}\Re' = U^{-1}U' + 3U^{-1} + 2\xi^{-1}, \tag{6.2.7}$$

$$-\Re^{-1}P' = UU' + (\lambda + 1)U + \lambda\xi, \tag{6.2.8}$$

$$P^{-1}P' - \gamma \Re^{-1}\Re' = -2\lambda U^{-1} \tag{6.2.9}$$

provided

$$\lambda = R\dot{R}^{-2}\ddot{R} \tag{6.2.10}$$

is a constant. This requires that

$$R(t) = \text{const} \cdot (1 - t/t_c)^\alpha, \quad \alpha = \frac{1}{1-\lambda}, \tag{6.2.11}$$

where α and t_c are constants.

The strong shock conditions

$$p(R,t) = \frac{2}{\gamma+1}\rho_0\dot{R}^2, \quad \rho(R,t) = \frac{\gamma+1}{\gamma-1}\rho_0,$$

$$u(R,t) = \frac{2}{\gamma+1}\dot{R}, \tag{6.2.12}$$

in the light of (6.2.4), (6.2.6), and (6.2.11), become

$$P(1) = \frac{2}{\gamma+1}, \quad \Re(1) = \frac{\gamma+1}{\gamma-1}, \quad U(1) = \frac{1-\gamma}{1+\gamma}. \tag{6.2.13}$$

It is assumed that the shock radius is given by $\xi = 1$, that is, $R(t) = \text{const.} (1 - t/t_c)^\alpha$. The exponent α must be found by solving (6.2.7)–(6.2.9) analytically or numerically subject to (6.2.13). It is not difficult to write the equations (6.2.7) and (6.2.9) in the 'integrated' form

$$\frac{P(\xi)}{P(1)} = \left(\frac{U(1)}{\xi^2 U(\xi)}\right)^\gamma \sigma^{2\lambda + 3\gamma}, \tag{6.2.14}$$

$$\frac{\Re(\xi)}{\Re(1)} = \frac{U(1)}{\xi^2 U(\xi)} \sigma^3, \tag{6.2.15}$$

implying the relation

$$\left(\frac{P(\xi)}{P(1)}\right)^3 = \left(\frac{\Re(\xi)}{\Re(1)}\right)^{2\lambda + 3\gamma} \left(\frac{\xi^2 U(\xi)}{U(1)}\right)^{2\lambda}, \tag{6.2.16}$$

where

$$\sigma(\xi) = \exp\left(-\int_1^\xi U^{-1}(\xi')d\xi'\right), \quad \sigma(1) = 1, \tag{6.2.17}$$

or

$$\sigma'(\xi) = -\sigma(\xi)U^{-1}(\xi). \tag{6.2.18}$$

If we use the RH conditions (6.2.13), equations (6.2.14) and (6.2.15) become

$$P(\xi) = \frac{2}{\gamma + 1}\left(\frac{1 - \gamma}{(1 + \gamma)\xi^2 U(\xi)}\right)^\gamma \sigma^{2\lambda + 3\gamma}, \tag{6.2.19}$$

$$\Re(\xi) = -\frac{\sigma^3}{\xi^2 U(\xi)}, \tag{6.2.20}$$

implying the relation

$$P(\xi) = \frac{2}{\gamma + 1}\left[\frac{\gamma - 1}{\gamma + 1}\Re(\xi)\right]^\gamma \sigma^{2\lambda}. \tag{6.2.21}$$

Since $\Re(\xi)$ and $P(\xi)$ can be expressed in terms of $U(\xi)$ and σ, (6.2.8) involves U, U' and σ only. Substituting (6.2.14) and (6.2.15) into (6.2.8) and simplifying we get

$$\sigma^{2\lambda + 3\gamma - 3} = \frac{UU' + (\lambda + 1)U + \lambda\xi}{\gamma\xi U' + 2\gamma U + (2\lambda + 3\gamma)\xi}$$
$$\times U^\gamma \xi^{2\gamma - 1} \Re(1) P^{-1}(1) U(1)^{1 - \gamma}. \tag{6.2.22}$$

Differentiating (6.2.22) we have

$$-(2\lambda + 3\gamma - 3)U^{-1} = \frac{UU'' + U'^2 + (\lambda + 1)U' + \lambda}{UU' + (\lambda + 1)U + \lambda\xi}$$
$$-\frac{\gamma\xi U'^2 + 3\gamma U' + 2\lambda + 3\lambda}{\gamma\xi U' + 2\gamma U + (2\lambda + 3\lambda)\xi}$$
$$+\gamma U^{-1}U' + (2\gamma - 1)\xi^{-1}. \tag{6.2.23}$$

If we now introduce the variables x and y via

$$x = U'(\xi), \quad y = \xi^{-1}U(\xi), \tag{6.2.24}$$

then

$$\frac{dx}{d\xi} = U'', \quad \frac{dy}{d\xi} = \xi^{-1}(x - y),$$

and so

$$\frac{dy}{dx} = \xi^{-1}(x - y)(U''(\xi))^{-1}. \tag{6.2.25}$$

The second derivative $U''(\xi)$ can be eliminated from (6.2.25) with the help of (6.2.23). The former therefore becomes

$$\frac{dy}{dx} = y(y - x)\frac{F(y; \lambda)}{G(x, y; \lambda)}, \tag{6.2.26}$$

where

$$F(y; \lambda) = 2\gamma y^2 + (2\lambda + 2\gamma - \gamma\lambda)y - \gamma\lambda, \tag{6.2.27}$$

and

$$\begin{aligned}
G(x, y; \lambda) = \quad & y(\gamma x + 2\gamma y + 2\lambda + 3\gamma)[x^2 + (\lambda + 1)x + \lambda] \\
& + [xy + (\lambda + 1)y + \lambda]\{(\gamma x + 2\gamma y + 2\lambda + 3\gamma) \\
& \times [\gamma x + (2\gamma - 1)y + 2\lambda + 3\gamma - 3] \\
& - y(3\gamma x + 2\lambda + 3\gamma)\}.
\end{aligned} \tag{6.2.28}$$

The curve $G(x, y; \lambda) = 0$ intersects the straight line $y = x$ at the four points

$$x = y = -1, -1 - \frac{2\lambda}{3\gamma}, -1 - \frac{2(\lambda - 1)}{3\gamma - 1}, -\lambda. \tag{6.2.29}$$

As $\xi \to \infty$, the reduced velocity $U_1(\infty)$ must vanish; therefore,

$$x(\infty) = \lim_{\xi \to \infty} \frac{d}{d\xi}[U_1(\xi) - \xi] = -1, \tag{6.2.30}$$

$$y(\infty) = \lim_{\xi \to \infty} \frac{U_1(\xi) - \xi}{\xi} = -1, \tag{6.2.31}$$

see (6.2.24). Also, at the shock $\xi = 1$, we have

$$x(1) = -\frac{6(\gamma + 1)\lambda + \gamma^2 + 10\gamma + 1}{(\gamma + 1)^2}, \quad y(1) = \frac{1 - \gamma}{1 + \gamma}, \tag{6.2.32}$$

see (6.2.6), (6.2.22), (6.2.24) and (6.2.13). It was observed in section 6.1 that, for some values of γ, the reduced pressure behind the shock first increases and then decreases. In fact, it follows from (6.2.8) and the integral (6.2.14) that the slope $P'(\xi)$ at the shock $\xi = 1$ is positive if $\gamma < 2 + \sqrt{3}$. It therefore follows that the reduced pressure must have a maximum at some finite positive value $\xi = \xi_m$ where $P'(\xi_m) = 0$. We can also deduce from (6.2.14), (6.2.15) and (6.2.8) by using the expressions for x and y in (6.2.24) that, at the maximum $P'(\xi) = 0$, we have

$$xy + (\lambda + 1)y + \lambda = 0, \qquad (6.2.33)$$

$$\gamma x + 2\gamma y + 2\lambda + 3\gamma = 0 \qquad (6.2.34)$$

or

$$y^2 + \frac{2\lambda + 2\gamma - \lambda\gamma}{2\gamma}y - \frac{\lambda}{2} = 0. \qquad (6.2.35)$$

Since there is only one pressure maximum behind the shock, (6.2.35) implies that

$$\lambda = \lambda_m = -\frac{2\gamma}{(\sqrt{\gamma} + \sqrt{2})^2}. \qquad (6.2.36)$$

From (6.2.33)–(6.2.34) we have the corresponding co-ordinates of the maximum as

$$y_m = -(-\tfrac{1}{2}\lambda_m)^{1/2}, \quad x_m = \lambda_m(2/\gamma)^{1/2} - 1. \qquad (6.2.37)$$

Thus the exponent λ in (6.2.36) depends only on γ; it does not depend on the strength of the shock and its manner of excitation. It may now be verified that both numerator and denominator on the RHS of (6.2.26) vanish when (6.2.33) and (6.2.34) are satisfied. Therefore, the integral curve of (6.2.26) passes through the point (x_m, y_m) and the slope dy/dx remains finite there. We also infer from (6.2.24) and (6.2.37) that U_m at this point is given by

$$U_m = -\xi_m(-\tfrac{1}{2}\lambda_m)^{1/2}. \qquad (6.2.38)$$

As γ increases, the similarity variable $\xi = \xi_m$ (where the maximum pressure occurs) decreases until at $\gamma = 2 + \sqrt{3}$, $\lambda_m = -2/3$, we have $y_m = y(1)$ (see (6.2.32), (6.2.36) and (6.2.37)). Thus, the maximum pressure for this value of γ occurs at the shock front where

$$y_m = y(1) = -\left(-\tfrac{1}{2}\lambda_m\right)^{1/2} = -1/\sqrt{3}. \qquad (6.2.39)$$

This statement is not easy to derive from numerical results.

Brushlinskii and Kazhdan (1963) showed that there is a whole interval of possible α values corresponding to each γ and conjectured that the unique solution corresponds to the smallest value of α. Butler (1954) had encountered the same situation in his investigations.

From the analysis of Fujimoto and Mishkin (1978) it follows that dy/dx as given by (6.2.26) is nonsingular when (6.2.33) and (6.2.34) are both satisfied. The argument of existence of a point of maximum pressure and the

Table 6.1. Analytic and numerical values of self-similar exponent α for different values of γ (see (6.2.11)) (Fujimoto and Mishkin, 1978).

γ	Analytic values of α (Fujimoto and Mishkin (1978))	Numerical values of α (Lazarus and Richtmyer (1977))
1.0	0.749	
1.1	0.734	0.769
1.4	0.707	0.717
5/3	0.687	0.688
2.0	0.667	0.667
3.0	0.623	0.636
$2 + \sqrt{3}$	0.600	0.625
6	0.562	0.610
∞	0.500	0.588

analyticity arguments proffered by Butler (1954) lead to the same unique solution (see also reference to I.M. Gelfand in Brushlinskii and Kazhdan (1963)).

The above arguments hold only for $\gamma \leq 2 + \sqrt{3}$. For higher values of γ, other arguments must be used. Table 6.1 gives the values of α as obtained by the theory presented here and the direct numerical solution of the eigenvalue problem (see section 6.1). The agreement is good except when γ is close to 1. For larger values of γ, the numerical solution and the analysis presented here do not agree.

Lazarus (1980) criticised the above work as 'erroneous'. However, in a rejoinder, Mishkin (1980) rebutted the charge and showed that there was no logical error in the analysis of Fujimoto and Mishkin (1978).

6.3 Converging Shock Waves Caused by Spherical or Cylindrical Piston Motions

It is natural to enquire how the converging shock may be generated. It may then be related to its asymptotic behaviour near the center or axis of symmetry (see sections 6.1 and 6.2). This aspect was analysed by Van Dyke and Guttman (1982) who treated the entire problem analytically, requiring however, in the end, an efficient computation of the series solution they obtained. Imagine a spherical (or cylindrical) container of initial radius R_0, filled with a perfect gas at rest with density ρ_0 and adiabatic constant γ. Let the container move at $t = 0$ with a very large constant velocity V, causing a strong shock of radius $R = R(t)$. This shock rushes ahead to the center (axis).

The equations of motion governing the flow behind the shock are

$$\rho_t + (\rho v)_r + j\frac{\rho v}{r} = 0, \qquad (6.3.1)$$

$$v_t + vv_r + \frac{1}{\rho}p_r = 0, \qquad (6.3.2)$$

$$\left(\frac{\partial}{\partial t} + v\frac{\partial}{\partial r}\right)(p\rho^{-\gamma}) = 0, \qquad (6.3.3)$$

with the usual notation for density, pressure and particle velocity. $j = 1, 2$ for cylindrical and spherical symmetry, respectively. The Rankine-Hugoniot conditions at the strong shock $r = R(t)$ are

$$v = \frac{2}{\gamma+1}\dot{R}, \qquad (6.3.4)$$

$$\rho = \frac{\gamma+1}{\gamma-1}\rho_0, \qquad (6.3.5)$$

$$p = \frac{2}{\gamma+1}\rho_0\dot{R}^2. \qquad (6.3.6)$$

The boundary condition at the piston is

$$v = -V \quad \text{at} \quad r = R_0 - Vt, \qquad (6.3.7)$$

describing the motion of the piston inward with initial radius R_0 at $t = 0$. Van Dyke and Guttman (1982) used this transformation, measuring the distance r inward from the original radius R_0. The inward particle velocity u equals $-v$. We may also introduce the variable, $x = R_0 - r$. The basic assumption in this analysis is that the initial motion of the piston may be considered planar so that the shock thus produced moves with the speed $\frac{1}{2}(\gamma+1)V$ (see (6.3.4)), where V is the constant speed of the gas between the piston and the shock. The other quantities in this region are $\rho = \rho_0(\gamma+1)/(\gamma-1)$ and $p = \frac{1}{2}(\gamma+1)\rho_0V^2$.

It is convenient to introduce the variable

$$\xi = \frac{2}{\gamma-1}\left(\frac{x}{Vt} - 1\right), \qquad (6.3.8)$$

which varies, in the planar case, from zero at the piston to unity at the basic position of the shock. We may also render the variables dimensionless by referring the length to R_0, speed to V, density to ρ_0, pressure to ρ_0V^2, and time to R_0/V. The system (6.3.1)–(6.3.3) now becomes

$$\left[1 - \left(1 + \frac{1}{2}(\gamma-1)\xi\right)t\right]$$
$$\times\left[\rho u_\xi + \left(u - 1 - \frac{1}{2}(\gamma-1)\xi\right)\rho_\xi + \frac{1}{2}(\gamma-1)t\rho_t\right]$$
$$= \frac{1}{2}(\gamma-1)jt\rho u, \qquad (6.3.9)$$

$$\rho\left(u - 1 - \frac{1}{2}(\gamma - 1)\xi\right)u_\xi + \frac{1}{2}(\gamma - 1)t\rho u_t + p_\xi = 0, \tag{6.3.10}$$

$$\left(u - 1 - \frac{1}{2}(\gamma - 1)\xi\right)(\rho p_\xi - \gamma p \rho_\xi)$$
$$+ \frac{1}{2}(\gamma - 1)t(\rho p_t - \gamma p \rho_t) = 0. \tag{6.3.11}$$

The boundary conditions (6.3.4)–(6.3.6) at the shock and (6.3.7) at the piston now become

$$u = \frac{2}{\gamma + 1}\dot{X}, \quad \rho = \frac{\gamma + 1}{\gamma - 1}, \quad p = \frac{2}{\gamma + 1}\dot{X}^2 \tag{6.3.12}$$

at $\xi = \frac{2}{\gamma - 1}\left[\frac{X(t)}{t} - 1\right]$ and

$$u = 1 \quad \text{at} \quad \xi = 0, \tag{6.3.13}$$

respectively. A basic assumption of the analysis is that the solution is analytic in time so that one may assume the shock position $X(t)$ in the form

$$X(t) = \sum_{n=1}^{\infty} X_n t^n. \tag{6.3.14}$$

The other flow variables behind the shock have the form

$$u = \sum_{n=1}^{\infty} U_n(\xi)t^{n-1}, \quad \rho = \sum_{n=1}^{\infty} R_n(\xi)t^{n-1}, \quad p = \sum_{n=1}^{\infty} P_n(\xi)t^{n-1}. \tag{6.3.15}$$

The lowest approximation in (6.3.15) is assumed to be that given by the plane piston motion as noted above:

$$U_1 = 1, \quad R_1 = \frac{\gamma + 1}{\gamma - 1}, \quad P_1 = \frac{1}{2}(\gamma + 1), \quad X_1 = \frac{1}{2}(\gamma + 1). \tag{6.3.16}$$

Substituting (6.3.15) into (6.3.9)–(6.3.11) and equating like powers of t on both sides, we get the following system of ODEs for the first order terms:

$$\frac{\gamma + 1}{\gamma - 1}U_2' - \frac{1}{2}(\gamma - 1)\xi R_2' + \frac{1}{2}(\gamma - 1)R_2 = \frac{1}{2}(\gamma + 1)j, \tag{6.3.17}$$

$$-\xi U_2' + U_2 + \frac{2}{\gamma + 1}P_2' = 0, \tag{6.3.18}$$

$$\xi\left(P_2' - \frac{1}{2}\gamma(\gamma - 1)R_2'\right) - \left(P_2 - \frac{1}{2}\gamma(\gamma - 1)R_2\right) = 0. \tag{6.3.19}$$

The substitution of first of (6.3.15) into the boundary condition (6.3.13) at the piston gives

$$U_n(0) = 0 \quad \text{for all} \quad n > 1. \tag{6.3.20}$$

The boundary conditions (6.3.12) at the shock $\xi = 1$ are satisfied if we put (6.3.14) and (6.3.15) therein and equate like powers of t on both sides. The second order BCs are

$$U_2(1) = \frac{4}{\gamma+1}X_2, \quad R_2(1) = 0, \quad P_2(1) = 4X_2. \tag{6.3.21}$$

The solution of (6.3.17)–(6.3.19), subject to (6.3.21) and $U_2(0) = 0$, is found to be

$$U_2 = \frac{\gamma(\gamma-1)}{2(2\gamma-1)}j\xi, \quad R_2 = \frac{\gamma+1}{2\gamma-1}j(1-\xi),$$

$$P_2 = \frac{\gamma(\gamma+1)(\gamma-1)}{2(2\gamma-1)}j, \quad X_2 = \frac{\gamma(\gamma+1)(\gamma-1)}{8(2\gamma-1)}j. \tag{6.3.22}$$

Van Dyke and Guttman (1982), extrapolating from first and second order forms of the solution, wrote the nth order solution in the form of polynomials in ξ of degree $n-1$:

$$U_n(\xi) = \sum_{k=2}^{n} U_{nk}\xi^{k-1}, \quad R_n(\xi) = \sum_{k=1}^{n} R_{nk}\xi^{k-1}, \quad P_n(\xi) = \sum_{k=1}^{n} P_{nk}\xi^{k-1}.$$

$$\tag{6.3.23}$$

We substitute (6.3.23) into the system of ODEs for $U_n(\xi)$, $R_n(\xi)$, and $P_n(\xi)$ obtained from (6.3.9)–(6.3.11) via (6.3.15) and equate like powers of ξ. We thus obtain for each approximation a system of $3n$ linear algebraic equations for the coefficients U_{nk}, R_{nk}, P_{nk} and X_n, whose nonhomogeneous terms depend on all previous approximations. The conditions at the shock are

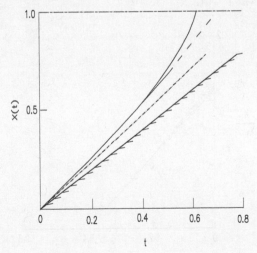

Figure 6.4 The shock locus for a spherical converging shock for $\gamma = 7/5$. Dotted lines represent a one term approximation to the shock wave; dashed line is a three term approximation; thick line is the full solution. The bottom line represents the path of the piston (Van Dyke and Guttman, 1982).

obtained by transferring them to the basic position $\xi = 1$ by Taylor series expansion. The trajectory of the shock, to third order in t, is found to be

$$X(t) = \frac{1}{2}(\gamma+1)t + \frac{\gamma(\gamma+1)(\gamma-1)}{8(2\gamma-1)}jt^2$$

$$+\frac{(\gamma+1)(\gamma-1)}{48(7\gamma-5)}\Big[(\gamma+1)(3\gamma+1)j$$

$$+\frac{\gamma(13\gamma^3-21\gamma^2+13\gamma-1)}{(2\gamma-1)^2}j^2\Big]t^3$$

$$+\cdots\cdots. \tag{6.3.24}$$

This shock locus is shown in Figure 6.4 for a spherical converging shock for $\gamma = 7/5$. The numerical results thus obtained for the cylindrical shock for $\gamma = 7/5$ agree closely with the direct numerical integration of the governing system of ODEs to this order by Lee (1968). Van Dyke and Guttman (1982) carried out the expansion (6.3.24) for $\gamma = 3$ for a spherical piston to fifth order and further confirmed the accuracy of this procedure. To obtain more accurate results, Van Dyke and Guttman (1982) wrote a computer program for the general term and computed the results to 40th approximation; they claim an accuracy to 14 figures. They also used Domb and Sykes (1957) method to estimate the radius of convergence of the series (6.3.24). Thus, writing

$$X(t) = \Sigma X_n t^n \sim A_1 \left(1 - t/t_c\right)^{\alpha_1} \quad \text{as} \quad t \to t_c, \tag{6.3.25}$$

one has

$$\frac{X_n}{X_{n-1}} \sim \frac{1}{t_c}\left(1 - \frac{1+\alpha_1}{n}\right) \quad \text{as} \quad n \to \infty. \tag{6.3.26}$$

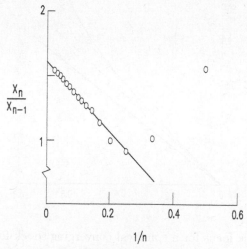

Figure 6.5 X_n/X_{n-1} versus $1/n$ with the value of $\alpha_1 = 0.717$ of Guderley's exponent for spherical symmetry with $\gamma = 7/5$, see (6.3.26) (Van Dyke and Guttman, 1982).

Table 6.2. The history of evaluation of Guderley's similarity exponent α_1 (Van Dyke and Guttman, 1982).

	Spherical, $\gamma = 7/5$	Spherical, $\gamma = 5/3$	Spherical, $\gamma = 3$	Cylindrical, $\gamma = 7/5$
Guderley (1942)	0.717	–	–	0.834
Butler (1954)	0.717173	0.688377	–	0.835217
Stanyukovich (1960)	0.717	–	0.638	0.834
Brushlinskii & Kazhdan (1963)	0.7170	0.68838	0.6364	–
Welsh (1967)	0.717174	0.688377	0.636411	0.835323
Goldman (1973)	–	0.688377	–	–
Lazarus & Richtmyer (1977)	0.71717450	0.68837682	0.63641060	0.83532320
Fujimoto & Mishkin (1978)	0.707	0.687	0.623	–
Mishkin & Fujimoto (1978)	–	–	–	0.828

Figure 6.5 shows that, for spherical symmetry and $\gamma = 7/5$, a linear fit in $1/n$ with the value $\alpha_1 = 0.717$ of Guderley's exponent gives $1/t_c \sim 1.61$ or $t_c \sim 0.62$ to graphical accuracy. All these values were later refined to higher accuracy for various values of γ for both spherical and cylindrical symmetries. The history of evaluation of Guderley's similarity exponent α_1 is given in Table 6.2. The value 0.707 of Fujimoto and Mishkin for $\gamma = 7/5$ for spherical geometry, reported in section 6.2, differs significantly from the accurate value 0.717174 obtained by several other authors.

Following the conjecture of Guderley (1942), Van Dyke and Guttman (1982) sought the following expansion for the radius of the shock wave

$$R(t) = 1 - X(t) \sim \sum_{i=1} A_i \left(1 - t/t_c\right)^{\alpha_i}, \qquad (6.3.27)$$

using the method of Padé approximants. The values of A_i and α_i, $i = 1, 2, 3$, for $\gamma = 7/5, 5/3, 3$ for spherical symmetry and for $\gamma = 7/5$ for cylindrical symmetry are listed in Table 6.3. The radius of the converging shock is given to an accuracy of $1/2$ percent by the first three terms (see (6.3.14)). Other piston motions could lead to the same asymptotic similarity solution of the converging shock. An analysis proving this statement remains to be carried out.

We conclude this section by summarising an interesting review paper by Lazarus (1981). He laid stress on the role of nonanalytic solutions which involve simultaneous arrival at the origin of two (or more) discontinuities, for example, a shock and a discontinuous pressure gradient, or of one discontinuity and a point of nonanalytic continuity. These solutions may seem artificial. Lazarus (1981) asserted that this view is not right since, for self-similar solutions, all the information contained in the solution arrives simultaneously at the origin. In some of the new solutions found by Lazarus, an arbitrary number of secondary shocks are possible; the physical nature

Table 6.3. Exponents $\alpha_i (i = 1, 2, 3)$ and amplitudes $A_i (i = 1, 2, 3)$ in Guderley's local expansion (6.3.27) for spherical and cylindrical converging shocks corresponding to $\gamma = 7/5, 5/3, 3$ (Van Dyke and Guttman, 1982).

Geometry	γ	α_1	α_2	α_3	A_1	A_2	A_3
Spherical	7/5	0.7171745	2.045	3.4	0.981706	0.0140	0.007
Spherical	5/3	0.6883768	1.885	3.1	0.989732	0.0055	0.006
Spherical	3	0.636411	1.638	2.5	1.016952	−0.0244	0.01
Cylindrical	7/5	0.835324	2.033	3	0.983865	0.0133	0.01

of the previously rejected partial solutions is discussed. Many (unresolved) questions are also posed. In particular it is suggested that the asymptotic approach (or nonapproach) to self-similar solutions obtained by direct numerical integration of partial differential equations needs more careful analysis since the evidence available for approach to a unique self-similar solution is not convincing.

Chapter 7

Spherical Blast Waves Produced by Sudden Expansion of a High Pressure Gas

7.1 Introduction

We have so far modelled the blast wave in several idealised ways. Each of these models represents reality in certain space-time regimes and describes some physical aspect(s) of the phenomenon. For example, the Taylor-Sedov self-similar solution is an extremely good descriptor of the initial stages of a very strong blast wave but begins to depart from reality as the shock moderates to a finite strength; Sakurai's (1953) extension gives a solution which is valid for some further time and distance. Similarly, the piston motions describe blast waves for which the energy released is not constant. Taylor-Sedov solution comes out as a special case with constant energy of the blast wave. Bethe's theory (1942), as also that of Whitham (1950), deals with weak explosions. Brinkley and Kirkwood (1947) and Sachdev (1971, 1972) give a local theory of the blast wave. It describes how the shock decays all the way to a sound wave; it does not give details of the flow behind the shock.

A different, more realistic, but also mathematically more complicated, model was proposed by McFadden (1952) and Friedman (1961). Each of

195

these investigators assumed that at $t = 0$, a unit sphere containing a perfect gas at a high pressure is allowed to expand suddenly into a homogeneous atmosphere, referred to as air. This is an analogue of a plane shock tube problem where, additionally, the spherical term must also be included. The subsequent behaviour for $t > 0$, the 'equalisation', may be described in the space-time diagram as follows. The region (0) is the undisturbed air; the air which has been overtaken by the main blast wave is contained in region (1). These two regions are separated by the main shock. There is an interface, a contact discontinuity, which separates the (hot) air in region (1) from the gas in region (2). The latter is a nearly uniform region outside the main expansion which itself spans region (3); this region is bounded by its head, adjoining the uniform gas region (4) and a tail which separates it from the nearly uniform region (2) outside of the main expansion region. Both McFadden (1952) and Friedman (1961) essentially deal with the same model, but their analysis is quite distinct. The gas-sphere in the former is at a relatively lower pressure so that the phenomenon of secondary shock is not observed. In the case of expansion of a higher pressure gas sphere considered by Friedman (1961), the secondary shock is clearly seen. This shock is absent in the one-dimensional shock-tube problem since the main shock and the expansion come into an instantaneous equilibrium, being separated by a region of uniform pressure and velocity. Physically, the high pressure gas passing through a spherical rarefaction wave must expand to lower pressures than those reached through an equivalent one-dimensional expansion, clearly due to the increase in volume. Therefore, the pressures at the tail of the rarefaction wave are lower than those transmitted by the main shock and a compression or a secondary shock must be inserted to connect these two phases. McFadden (1952), however, considered the case when the pressure difference referred to above is not severe so that a weak discontinuity or a characteristic replaces the secondary shock.

The main aim of McFadden's (1952) analysis was to get an initial ($t \sim 0$) analytic behaviour of the blast wave where the initial discontinuities are smeared and the flow may be computed by the numerical methods popular in those years. The basic idea was to write a series solution in time with coefficients functions of an appropriate 'similarity' variable, suggested by the solution of one-dimensional shock tube problem. The series solutions in each of the domains were appropriately matched to those in other regions; the loci of the dividing surfaces were also determined to first order in time.

The analysis of Friedman (1961) is entirely different; he was probably not aware of the work of McFadden (1952). Here the rarefaction wave is found by a perturbation of the corresponding plane rarefaction wave. A considerable use is made of the approximate shock wave theory of Chisnell (1957), Chester(1954, 1960) and Whitham (1958) to determine both the secondary shock and the main shock. The solution, though explicit, is quite intuitive and approximate.

It is the purpose of this chapter to bring out the analytic features of these valiant approches, to vindicate the power of analysis for this very complicated problem. Section 7.2 deals with the McFadden (1952) approach while section 7.3 corrects and corroborates the work of Friedman (1961).

7.2   Expansion of a High Pressure Gas into Air:

A Series Solution

One of the earliest attempts to simulate a blast wave, which is not too strong, is due to McFadden (1952). Here, the point explosion hypothesis and the assumption that the energy of the blast wave remains constant are both dispensed with. The blast is also not assumed to be weak. Instead it is envisioned that a unit sphere (in nondimensional variables), containing a perfect gas at a (uniform) high pressure, is allowed to expand suddenly at $t = 0$ into a homogeneous atmosphere. The medium in the sphere is called 'gas' while that outside is referred to as 'air'. After the 'diaphragm' is destroyed, the gas rushes outward compressing the air around it. The flow for $t > 0$ may be described succinctly in the (x, t) plane (see Figure 7.1). Region A is the undisturbed gas. Region B is a rarefaction wave which is bounded by its 'head', a straight characteristic on the left, and its 'tail' on the right. The latter, another characteristic, adjoins region C, which is a rarefied gas moving outward. This rarefied gas and the air overtaken by the main shock are separated by an interface, a contact discontinuity across which pressure and particle velocity are continuous but other variables get a jump. The region D contains compressed air, overtaken by the main shock, and is bounded by the contact discontinuity and the main shock. The gas

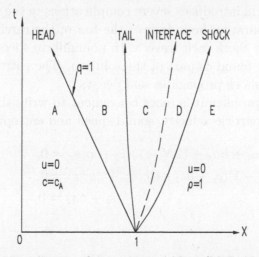

Figure 7.1 Space–time diagram for a spherical blast (McFadden, 1952).

sphere is not envisaged to be at a very high pressure; thus, it is assumed that no secondary shock is formed. This phenomenon was subsequently discussed by Friedman (1961) and is detailed in the next section. An earlier study due to Wecken (1950) had indicated the formation and strengthening of the secondary shock.

McFadden (1952) was interested in an analysis which would provide an initial solution to continue the computation of the problem by numerical methods then available, without the need to tackle discontinuities in the initial data. He wrote out a series form of the solution for particle velocity, pressure and entropy in powers of (nondimensional) time with coefficients dependent on a slope co-ordinate which is a combination of space and time co-ordinates and which appears prominently in the solution of the planar form of this problem (the shock tube problem). Indeed, the zeroth order solution in the series is simply the solution of the shock tube problem. McFadden (1952) found first order correction for various regions shown in Figure 7.1 and explained how geometry affected the solutions of the shock tube problem.

Here we indicate how to generalise the work of McFadden (1952) to write a series solution (different in each region) with an arbitrary number of terms, which satisfies appropriate boundary conditions on each of the boundaries shown in Figure 7.1. The solution is given explicitly up to two terms. The series is summed up as it is and by the use of Padé approximation, and is shown to converge for different times $t = t_c$ for different values of γ. The latter is assumed to be constant in the entire flow.

We remark that the mathematical approach here is essentially the same as was used much later by Van Dyke and Guttman (1982) for the converging shock wave arising from a spherical or cylindrical piston motion. The series solution was shown to converge to Guderley's self-similar solution holding near the focusing (see section 6.3 for a detailed discussion).

The spherical term introduces severe complications in the flow and hence its analysis: the separating boundaries—the tail of the rarefaction, the interface and the main shock each travel with nonuniform speeds and so their trajectories must be found as part of the solution. The entropy behind the shock varies as the shock propagates and decays.

For the present problem it is more convenient to write the equations of motion in terms of particle velocity, sound speed and entropy:

$$u_t + uu_x + (2N - 1)cc_x - c^2 s_x = 0, \tag{7.2.1}$$
$$(2N - 1)(c_t + uc_x) + c(u_x + 2u/x) = 0, \tag{7.2.2}$$
$$s_t + us_x = 0, \tag{7.2.3}$$

where

$$s = S/c_v\gamma(\gamma - 1), \quad p = c^{2N+1}\exp(-\gamma s),$$

$$\rho = \gamma c^{2N-1} \exp(-\gamma s), \tag{7.2.4}$$

$$N = \frac{1}{2}\frac{(\gamma+1)}{\gamma-1};$$

where S is the entropy per unit mole. The equation of state is chosen in the form

$$p\rho^{-\gamma} = \gamma^{-\gamma} \exp(S/c_v), \tag{7.2.5}$$

so that $S = 0$ when $p = 1$ and $\rho = \gamma$. The other dependent variables p, ρ, and u have the usual meaning. The distance x is measured from the geometric center and t is the time. The system of equations (7.2.1)–(7.2.3) must be solved in each of the regions A to E, shown in Figure 7.1.

If the conditions in the high pressured gas sphere, $0 < x < 1$, are taken to be $u = 0$, $c = c_A$, $s = s_A$, where c_A and s_A are constant, and those in the air ahead, $x > 1$, as $u = 0$, $c = 1$ and $s = 0$, these respectively constitute the solutions in the regions A and E ($s = 0$ is actually normalised by subtracting from it some value $s = s_0$, say).

We now turn to the conditions to be imposed on the boundaries between these constant regions. The head of the rarefaction wave,

$$x_H = 1 - c_A t, \tag{7.2.6}$$

is a characteristic and permits a jump in the derivatives of u and c across it; the functions u, c, and s must however be continuous. Thus, when $x = x_H$, we have on the head of the rarefaction wave the conditions

$$u_B = 0, \quad c_B = c_A, \quad s_B = s_A, \tag{7.2.7}$$

until it reaches the center. This imposes the restriction $0 < t < 1/c_A$ (see (7.2.6)).

The tail of the rarefaction wave is assumed to be a characteristic moving inward relative to the particles; it is not a straight line. Its actual slope may be negative or positive. McFadden (1952) made this assumption in consonance with the shock tube solution. In actual practice, there is possibility of a weak shock developing and moving inward to collapse at the center. This aspect of the problem remains to be incorporated in the present analysis. Here we continue to use McFadden's assumptions.

The interface between the rarefied gas and the compressed air is a contact surface. It is a particle path. Therefore, the particle velocity must assume the same value as the interface is approached from both sides. The interface moves with a finite speed (except at $t = 0$); the pressure and the particle velocity across it are continuous. Thus, denoting the interface by $x = x_I(t)$, we have, on this moving front,

$$u_C = c_D, t \geq 0, \quad p_C = p_D, t > 0. \tag{7.2.8}$$

The other variables, namely, density, sound speed, and entropy suffer a jump across $x = x_I(t)$.

With nondimensionalised values of the variables in the air ahead of the main shock, namely, $p = 1$, $c = 1$, $u = 0$, the Rankine-Hugoniot conditions give the following relations across the shock, $x_s = x_s(t)$, $t > 0$ in terms of the shock velocity $U(t)$:

$$\begin{aligned}
u_D &= (2N-1)(U^2-1)/2NU, \\
c_D^2 &= [(2N+1)U^2-1][U^2+(2N-1)]/4N^2U^2, \qquad (7.2.9) \\
p_D &= [(2N+1)U^2-1]/2N.
\end{aligned}$$

We introduce the independent variables

$$q = (1/2N)[(2N-1)+(1-x)/c_A t] \quad \text{and} \quad y = c_A t \qquad (7.2.10)$$

into the basic system of equations (7.2.1)–(7.2.3) since q appears prominently in the solution of the shock tube problem. The transformed system becomes

$$2Nyc_A u_y + c_A\{(2N-1)-2Nq\}u_q - uu_q - (2N-1)cc_q + c^2 s_q = 0, \quad (7.2.11)$$

$$\begin{aligned}
&2N(2N-1)c_A\{1+(2N-1-2Nq)y\}yc_y \\
&+c_A\{(2N-1)-2Nq\}\{(2N-1)+(2N-1)y((2N-1)-2Nq)c_q \\
&-(2N-1)\{1+((2N-1)-2Nq)y\}uc_q \\
&-\{1+((2N-1)-2Nq)y\}cu_q + 4Nyc_q = 0, \qquad (7.2.12)
\end{aligned}$$

$$2Nyc_A s_y + c_A\{(2N-1)-2Nq\}s_q - us_q = 0, \qquad (7.2.13)$$

We seek a series solution of the system (7.2.11)–(7.2.13) in each of the regions B, C, and D for $0 < t < 1/c_A$, that is, for $0 < y < 1$. Thus, we write

$$\begin{aligned}
u(q,y) &= \sum_{n=0}^{\infty} u_n(q)y^n, \quad c(q,y) = \sum_{n=0}^{\infty} c_n(q)y^n, \\
s(q,y) &= \sum_{n=0}^{\infty} s_n(q)y^n, \qquad (7.2.14)
\end{aligned}$$

where the coefficients $u_n(q)$, $c_n(q)$, and $s_n(q)$ are assumed to be sufficiently smooth. Substituting (7.2.14) into (7.2.11)–(7.2.13), we get the zeroth order system as

$$[c_A\{(2N-1)-2Nq\}-u_0]u_0' - (2N-1)c_0 c_0' + c_0^2 s_0' = 0, \qquad (7.2.15)$$

$$-c_0 u_0' + (2N-1)\{[c_A[(2N-1)-2Nq]-u_0\}c_0' = 0, \qquad (7.2.16)$$

$$[c_A\{(2N-1)-2Nq\}-u_0]s_0' = 0. \qquad (7.2.17)$$

We may observe that the zeroth order system of ODEs is nonlinear while all higher order systems are linear and inhomogeneous. We write the first order system later (see (7.2.56)–(7.2.58)).

We must now find the boundary conditions, to different orders, across different interfaces to use them in conjunction with systems of different orders.

The q-co-ordinates, $q = q_T(y)$, $q = q_I(y)$, and $q = q_S(y)$ of the unknown boundaries—the tail of the rarefaction wave, the contact front and the shock—are written as series in time:

$$q_T = \sum_{n=0}^{\infty} q_{T_n} y^n, \quad q_I = \sum_{n=0}^{\infty} q_{I_n} y^n, \quad q_S = \sum_{n=0}^{\infty} q_{S_n} y^n, \qquad (7.2.18)$$

where q_{T_0}, q_{I_0}, and q_{S_0} are related to the slopes of the respective paths at $t = 0$ in view of (7.2.10).

At the head of the rarefaction wave (7.2.6), $q = 1$ (see (7.2.10)), the boundary conditions (7.2.7) by virtue of (7.2.14) become

$$u_0(1) = 0, \quad c_0(1) = c_A, \quad s_0(1) = s_A, \qquad (7.2.19)$$

and

$$u_n(1) = 0, \quad c_n(1) = 0, \quad s_n(1) = 0, \quad n \geq 1. \qquad (7.2.20)$$

These conditions suffice to determine uniquely the series solution in the region B.

Now we proceed to derive conditions across the tail of the rarefaction wave which is a negative characteristic with the slope

$$\frac{dx}{dt} = u - c. \qquad (7.2.21)$$

Substituting (7.2.14) and (7.2.18$_1$) into (7.2.21), using (7.2.10) for $q = q_T$, $x = x_T$, etc. and equating different powers of y on both sides, we have

$$u_0(q_{T_0}) - c_0(q_{T_0}) = c_A\{(2N - 1) - 2N q_{T_0}\}, \qquad (7.2.22)$$

$$u_1(q_{T_0}) - c_1(q_{T_0}) = -2N c_A q_{T_1}, \text{etc.} \qquad (7.2.23)$$

It suffices to consider the continuity of the positive Riemann invariant $u + (2N - 1)c$ across the tail to carry the information from the region B to the region C. Introducing in this expression the expansions (7.2.14) along the tail $q = q_T$ (see (7.2.18)) and equating coefficients of different powers of y on both sides, we have

$$[u_0(q_{T_0}) + (2N - 1)c_0(q_{T_0})]_B = [u_0(q_{T_0}) + (2N - 1)c_0(q_{T_0})]_C, \qquad (7.2.24)$$

$$[u_1(q_{T_0}) + (2N - 1)c_1(q_{T_0})]_B = [u_1(q_{T_0}) + (2N - 1)c_1(q_{T_0})]_C, \text{etc.} \qquad (7.2.25)$$

The locus of the contact discontinuity $dx/dt = u$ is found in the same manner as the tail of the expansion wave (see (7.2.21)–(7.2.23)). The result is

$$u_0(q_{I_0}) = c_A[(2N - 1) - 2Nq_{I_0}], \qquad (7.2.26)$$

$$u_1(q_{I_0}) = -4Nc_A q_{I_1}, \text{etc.} \qquad (7.2.27)$$

At the contact surface the conditions (7.2.8) apply. Therefore, using the expansions (7.2.14) in (7.2.8), evaluating them at $q = q_I$ (see (7.2.18)) and equating powers of y on both sides, we get

$$[u_0(q_{I_0})]_C = [u_0(q_{I_0})]_D, \qquad (7.2.28)$$

$$[u_1(q_{I_0})]_C = [u_1(q_{I_0})]_D, \text{etc.} \qquad (7.2.29)$$

$$[p_0(q_{I_0})]_C = [p_0(q_{I_0})]_D, \qquad (7.2.30)$$

$$[p_1(q_{I_0})]_C = [p_1(q_{I_0})]_D, \text{etc.} \qquad (7.2.31)$$

Finally, we consider the conditions at the shock front. If the shock law is assumed in the form

$$U = \sum_{n=0}^{\infty} U_n y^n \qquad (7.2.32)$$

or

$$\frac{dx_S}{dt} = \sum_{n=0}^{\infty} U_n y^n, \qquad (7.2.33)$$

then, putting $q = q_S$ (see (7.2.18)) and $x = x_S$ in the definition (7.2.10) of q, we obtain, as for the tail of the rarefaction wave, the following relation:

$$U_0 = c_A\{(2N - 1) - 2Nq_{S_0}\}, \qquad (7.2.34)$$

$$U_1 = -4Nc_A q_{S_1}, \quad \text{etc.} \qquad (7.2.35)$$

Substituting the expansions (7.2.14) and (7.2.18$_3$) into the Rankine-Hugoniot conditions (7.2.9) and equating coefficients of equal powers of y on both sides, we get

$$u_{D_0}(q_{S_0} + 0) = \frac{2N - 1}{2N}\left(\frac{U_0^2 - 1}{U_0}\right), \qquad (7.2.36)$$

$$u_{D_1}(q_{S_0} + 0) = \frac{2N - 1}{2N}\left(\frac{U_0^2 + 1}{U_0^2}\right)U_1, \qquad (7.2.37)$$

$$c_{D_0}^2(q_{S_0} + 0) = \frac{\{(2N + 1)U_0^2 - 1\}(U_0^2 + (2N - 1))}{4N^2 U_0^2}, \qquad (7.2.38)$$

$$c_{D_1}(q_{S_0} + 0) = \frac{\{(2N + 1)U_0^4 + (2N - 1)\}U_1}{4N^2 c_{D_0} U_0^3}, \qquad (7.2.39)$$

$$c_{D_0}^{2N+1}(q_{S_0}+0)e^{-\gamma s_0} = \frac{(2N+1)U_0^2 - 1}{2N}, \tag{7.2.40}$$

$$S_{D_1}(q_{S_0}) = \left\{ \frac{(2N-1)c_{D_1}}{c_{D_0}} - \frac{(2N-1)2U_0}{(2N+1)U_0^2 - 1} \right\} U_1, \text{etc.} \tag{7.2.41}$$

Now we turn to explicit forms of the zeroth and first order solutions which satisfy appropriate boundary conditions at the boundaries of regions B, C, and D. From (7.2.17) it follows that $s_0' = 0$. Thus,

$$s_0 = s_A, \quad 1 \geq q \geq q_{I_0} \tag{7.2.42}$$

in the entire gaseous region A. The constant value of entropy in region D may be denoted by

$$s_0 = C_0, \quad q_{I_0} > q > q_{S_0}, \tag{7.2.43}$$

where $C_0 \neq s_A$. To this order, the entropy is constant also in the regions B and C. Putting $s_0' = 0$ in equations (7.2.15)–(7.2.16) and combining them suitably we obtain

$$\{c_A[(2N-1) - 2Nq] - (u_0 \pm c_0)\}[u_0' \pm (2N-1)c_0'] = 0. \tag{7.2.44}$$

These are easily checked to be equations for plane isentropic flow in the variable q. Of the four possibilities, three give a solution linear in q while the fourth leads to a constant solution. The linear solution compatible with the boundary condition (7.2.19) at the head of the rarefaction wave is

$$u_0 = c_A(2N-1)(1-q), \quad c_0 = c_A q, \quad 1 \geq q \geq q_{T_0}. \tag{7.2.45}$$

This is a centered simple wave. The only solution compatible with the zeroth order boundary conditions in regions C and D is a constant solution (see (7.2.28), (7.2.36),(7.2.38) and (7.2.40)). Since this solution must match (7.2.45) at $q = q_{T_0}$, we have

$$u_0 = c_A(2N-1)(1-q_{T_0}), \quad c_0 = c_A q_{T_0}, \quad q_{T_0} \geq q \geq q_{I_0}. \tag{7.2.46}$$

In region D, we denote the constant solution by A_0 and B_0:

$$u_0 = A_0, \quad c_0 = B_0, \quad q_{I_0} > q > q_{S_0}. \tag{7.2.47}$$

We match the solution (7.2.43) and (7.2.47) in region D to (7.2.42) and (7.2.46) in region C via the boundary conditions expressing the continuity of particle velocity and pressure across the contact front. We obtain

$$A_0 = c_A(2N-1)(1-q_{T_0}), \tag{7.2.48}$$

$$B_0^{2N+1} \exp(-\gamma C_0) = (c_A q_{T_0})^{2N+1} \exp(-\gamma S_A). \tag{7.2.49}$$

Using the zeroth order solution (7.2.43) and (7.2.47) in the zeroth order R-H conditions (7.2.36), (7.2.38), and (7.2.40), we get

$$A_0 = (2N - 1)(U_0^2 - 1)/2NU_0, \qquad (7.2.50)$$

$$B_0^2 = [(2N + 1)U_0^2 - 1][U_0^2 + (2N - 1)]/4N^2U_0^2, \qquad (7.2.51)$$

$$B_0^{2N+1} \exp(-\gamma C_0) = [(2N + 1)U_0^2 - 1]/2N. \qquad (7.2.52)$$

The system of equations (7.2.48)–(7.2.52) must be solved for the five constants A_0, B_0, C_0, q_{T_0} and U_0. This problem can be reduced to the solution of the equation

$$[c_A(2N - 1)(1 - q_{T_0})]^2 = \frac{(2N - 1)^2}{2N + 1} \frac{(p_A q_{T_0}^{2N+1} - 1)^2}{2Np_A q_{T_0}^{2N+1} + 1} \qquad (7.2.53)$$

for q_{T_0}, where $p_A = c_A^{2N+1} \exp(-\gamma s_A)$ is the pressure in the (undisturbed) gas region A. Once q_{T_0} is determined, other unknowns may be found from (7.2.48)–(7.2.52). The constant q_{s_0} in the series representation of the shock locus (7.2.18) may be found by substituting U_0 from the above into (7.2.34). From (7.2.26) and (7.2.47), we have

$$A_0 = c_A[(2N - 1) - 2Nq_{I_0}]. \qquad (7.2.54)$$

Eliminating A_0 between (7.2.48) and (7.2.54), we get a relation between q_{I_0} and q_{T_0}:

$$q_{I_0} = (2N - 1)q_{T_0}/2N. \qquad (7.2.55)$$

The solution thus found corresponds to the plane shock tube problem and forms the basis for the series solution presented here. Higher order terms give the effect of geometry.

To obtain the first order system, we consider the solution of this system in each region by substituting the respective zeroth order solution on the right hand sides.

For the rarefaction region B, if we use the zeroth order solution (7.2.42) and (7.2.45), we get the following inhomogeneous system of ODEs:

$$qu_1' - (4N - 1)u_1 + (2N - 1)qc_1' + (2N - 1)c_1 = c_A q^2 s_1', \qquad (7.2.56)$$

$$qu_1' + (2N - 1)u_1 + (2N - 1)qc_1' - (4N^2 - 1)c_1 = 4c_A N(2N - 1) \\ \times q(1 - q), \qquad (7.2.57)$$

$$qs_1' - 2Ns_1 = 0. \qquad (7.2.58)$$

The system (7.2.56)–(7.2.58) must be solved subject to the boundary conditions (7.2.20) with $n = 1$ at the head of the wave. The solution $s_1 = s_{10}q^{2N}$

of (7.2.58), with s_{10} a constant, shows that $s_1 = 0$ since the entropy is continuous across the head of the rarefaction wave. The solution of the above system in region B has different forms depending on the value of N.

For $N \neq 1, 2$, we have

$$u_1 = -\frac{c_A(2N-1)q}{2}\frac{2(N-2) - 3(N-1)q + (N+1)q^{N-1}}{(N-1)(N-2)}, \qquad (7.2.59)$$

$$c_1 = -\frac{c_A q}{2}\frac{2(N-2)(2N-1) - (N-1)(4N-3)q + (3N-1)q^{N-1}}{(N-1)(N-2)}, \qquad (7.2.60)$$

$$s_1 = 0, \quad 1 \geq q > q_T. \qquad (7.2.61)$$

For $N = 3, 4, 5$, u_1 and c_1 are polynomials of degree N in q. For $N = 1, 2$, the solution involves logarithmic terms. For $N = 2$, which corresponds to $\gamma = 5/3$, the solution has the form

$$u_1 = -(3c_A q/2)[3q \log q + 2(1-q)], \qquad (7.2.62)$$

$$c_1 = -(c_A q/2)[5q \log q + 6(1-q)], \qquad (7.2.63)$$

$$s_1 = 0, \quad 1 \geq q > q_T. \qquad (7.2.64)$$

To get the first order solution in region C, we use the corresponding zeroth order solution (7.2.42) and (7.2.46). We thus have

$$[(2N-1)q_{T_0} - 2Nq]u_1' + 2Nu_1 - (2N-1)q_{T_0}c_1' + c_A q_{T_0}^2 s_1' = 0, \qquad (7.2.65)$$

$$-q_{T_0}u_1' + (2N-1)[(2N-1)q_{T_0} - 2Nq]c_1' + 2N(2N-1)c_1 = -4c_A N(2N-1)q_{T_0}(1 - q_{T_0}), \qquad (7.2.66)$$

$$[(2N-1)q_{T_0} - 2Nq]s_1' + 2Ns_1 = 0. \qquad (7.2.67)$$

The system (7.2.65)–(7.2.67) has the following general solution for $N \neq 0, 1/2$:

$$u_1/c_A = k_1[(N-1)q_{T_0} - Nq] - k_2 N[q_{T_0} - q] \\ -k_3 q_{T_0}^2/2N, \qquad (7.2.68)$$

$$c_1/c_A = \frac{k_1[(N-1)q_{T_0} - Nq] - k_2 N[q_{T_0} - q]}{2N - 1} \\ -2q_{T_0}(1 - q_{T_0}), \qquad (7.2.69)$$

$$s_1 = -(k_3/2N)[(2N-1)q_{T_0} - 2Nq], \quad q_T > q > q_I, \qquad (7.2.70)$$

where k_1, k_2, and k_3 are constants of integration.

Using (7.2.55) in (7.2.70), we infer that as $q \to q_{I_0}$, $s_1 \to 0$, for any value of k_3. Therefore, we have

$$s(q_I + 0, y) = s_A + O(y^2) \tag{7.2.71}$$

at the contact surface.

In fact, $s(q_I + 0, y) = s_A$ to all orders in regions B and C since there is no entropy jump across the head and the tail of the rarefaction wave.

The solution of the first order system of ODEs holding in region D is found in the same way as for region C; here the zeroth order solution is constant and is given by (7.2.43) and (7.2.47). The constant A_0 is 'partially' eliminated in terms of q_{I_0} with the help of (7.2.54). The result is

$$u_1 = \left(\frac{K_1}{2}\right)[2c_A N(q_{I_0} - q) - B_0] + \left(\frac{K_2}{2}\right)[2c_A N(q_{I_0} - q) + B_0]$$
$$\qquad - B_0^2 K_3/2c_A N, \tag{7.2.72}$$

$$c_1 = \frac{K_1[2c_A N(q_{I_0} - q) - B_0] - K_2[2c_A N(q_{I_0} - q) + B_0]}{2(2N-1)},$$

$$\qquad - \frac{2A_0 B_0}{c_A(2N-1)}, \tag{7.2.73}$$

$$s_1 = -K_3(q_{I_0} - q), \quad q_I > q > q_s, \tag{7.2.74}$$

where K_1, K_2, and K_3 are constants of integration. In the zeroth and first order solutions for regions B, C, and D, c_A and N are known from the statement of the problem while q_{T_0}, q_{I_0}, A_0 and B_0 have already been found from the zeroth order solution. We shall now investigate how to find the constants k_1, k_2, K_1, K_2, and K_3 from the boundary conditions corresponding to order one.

McFadden (1952) showed that, in fact, all the variables namely, velocity, sound speed and entropy can all be made continuous across the tail to first order in y. That is,

$$u(q_T + 0, y) - u(q_T - 0, y) = O(y^2),$$
$$c(q_T + 0, y) - c(q_T - 0, y) = O(y^2), \tag{7.2.75}$$
$$s(q_T + 0, y) - s(q_T - 0, y) = O(y^2),$$

It is easily seen from (7.2.70) that, if we choose $k_3 = 0$, $s_1 = 0$ and, therefore, entropy is continuous across the tail to first order. If we consider the negative Riemann invariant, namely, $u - (2N-1)c$ in region C and take the limit $q \to q_T - 0$, then, since $k_3 = 0$, $q_T - 0$ depends only on c_A, N and q_{T_0} to first order. The same can be shown to be true if we approach the tail of rarefaction wave from region B. Thus, the negative Riemann invariant is continuous across the tail to first order for any k_1 or k_2. Now, we consider the positive Riemann invariant $u + (2N - 1)c$ in regions B and C; it can be made continuous across the tail of the rarefaction wave provided

$$
\begin{aligned}
k_1 &= \frac{2N - 1}{(N - 1)(N - 2)}[(N - 2) - 2(N - 1)q_{T_0} + Nq_{T_0}^{N-1}], \quad N \neq 1, 2, \\
&= 6q_{T_0} \log q_{T_0} + 3(1 - q_{T_0}), \quad N = 2;
\end{aligned}
\tag{7.2.76}
$$

see (7.2.59)–(7.2.61) and (7.2.68)–(7.2.70). Here we have assumed that $k_3 = 0$. From the value of k_1 given by (7.2.76) and $k_3 = 0$, the solutions in regions B and C satisfy (7.2.75) for arbitrary k_2. The three shock conditions (7.2.37), (7.2.39) and (7.2.41) can be used to determine K_1, K_2 and K_3 in terms of U_1. The solution of order one is then substituted into the interface conditions (7.2.29) and (7.2.31). These together give U_1 and $u_0(q_{I_0} + 0)$.

The coefficients q_{T_1}, q_{I_1}, and q_{S_1} in the loci of the tail of the rarefaction wave, the contact discontinuity and the shock may now be found from (7.2.23), (7.2.27), and (7.2.35) as

$$
q_{T_1} = \frac{c_1(q_{T_0}) - u_1(q_{T_0})}{2c_A N},
\tag{7.2.77}
$$

$$
q_{I_1} = -\frac{u_1(q_{I_0} + 0)}{4c_A N},
\tag{7.2.78}
$$

$$
q_{S_1} = -\frac{U_1}{4c_A N}.
\tag{7.2.79}
$$

Thus the solutions (7.2.59)–(7.2.61), (7.2.68)–(7.2.70), and (7.2.72)–(7.2.74) in regions B, C, and D, respectively, and the loci of the boundaries, namely, the tail of the rarefaction wave, the interface, and the shock, have been completely found to order one.

The above procedure can be systematized as an algorithm to find higher order terms in the series solution (7.2.14).

It may be observed that the process of finding the higher order coefficients in the series (7.2.14) and (7.2.18) becomes rather unwieldy. The task of generating higher order terms satisfying appropriate boundary conditions in each of the regions was delegated to the computer. Twenty terms were generated in each of the regions and summed directly and by the use of Padé summation. The results were obtained by direct summation and by the use of Padé summation for $c_A t = 0.05, 0.30, 0.55$. The initial pressure ratio and the initial density ratio with $\gamma = 1.4$ were chosen to be 12.817 and 3.956, respectively. These are the same initial conditions as were used by McFadden (1952). The direct series sum and the Padé sum agree very well for $t \leq 0.2$. They begin to diverge thereafter. Padé summation extends the validity of the series solution to $t = 1/c_A$, the time up to which the present analysis holds. The series solution is significantly different from the first order solution of McFadden (1952) (see Yogi (1995) for details). The convergence of the series solution deserves further investigation.

The pressure decreases monotonically from the head of the rarefaction to the tail. Then it increases through the contact discontinuity to reach its maximum behind the shock. The density decreases from the head of the rarefaction to the tail. Then it begins to increase, gets a (positive) jump at the contact discontinuity and continues to grow until the shock. The particle velocity increases from its value 0 at the head of the rarefaction wave to its tail and then monotonically decreases through regions C and D. The sound speed behaves like the pressure from the head of the rarefaction to the contact discontinuity. It suffers a jump there and then decreases monotonically to the shock. The qualitative behaviour described above was also observed by Saito and Glass (1979) in their numerical solution. We may point out that Saito and Glass (1979) too did not envisage the presence of a secondary shock behind the main shock.

7.3 Blast Wave Caused by the Expansion of a High Pressure Gas Sphere: An Approximate Analytic Solution

In section 7.2 we studied the initial behaviour of a spherical blast as simulated by the sudden expansion of a uniform high pressure gas into the ambient air when the gas-pressure is not too high. The solution was sought in the form of series in time with coefficients functions of a 'similarity variable'. It was essentially a short time solution but, by an efficient use of the series, could be made to yield good results even for a finite time. The scheme of the series solution was such that the zeroth order term constituted the solution of the shock tube problem.

Essentially the same problem was treated subsequently by Friedman (1961). There was one major difference in the model though. The gas pressure was of such intensity that it gave rise to the phenomenon of a secondary shock, not treated by McFadden (1952). The latter author, however, does refer to the possibility of this shock, attributing it to an earlier investigation of Wecken (1950). McFadden (1952) also referred to some numerical evidence for the secondary shock. He himself studied the problem such that there was no secondary shock to first order approximation in the solution that he actually constructed.

In addition to treating the phenomenon of secondary shock, Friedman (1961) dealt with the blast wave problem by an approximate method quite distinct from that of McFadden (1952). The solution in the rarefaction region was found as a perturbation on the plane (shock tube) solution. The trajectory of the main shock and the formation and subsequent motion of the secondary shock were found by using an approximate technique, called CCW method (Chester (1954), Chisnell (1957) and Whitham (1958)). The contact surface was found explicitly by making use of these approximate results.

The analysis of Friedman (1961), though novel, is in error, as we shall show, since the approximate integration of some of the equations leads to (spurious) singularities for $\gamma = 5/3$ and $\gamma = 3$ and, therefore, to large values of the perturbation term rendering the perturbation scheme invalid. This therefore affects the entire solution of the problem. We shall first briefly describe Friedman's solution and show how more precise integration eliminates the errors in his analysis.

We now discuss the physical model and Friedman's approach in some detail. At time $t = 0$, a gas sphere (or cylinder) of radius x_0 under high internal pressure $p = p_4$, say, is surrounded by still air at pressure $p = p_0$, where $p_4 >> p_0$. As in section 7.2, the medium in the sphere is referred to as gas while that outside is air. For $t \geq 0$, an equalisation or explosion takes place giving rise to the following regions in the (x, t) plane (see Figure 7.2): (0) refers to the air which is not overtaken by the main shock, (1) is the compressed air enveloped by the main shock, (2) refers to the nearly uniform region outside the main expansion, (3) is the main expansion or rarefaction region, and (4) is the gas not yet disturbed by the centered expansion. The surface separating regions (1) and (2) is a contact discontinuity across which pressure and particle velocity are continuous while temperature, density and entropy suffer a jump.

The phenomenon of secondary shock which does not appear in the shock tube problem may be described as follows. The high pressure gas upon

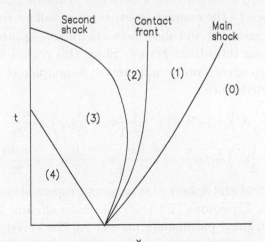

Figure 7.2 Explosion flow diagram (Friedman, 1961).

passing through a spherical rarefaction must expand to lower pressures than those reached through an equivalent one-dimensional expansion; this is caused by the increase in volume. This results in lower pressures at the tail of the rarefaction than the pressures transmitted by the main shock. A compression or secondary shock must be inserted to connect these two phases. Mathematically, this can be explained as follows. The centred expansion is described by a set of negative characteristics $(dx/dt = u - a)$ which, initially, point in the direction of decreasing x. However, later they turn around to the increasing x direction as the particle velocity increases. Negative characteristics also carry information after reflection from the shock. As the main shock propagates outward, it becomes weaker and the reflected negative characteristics incline more and more toward the decreasing x-direction. Thus the characteristics of the same family but arising from two different sources tend to meet and a shock must be inserted to make the flow compatible. This weak shock itself is determined by the following well-known result (see Courant and Friedrichs (1948)): the slope of a weak shock at each point is nearly equal to the average of the slopes of the incoming characteristics at that point. Friedman (1961) used this idea, developed earlier by Whitham (1952). This secondary shock ceases to be weak as it evolves. Friedman (1961) used the CCW approach, referred to earlier, to find the locus of this secondary shock; the flow ahead of this shock is nonuniform. The main shock between regions 0 and 1 was also found by the CCW method. For the trajectory of the contact front, Friedman (1961) derived a differential equation from the known flow properties on the characteristics coming from the main shock. The flow is assumed to be isentropic in the expansion region (3) and in the initial portions of the region (2) since here the secondary shock is weak. The shock-area rule (CCW method), where ever it has been used, correctly takes entropy changes into account via the shock conditions. In region (1) the entropy changes were ignored in following the flow properties from the main shock to the contact front; this may affect the solution.

We first treat region (3) and show how Friedman's approximation of the perturbation solution introduces errors. Since this region is isentropic, we may write equations of continuity and motion in nonplanar geometry in the following characteristic form:

$$\left(\frac{1}{\gamma - 1}a_t + \frac{1}{2}u_t\right) + (u + a)\left(\frac{1}{\gamma - 1}a_x + \frac{1}{2}u_x\right) + \frac{nua}{2x} = 0, \qquad (7.3.1)$$

$$\left(\frac{1}{\gamma - 1}a_t - \frac{1}{2}u_t\right) + (u - a)\left(\frac{1}{\gamma - 1}a_x - \frac{1}{2}u_x\right) + \frac{nua}{2x} = 0. \qquad (7.3.2)$$

$n = 1, 2$ for cylindrical and spherical symmetry, respectively, and γ is the ratio of specific heats. Equations (7.3.1)–(7.3.2) have already been normalised by using the dimensional parameters a_0 and x_0, the speed of sound in the undisturbed medium and the initial radius of the blast, respectively:

$$a = \bar{a}/a_0, \quad u = \bar{u}/a_0, \quad x = \bar{x}/x_0, \quad t = \bar{t}a_0/x_0. \qquad (7.3.3)$$

Here, the barred quantities are dimensional. Friedmann (1961) first wrote out the one-dimensional centered expansion for the plane case with $n = 0$ in (7.3.1)–(7.3.2):

$$u_1 = 2\mu r + (1 - \mu)\frac{x - 1}{t},$$

$$a_1 = \mu\left(2r - \frac{x - 1}{t}\right), \quad \mu = \frac{\gamma - 1}{\gamma + 1}, \quad r = \frac{1}{\gamma - 1}a_1 + \frac{1}{2}u_1 = \text{constant},$$

$$s = \frac{1}{\gamma - 1}a_1 - \frac{1}{2}u_1 = (1 - 2\mu)r - (1 - \mu)\frac{x - 1}{t}, \tag{7.3.4}$$

where r and s are the Riemann invariants. For the plane centered expansion wave, r is constant throughout the region (3). Friedman (1961) sought solution of (7.3.1) and (7.3.2) with $n = 1, 2$ in the form

$$u = u_1 + u_2, \quad a = a_1 + a_2,$$

$$R = \frac{1}{\gamma - 1}a_2 + \frac{1}{2}u_2, \quad S = \frac{1}{\gamma - 1}a_2 - \frac{1}{2}u_2, \tag{7.3.5}$$

where the subscript '2' denotes perturbation quantities due to the nonplanar geometrical terms. The perturbation in Riemann invariants are denoted by capital letters, R and S. Substituting (7.3.5) in (7.3.1) and retaining only first order terms in R, we have

$$R_t + (u_1 + a_1)R_x + \frac{nu_1a_1}{2x} = 0 \tag{7.3.6}$$

which, in the characteristic form, is

$$\frac{dt}{t} = \frac{dx}{4\mu rt + (1 - 2\mu)(x - 1)} = -2x\frac{dR}{ntu_1a_1}. \tag{7.3.7}$$

The first of (7.3.7) written with x as the dependent variable can be solved to yield

$$\frac{x - 1}{t} - 2r = -Kt^{-2\mu}, \tag{7.3.8}$$

or using the expression for a_1 from (7.3.4),

$$a_1t^{2\mu} = K\mu, \tag{7.3.9}$$

where K is a constant for each characteristic. Eliminating x from the second of (7.3.7) with the help of (7.3.8) we obtain an equation for $R = R(t)$ along the positive characteristic:

$$\frac{dR}{dt} = -\frac{n\mu K[2r - (1 - \mu)Kt^{-2\mu}]t^{-2\mu}}{2[1 + t(2r - Kt^{-2\mu})]}. \tag{7.3.10}$$

Friedman (1961) approximated the denominator in (7.3.10) by 1, implying thus that the characteristics remain close to $x = 1$ (see (7.3.8)). This, he averred, holds except for blasts of high intensity. Thus, (7.3.10) reduces to

$$\frac{dR}{dt} = -\frac{n\mu K}{2}[2r - (1 - 2\mu)Kt^{-2\mu}]t^{-2\mu}. \tag{7.3.11}$$

The solution of (7.3.6), therefore, is

$$R = -\frac{nK\mu t}{2}\left[\frac{2r}{1 - 2\mu}t^{-2\mu} - \frac{1 - \mu}{1 - 4\mu}Kt^{-4\mu}\right] + f(a_1 t^{2\mu}), \tag{7.3.12}$$

where $f(a_1 t^{2\mu})$ is a 'function' of integration which is constant along the characteristic (7.3.7) (see (7.3.9)). It is clear from (7.3.12) that R blows up when either $\mu = 1/2$ or $\mu = 1/4$, that is, when $\gamma = 3, 5/3$. Moreover, the factor $1 - 2\mu$ would also be small in the range $1 < \gamma < 5/3$. Thus, the approximation $x \sim 1$ introduces considerable errors. One must solve (7.3.10) exactly to eliminate these inaccuracies which also affect subsequent analysis.

Equation (7.3.10) can be written more simply as

$$\frac{dR}{dT} = \frac{A + BT}{CT^m + DT + E}, \quad T = t^{-1/m}, \tag{7.3.13}$$

where

$$
\begin{aligned}
m &= \frac{1}{2\mu} = \frac{1}{2}\frac{\gamma + 1}{\gamma - 1}, \\
A &= 2nK\gamma, \\
B &= -\frac{nK^2(2m - 1)}{2m}, \\
C &= 4, \quad D = -4K, \quad E = 8\gamma.
\end{aligned} \tag{7.3.14}
$$

For integral values of $m = 1, 2, 3, 4$, which correspond to $\gamma = 3, 5/3, 7/5, 9/7$, respectively, it is possible to integrate (7.3.13) in a closed form. We write the solution for $\gamma = 5/3$ for which R in (7.3.12) blows up.

For $\gamma = 5/3$, $m = 2$, and $\mu = 1/4$, and the exact solution of (7.3.6 is

$$
R = \begin{cases}
\dfrac{\log\left[\left(\dfrac{2t^{-1/2} - K - \sqrt{K^2 - 8\gamma}}{2t^{-1/2} - K + \sqrt{K^2 - 8\gamma}}\right)^{\left(8\gamma - \frac{3}{2}K^2\right)\frac{nK}{16\sqrt{K^2 - 8\gamma}}}\right]}{(t^{-1} - Kt^{-1/2} + 2\gamma)^{\frac{3nK^2}{32}}} + f(a_1 t^{1/2}), \quad K^2 > 8\gamma, \\[2em]
-\dfrac{3nK^2}{32}\log(t^{-1} - Kt^{-1/2} + 2\gamma) + \dfrac{nK}{8}\dfrac{\left(8\gamma - \frac{3}{2}K^2\right)}{\sqrt{8\gamma - K^2}}\tan^{-1}\dfrac{2t^{-1/2} - K}{\sqrt{8\gamma - K^2}} \\
\hspace{6cm} + f(a_1 t^{1/2}), \quad K^2 < 8\gamma, \\[1em]
-\dfrac{3nK^2}{32}\log(t^{-1} - Kt^{-1/2} + 2\gamma) - \dfrac{2}{2t^{-1/2} - K} + f(a_1 t^{1/2}), \quad K^2 = 8\gamma.
\end{cases} \tag{7.3.15}
$$

Friedman (1961) argued that since his entire analysis was approximate, this kind of complicated form was not called for. This argument is fallacious for $\gamma = 5/3$, as we pointed out earlier, since R blows up for this value of γ. For $\gamma = 1.4$, however, the integral (7.3.12), though approximate, may qualitatively be correct. Following Friedman (1961), this is the case we shall discuss specifically though we carry out the analysis for general γ. We may write (7.3.12) in the form

$$R = \frac{nt}{2}\left[\frac{(1-\mu)}{\mu(1\quad 4\mu)}a_1^2 - \frac{2r}{1\quad 2\mu}a_1\right] + f(a_1 t^{2\mu}) \qquad (7.3.16)$$

if we eliminate K therein with the help of (7.3.9). The function $f(a_1 t^{2\mu})$, constant along each positive characteristic, and the constant r are obtained by using the continuity of $(\gamma-1)^{-1}a+\frac{1}{2}u$ across the boundary characteristic between regions (3) and (4). Since $u_4 = 0$ and a_4 is constant in the region (4), the rarefaction front is given by

$$x - 1 = -a_4 t. \qquad (7.3.17)$$

The value of the Riemann invariant in region (4) is

$$r = \frac{1}{\gamma - 1}a_4. \qquad (7.3.18)$$

Since $R = 0$ in region 4, we have

$$f(a_4 t^{2\mu}) = -\frac{nt}{2}\left[\frac{1-\mu}{\mu(1-4\mu)}a_4^2 - \frac{2r}{1-2\mu}a_4\right]. \qquad (7.3.19)$$

Therefore, we may easily find that

$$f(a_1 t^{2\mu}) = -\frac{nt}{2}\left(\frac{a_1}{a_4}\right)^{1/2\mu}\left[\frac{1-\mu}{\mu(1-4\mu)}a_4^2 - \frac{2r}{1-2\mu}a_4\right]. \qquad (7.3.20)$$

We may obtain the Reimann invariant in region (3) with the help of (7.3.16) and (7.3.20) as

$$\begin{aligned}
\frac{1}{\gamma - 1}a_3 + \frac{1}{2}u_3 &= \left(\frac{1}{\gamma - 1}a_4 + 0\right) + R, \\
&= \frac{1}{\gamma - 1}a_4 + R, \\
&= \frac{1}{\gamma - 1}a_4 + tH(a_1), \qquad (7.3.21)
\end{aligned}$$

where

$$\begin{aligned}
H(a_1) &= \frac{n}{2}\left\{\frac{1-\mu}{\mu(1-4\mu)}[a_1^2 - (\frac{a_1}{a_4})^{1/2\mu}a_4^2] - \frac{2r}{1-2\mu}[a_1 - (\frac{a_1}{a_4})^{1/2\mu}a_4]\right\}, \\
a_1 &= \mu(2r - \frac{x-1}{t}). \qquad (7.3.22)
\end{aligned}$$

Using the perturbation form (7.3.5) in the negative characteristic form
(7.3.2) and retaining the lowest order terms we get an equation for the
perturbation S in the other Riemann invariant, namely s,

$$tS_t + (x-1)S_x + [(2\mu - 1)R + S] + \frac{nu_1a_1t}{2x} = 0, \qquad (7.3.23)$$

where $R = tH(a_1)$ and $H(a_1)$ is given by (7.3.22). The characteristic form
of (7.3.23) is

$$\frac{dt}{t} = \frac{dx}{x-1} = \frac{dS}{-[(2\mu-1)R + S] - \frac{nu_1a_1t}{2x}}. \qquad (7.3.24)$$

The first of (7.3.24) gives the characteristics, $x - 1 = Lt$, where L is a
constant along each characteristic. Equations (7.3.24) can be combined to
yield

$$\frac{d}{dt}(tS) = (1 - 2\mu)R - \frac{nu_1a_1t}{2(1+Lt)} = (1 - 2\mu)tH(a_1) - \frac{nu_1a_1t}{2(1+Lt)}. \qquad$$
$$(7.3.25)$$

Since u_1, a_1 and L are constant along a negative characteristic, (7.3.25) can
be integrated to yield

$$S = (1 - 2\mu)\frac{t}{2}H(a_1) - \frac{nu_1a_1}{2L}\left[1 - \frac{\log(1 + Lt)}{Lt}\right]. \qquad (7.3.26)$$

Recalling that the (negative) Riemann invariant is given by $s + S$, we use
(7.3.4) and (7.3.26) to obtain

$$s + S = \frac{1}{\gamma - 1}a_3 - \frac{1}{2}u_3 = (1 - 2\mu)r - (1 - \mu)\frac{x-1}{t} + \frac{(1 - 2\mu)}{2}tH(a_1)$$
$$-\frac{nu_1a_1t}{2(x-1)^2}[x - 1 - \log x], \qquad (7.3.27)$$

where we have put $L = (x - 1)/t$.

In the above we have used $x = 1 + Lt$ as the lowest order equation for
the negative characteristics. We employ the above results to obtain more
accurate slope of the characteristics $u_3 - a_3$ and hence their loci.

Thus, combining (7.3.21) and (7.3.27) suitably we have

$$\frac{dx}{dt} = u_3 - a_3 = \frac{x-1}{t} + \frac{1 - 2\mu}{2 - 2\mu}tH(a_1) + \frac{nu_1a_1t}{2(1-\mu)(x-1)^2}[x - 1 - \log x]. \qquad$$
$$(7.3.28)$$

To solve (7.3.28), we again let $x = 1 + Lt$ on its RHS and assume that
u_1, a_1, and L are all constant along the negative characteristics. This gives

$$\frac{dx}{dt} = L + \frac{1 - 2\mu}{2 - 2\mu}tH(a_1) + \frac{nu_1a_1}{2(1-\mu)L}\left[1 - \frac{\log(1 + Lt)}{Lt}\right]. \qquad (7.3.29)$$

On integration of (7.3.29), we have

$$x = 1 + Lt + \frac{(1-2\mu)t^2}{4(1-\mu)}H(a_1) + \frac{nu_1a_1}{2(1-\mu)L}\int\left[1 - \frac{\log(1+Lt)}{Lt}\right]dt.$$

(7.3.30)

An approximate evaluation of the integral in (7.3.30) gives

$$x = 1 + Lt + \frac{(1-2\mu)t^2}{4(1-\mu)}H(a_1) + \frac{nu_1a_1}{2(1-\mu)L}\left[1 - \frac{\log(1+Lt)}{Lt}\right]\frac{t}{2}.$$

(7.3.31)

or putting $L = (x-1)/t$ in the last term in (7.3.31) we have

$$x = 1 + Lt + \frac{(1-2\mu)t^2}{4(1-\mu)}H(a_1) + \frac{nu_1a_1t^2}{4(1-\mu)(x-1)}\left[1 - \frac{\log(x)}{x-1}\right].$$

(7.3.32)

where $H(a_1)$ is given by (7.3.22).

We turn now to the other end of the flow, namely, the main shock. Friedman (1961), as we remarked earlier, followed CCW approach to obtain the locus of this shock. We briefly describe this approach. Whitham (1958), 'rather illogically', proposed that, to get the trajectory of a forward moving shock, one may write the compatibility condition holding along the positive characteristic and substitute the Rankine-Hugoniot conditions on this differential relation. This leads to an ODE for the Mach number with distance as the independent variable. This ODE may be solved in a closed form or integrated numerically with an appropriate initial condition. This approach makes sense if the shock is weak and therefore may be approximated by a characteristic close to it. That it works 'reasonably' even for a strong shock is surprising. Whitham (1958) also attempted to explain why this approach works. There have been several other attempts to critically examine and improvise upon this technique. Essentially, it works when there are no major effects catching up with the shock; this is true, for example, where self-similar solutions of the second kind hold (see section 6.1 for converging shocks).

Writing again the given flow equations with pressure and density as the dependent variables, we have

$$p_t + \rho a u_t + (u+a)(p_x + \rho a u_x) + \frac{n\rho a^2 u}{x} = 0, \qquad (7.3.33)$$

$$p_t - \rho a u_t + (u-a)(p_x - \rho a u_x) + \frac{n\rho a^2 u}{x} = 0. \qquad (7.3.34)$$

These equations hold along the characteristic directions, $u+a$ and $u-a$, respectively, whether the flow is isentropic or not. In the present case, it is nonisentropic behind the shock.

Since the positive characteristics meet the main shock, Whitham's rule requires their use. Thus, we have the following compatibility relation holding along $\frac{dx}{dt} = u + a$:

$$dp + \rho a\, du = -\frac{n\rho a^2 u}{u+a}\frac{dx}{x}. \tag{7.3.35}$$

The Rankine-Hugoniot conditions are

$$u = \frac{2a_0}{\gamma+1}(M - M^{-1}), \quad p = \frac{p_0}{\gamma+1}(2\gamma M^2 - \gamma + 1),$$

$$\rho = \frac{\rho_0(\gamma+1)M^2}{(\gamma-1)M^2+2}, \quad a^2 = \frac{a_0^2(2\gamma M^2 - \gamma + 1)\{M^2(\gamma-1)+2\}}{(\gamma+1)^2 M^2}, \tag{7.3.36}$$

where $M = U/a_0$, U is the shock velocity, and '0' refers to the undisturbed conditions in the region (0). Substituting (7.3.36) into (7.3.35) we have the following differential relation holding along the shock, $x_m = x_m(t)$:

$$-n\frac{dx_m}{x_m} = dM\left\{\frac{4M}{2\gamma M^2 - \gamma + 1} + \frac{2(M^2+1)}{M\{[2\gamma M^2 - \gamma + 1][(\gamma-1)M^2+2]\}^{1/2}}\right.$$

$$\left. +\frac{2M}{M^2-1}\left(\frac{(\gamma-1)M^2+2}{2\gamma M^2 - \gamma + 1}\right)^{1/2} + \frac{M^2+1}{M(M^2-1)}\right\}$$

or

$$\frac{dM}{dt_m} = -\frac{M}{n}x_m\left\{\frac{4M}{2\gamma M^2 - \gamma + 1} + \frac{2(M^2+1)}{M\{[2\gamma M^2 - \gamma + 1][(\gamma-1)M^2+2]\}^{1/2}}\right.$$

$$\left. +\frac{2M}{M^2-1}\left(\frac{(\gamma-1)M^2+2}{2\gamma M^2 - \gamma + 1}\right)^{1/2} + \frac{M^2+1}{M(M^2-1)}\right\}^{-1}$$

$$\equiv -\frac{Mx_m}{n}F(M). \tag{7.3.37}$$

Curiously, this complicated equation can be solved in a closed form:

$$(x_m)^n[\frac{\{2\gamma M^2 - \gamma + 1\}^{1/2} - \{(\gamma-1)M^2+2\}^{1/2}}{M}]^2[\{(\gamma-1)(2\gamma M^2$$

$$-\gamma+1)\}^{1/2} - \{2\gamma[(\gamma-1)M^2+2]\}^{1/2}]^{\sqrt{2\gamma/(\gamma-1)}}(2\gamma M^2 - \gamma + 1)^{1/\gamma}$$

$$\times \exp[\frac{1}{\{2(\gamma-1)\}^{1/2}}\sin^{-1}\{\frac{2(2\gamma M^2 - \gamma + 1) - (\gamma-1)[(\gamma-1)M^2+2]}{M^2(\gamma+1)^2}\}]$$

$$= \text{const.} \tag{7.3.38}$$

The (dimensionless) time may now be obtained from the relation

$$M\, dt_m = dx_m, \tag{7.3.39}$$

where dx_m is defined by (7.3.37). If we know the Mach number of the initial shock produced by the detonation at time $t = t_0$, say, equations (7.3.37) and (7.3.39) may be integrated simultaneously to give (x_m, t_m) as functions of Mach number. Specifically, if the high pressure gas has the initial pressure p_4 and sound speed a_4 while the outside atmospheric conditions are p_0 and a_0, the one-dimensional (shock tube) theory gives relation

$$\frac{p_4}{p_0} \left\{ 1 - \mu \frac{a_0}{a_4} (M - M^{-1}) \right\}^{(\mu+1)/\mu} = (1 + \mu)M^2 - \mu \qquad (7.3.40)$$

for the initial shock Mach number.

To get the flow between the main shock and the contact discontinuity, we need the loci of negative characteristics reflected from the main shock. Friedman (1961) argued that, since the distance between the shock and contact discontinuity is relatively short, the slope of each negative characteristic, $u_1 - a_1$, may be assumed to be constant and computed from the values of u_1 and a_1 immediately behind the main shock. This is tantamount to ignoring the effects of entropy changes and three dimensionality behind the shock. Changes in the shock strength are appropriately accounted for. Thus the negative characteristics are simply

$$x = x_m + w_1(M)(t - t_m), \qquad (7.3.41)$$

where $w_1(M) = u - a$ is evaluated at the point (x_m, t_m) on the main shock. Thus, (7.3.37), (7.3.39) and (7.3.41) define the main shock and negative characteristics behind the shock as functions of the shock Mach number M.

Now we turn to the determination of the trajectory of the contact discontinuity. In the nature of the analysis carried out by him, Friedman (1961) made several further assumptions to accomplish this task. He made use of the solution in domain (3), ignoring the presence of the secondary shock; he also assumed that the characteristics coming from the main shock were straight lines. The conditions at the contact surface—continuity of pressure and particle velocity—were satisfied. The path of the contact surface was found by observing that it was a particle line.

Since the contact front moves with the local particle velocity u, the positive characteristics with slope $u + a$ meet it from region (2) while the negative characteristics with slope $u - a$ intersect it from region (1). The Riemann invariant $(\gamma - 1)^{-1}a + (1/2)u$ is computed from (7.3.21), ignoring the entropy jump across the weak secondary shock separating regions (2) and (3). Introducing the notation

$$Q = (\gamma - 1)^{-1}a + \frac{1}{2}u, \qquad (7.3.42)$$

and $w = u - a$, as before, we observe that, since particle velocity is continuous across the contact front, we have in terms of Q and w,

$$Q_2 + \frac{1}{\gamma - 1}w_2 = Q_1 + \frac{1}{\gamma - 1}w_1$$

or

$$w_2 - w_1 = -(\gamma - 1)(Q_2 - Q_1). \tag{7.3.43}$$

The contact front is also a particle line, therefore, the entropy along it is constant. Thus, we have

$$\frac{p}{p_i} = \left(\frac{a}{a_i}\right)^{2\gamma/(\gamma-1)}, \tag{7.3.44}$$

where the subscript i denotes the initial state. Since the pressure is continuous across this front, we have from (7.3.43), (7.3.44) and the definition of w, the relation

$$\left(\frac{a}{a_i}\right)_2 = \left(\frac{a}{a_i}\right)_1 \tag{7.3.45}$$

or

$$2Q_2 - w_2 = \frac{a_{i2}}{a_{i1}}(2Q_1 - w_1). \tag{7.3.46}$$

Eliminating Q_1 from (7.3.43) and (7.3.46), we have the slope w_2 of negative characteristics in region (2) in terms of known quantities:

$$w_2 = \frac{\frac{\gamma+1}{\gamma-1}\frac{a_{i2}}{a_{i1}}w_1 + 2\left(1 - \frac{a_{i2}}{a_{i1}}\right)Q_2}{1 + \left(\frac{2}{\gamma-1}\right)\frac{a_{i2}}{a_{i1}}}. \tag{7.3.47}$$

This slope, however, involves the functions $w_1 = w_1(M)$ and $Q_2 = Q(x,t)$ (see (7.3.41) and (7.3.42)). We must determine the (x,t) co-ordinates of the contact front in terms of the parameter M. For this purpose, we first express a_1 in terms of w_1 and Q_2. Since the particle velocity is continuous across the contact front, we have

$$2Q_2 - w_1 = \frac{2}{\gamma - 1}a_2 + a_1. \tag{7.3.48}$$

Combining this relation with (7.3.47), we have

$$a_1 = \frac{2Q_2 - w_1}{1 + \left(\frac{2}{\gamma-1}\right)\frac{a_{i2}}{a_{i1}}}. \tag{7.3.49}$$

We may write the equation of the contact front as $x_c = C(t_c)$. The negative characteristic (7.3.41) from the main shock meets the contact front at

$$C(t_c) = x_m + w_1(M)(t_c - t_m). \tag{7.3.50}$$

Therefore,

$$\frac{dC}{dt_c} = w_1 + \frac{dM}{dt_c}\{x'_m + w'_1(t_c - t_m) - w_1 t'_m\}, \tag{7.3.51}$$

where prime denotes differentiation with respect to M. Also, at the contact front, we have the speed

$$\frac{dC}{dt_c} = \frac{dx_c}{dt_c} = u. \tag{7.3.52}$$

Using (7.3.51), (7.3.52) and the relation $w_1 = u_1 - a_1$, we have

$$a_1 = \frac{dM}{dt_c}\{x_m' + w_1'(t_c - t_m) - w_1 t_m'\}. \tag{7.3.53}$$

Eliminating a_1 from (7.3.49) and (7.3.53), we have an ODE relating t_c and M:

$$\frac{dt_c}{dM} = \frac{\{1 + \left(\frac{2}{\gamma-1}\right)\frac{a_{i_2}}{a_{i_1}}\}\{x_m' + w_1'(t_c - t_m) - w_1 t_m'\}}{2Q_2 - w_1}. \tag{7.3.54}$$

This equation, together with

$$x_c = x_m + w_1(M)(t_c - t_m), \tag{7.3.55}$$

defines the co-ordinates (x_c, t_c) of the contact front as functions of the parameter M. We thus have the locus of the negative characteristics in region (2) as

$$x = x_c + (t - t_c)w_2, \tag{7.3.56}$$

where w_2 is given by (7.3.47); t_c and x_c are obtained from (7.3.54) and (7.3.55).

With the above information from regions (2), (3) and (1), we may fit the secondary shock between regions (2) and (3). Friedman (1961) again made several assumptions to accomplish this. The secondary shock is formed by the intersection of negative characteristics from the main shock via the contact front and from the expansion region. The former tend to point more and more towards the decreasing x-direction due to weakening of the main shock, causing the characteristic slope $u_1 - a_1$ to decrease. The characteristics from the expansion region fan into the increasing x-direction. During the early phase the secondary shock is weak and, therefore, Whitham's (1952) rule may be justifiably used to find its locus.

The negative characteristics from the expansion fan and the main shock are given by (7.3.30) and (7.3.56), respectively:

$$x = 1 + Lt + f(x, t), \tag{7.3.57}$$
$$x = x_c + w_2(t - t_c), \tag{7.3.58}$$

where $f(x, t)$ represents other terms appearing in (7.3.30). t_c, x_c, and w_2 are functions of the main shock Mach number and are given by (7.3.54), (7.3.55), and (7.3.47), respectively. If we write the path of the secondary shock as

$$x_s = 1 + S(t_s), \tag{7.3.59}$$

then, along this path we have, on using (7.3.57) and (7.3.58),

$$
\begin{aligned}
S(t_s) &= w_2(t_s - t_c) + x_c - 1 \\
&= Lt_s + f\{x_c + (t_s - t_c)w_2, t_s\}. \tag{7.3.60}
\end{aligned}
$$

With the known slopes of the characteristics (7.3.57) and (7.3.58), we may write the slope of the secondary shock as their average:

$$
\frac{dS}{dt_s} = \frac{1}{2}\left(w_2 + \frac{L + f_t}{1 - f_x}\right). \tag{7.3.61}
$$

Friedman (1961) also obtained this slope following another route. Letting the parameters, M, L, and the co-ordinate x_s to depend on t_s, he found the derivative dS/dt_s from the two expressions for the shock path in (7.3.60) and took their average:

$$
\frac{dS}{dt_s} = \frac{1}{2}\left(w_2 + \frac{dM}{dt_s}[x'_c + w'_2(t_s - t_c) - w_2 t'_c] + \frac{L + f_t}{1 - f_x} + \frac{t_s(dL/dt)}{1 - f_x}\right). \tag{7.3.62}
$$

Equating (7.3.61) and (7.3.62) and simplifying, we get

$$
t\frac{dL}{dM} = -(1 - f_x)\{x'_c + w'_2(t_s - t_c) - w_2 t'_c\}. \tag{7.3.63}
$$

Eliminating L from this equation with the help of (7.3.60) we have

$$
\frac{dt_s}{dM} = \frac{2(1 - f_{x_s})(x'_c + w'_2(t_s - t_c) - w_2 t'_c)t_s}{x_c - 1 - w_2 t_c - f + (f_{t_s} + w_2 f_{x_s})t_s}.
$$

or

$$
\frac{dt_s}{dt_m} = \frac{2(1 - f_{x_s})[(x'_c + w'_2(t_s - t_c) - w_2 t'_c)]t_s}{x_c - 1 - w_2 t_c - f + (f_{t_s} + w_2 f_{x_s})t_s}\frac{dM}{dt_m}. \tag{7.3.64}
$$

Here, $f(x, t)$ and $f_x(x, t)$ are evaluated at $x_s = x_c + (t_s - t_c)w_2$, $t = t_s$. Integrating (7.3.64), we get a relation between t_s and M:

$$
\begin{aligned}
[x_c - 1 - t_c w_2]^2 &= t_s \int_{M_i}^{M} [x_c - 1 - t_c w_2]\left[-2w'_2 + \frac{d}{dM}\left(\frac{f}{t_s}\right)\right. \\
&\quad \left. + f_x/t_s\{x'_c + w'_2(t_s - t_c) - w_2 t'_c\}\right]dM. \tag{7.3.65}
\end{aligned}
$$

Here, M_i is the initial Mach number of the main shock. The secondary shock does not form at the the initial point of the fluid flow field. The time of formation of the secondary shock was found by Friedman (1961) by solving the integral equation (7.3.65) iteratively. For the specific problem solved by Friedman (1961), which we discuss towards the end of this section, the point of secondary shock formation was found to be $x_i = 1.14$, $t_i = 0.41$.

Equation (7.3.64) is solved in conjunction with

$$\frac{dM}{dt_m} = -\frac{M}{n}x_m F(M), \tag{7.3.66}$$

$$\frac{dx_m}{dt_m} = M, \tag{7.3.67}$$

$$\frac{dt_c}{dt_m} = \frac{\left\{1 + \left(\frac{2}{\gamma-1}\right)\frac{a_{i_2}}{a_{i_1}}\right\}\left\{x_m' + w_1'(t_c - t_m) - w_1 t_m'\right\}}{2Q_2 - w_1}\frac{dM}{dt_m} \tag{7.3.68}$$

(see (7.3.37), (7.3.39) and (7.3.54)) to obtain t_s, M, x_m and t_c as functions of t_m. x_c may then be found from (7.3.55). Equations (7.3.64) and (7.3.66)–(7.3.68) may be solved with the initial conditions $M = M_i$, $x_M = 1$, $t_{c_i} = 0$, $t_s = t_{s_i}$ to obtain M, x_m, t_c as functions of t_m, where M_i is the initial Mach number of the main shock. t_s is obtained from (7.3.64) for $t \geq t_{s_i}$.

The above technique for the initial motion of the secondary shock assumes that it is weak. It strengthens as it is carried outward by the expanding gases. Friedman (1961) again used the CCW approach to find the locus of the secondary shock. This procedure was adopted when the secondary shock begins to turn back towards the origin or when its strength $\overline{M} - 1$ becomes $O(1)$, which ever happens first; $\overline{M}$ is the Mach number of the secondary shock, given by $(u_3 - U_s)/a_3$. U_s is the velocity of the secondary shock.

In the present case we have an inward moving shock for which conditions ahead are given by a_3, p_3, and ρ_3, the flow in the expansion wave. The particle velocity across the shock now is

$$u = u_3 - \frac{2a_3}{\gamma+1}(\overline{M} - \overline{M}^{-1}). \tag{7.3.69}$$

The other conditions are

$$p = \frac{p_3}{\gamma+1}(2\gamma\overline{M}^2 - \gamma + 1), \tag{7.3.70}$$

$$\rho = \rho_3(\gamma+1)\overline{M}^2/[(\gamma-1)\overline{M}^2 + 2], \tag{7.3.71}$$

$$a^2 = \frac{a_3^2\overline{R}^2}{(\gamma+1)^2\overline{M}^2}, \tag{7.3.72}$$

where

$$\overline{R} = \{(2\gamma\overline{M}^2 - \gamma + 1)((\gamma-1)\overline{M}^2 + 2)\}^{1/2}.$$

The compatibility relation holding along the negative characteristic is

$$dp - \rho a\, du = -\frac{\rho a^2 u n}{u - a}\frac{dx}{x}. \tag{7.3.73}$$

Substituting the shock relations (7.3.69)–(7.3.72) into (7.3.73) we obtain the following relations holding along the secondary shock:

$$
\frac{d\overline{M}}{dx} = \left\{ \frac{\gamma+1}{2} \frac{\overline{M}}{\overline{R}} \left[\frac{a_{3t}/U_s + a_{3x}}{a_3} \right] - \left[\frac{\overline{M}^2 - 1}{\overline{R}} + \frac{1}{\gamma-1} \right] \left[\frac{a_{3t}/U_s + a_{3x}}{a_3} \right] \right.
$$
$$
\left. + \left[\frac{2a_3(\overline{M}^2 - 1) - (\gamma+1)\overline{M}u_3}{(\gamma+1)\overline{M}u_3 - 2a_3(\overline{M}^2 - 1) - \overline{R}} \right] \frac{n}{2x} \right\}
$$
$$
\bigg/ \left\{ \frac{2\overline{M}}{2\gamma\overline{M} - \gamma + 1} + \frac{\overline{M}^2 + 1}{\overline{M}\overline{R}} \right\}, \qquad (7.3.74)
$$

$$
\frac{dt}{dx} = U_s^{-1}, \quad U_s = u_3 - \overline{M}a_3. \qquad (7.3.75)
$$

The functions u_3, a_3 and their deivatives are obtained from (7.3.21) and (7.3.28). The system (7.3.74)–(7.3.75) is solved numerically starting from a point where x, t, and U_s (or $\overline{M}$) are prescribed.

As an example, Friedman analysed the case treated earlier numerically by Brode (1957) and experimentally by Boyer (1960). A sphere of (nondimensional) unit radius contains compressed gas at 22 atmospheres. It is surrounded by air at 1 atmosphere. The specific heat ratio γ of air and gas is assumed to be 1.4. Using plane shock tube theory, the initial strength of the main shock is found to be 1.846. The constant of integration of ODE in the Whitham's rule on the RHS of (7.3.38) is found to be 26.1 if these initial conditions are used. The initial conditions at $x = 1$ and $t = 0^+$ in different regions are given by $u_0 = 0$, $a_0 = 1$, $u_1 = 1.087$, $a_1 = 1.252$, $u_2 = 1.087$, $a_2 = 0.729$. The centered simple wave for $\gamma = 1.4$ is described

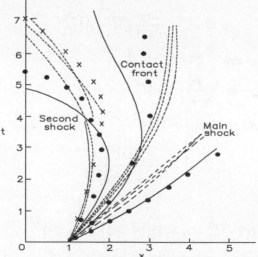

Figure 7.3 Experimental and theoretical spherical blast results: – – –, experimental (Boyer); ——, numerical integration (Brode); •, Friedman theory; ×, simplified secondary shock approximation (Friedman, 1961).

by $u_3 = \frac{5}{6}\left(1 + \frac{x-1}{t}\right)$, $a_3 = \frac{1}{6}\left(5 - \frac{x-1}{t}\right)$. It follows from (7.3.21), (7.3.28) and (7.3.32) that the flow in the region (3) for $\gamma = 1.4$ is given by

$$2.5a_3 + 0.5u_3 = 2.5 - \frac{t}{28.8}\left(5 - \frac{x-1}{t}\right)\left(1 + \frac{x-1}{t}\right)^2, \qquad (7.3.76)$$

$$u_3 - a_3 = \frac{x-1}{t} - \frac{1}{72}\left[\left(5 - \frac{x-1}{t}\right)\left(1 + \frac{x-1}{t}\right)t\right]$$

$$\times \left[\left(1 + \frac{x-1}{t}\right) - \frac{12}{x-1}\left(1 - \frac{\log x}{x-1}\right)\right], \qquad (7.3.77)$$

$$x = 1 + Lt - \frac{1}{144}\left[\left(5 - \frac{x-1}{t}\right)\left(1 + \frac{x-1}{t}\right)t^2\right]$$

$$\times \left[\left(1 + \frac{x-1}{t}\right) - \frac{12}{x-1}\left(1 - \frac{\log x}{x-1}\right)\right]. \qquad (7.3.78)$$

Figure 7.3 shows a comparison of results–experimental, numerical and by the simplified analysis of Friedman (1961). The main shock obtained by the CCW method is described quite well. Friedman (1961) points out the inadequecies of the numerical scheme of Brode (1957) and the experimental procedures of Boyer (1960). It is remarkable that, in spite of the highly simplified form of the analysis and neglect of entropy changes in different regimes, there is a reasonable qualitative agreement between the numerical solution and the analytic results of Friedman (1961). To obtain quantitatively correct results, the entire analysis of Friedman would have to be refurbished and made more rigorous. This is a formidable task, requiring considerable work.

Chapter 8

Numerical Simulation of

Blast Waves

8.1 Introduction

The rapid, almost exponential, growth of the power of the silicon computing chip, made available to the user at ever-decreasing cost, has made the computational approach a viable and practical alternative for a variety of problems in physics and engineering, in particular the complex nonlinear problems of fluid mechanics. Nearly a half century earlier, Von Neumann, the inventor of the modern electronic computing machine, carried out the first numerical calculations for the inviscid, nonlinear problem of gas dynamics involving shock waves. The concept of artificial viscosity, proposed first by Von Neumann and Richtmeyer (1950) and refined subsequently to a high degree of perfection by many others, has proved to be a powerful tool that made the numerical algorithms for such problems not only possible but also accurate, reliable and robust. The progress in the efficiency and accuracy of solution algorithms during the last five decades, keeping pace with the power of the computing chip, has proved beyond doubt the truth of Von Neumann's forecast in 1945: "Really efficient high speed computing devices may, in the field of nonlinear partial differential equations as well as in many other fields which are now difficult or are entirely denied of access, provide us with those heuristic hints which are needed in all parts of mathematics for genuine progress". Needless to say progress in numerical methods is equally dependent on the sharpening of the analytical tools of applied mathematics.

Goldstine and Von Neumann (1955) initiated one of the earliest numerical approaches to the explosion problem, assuming a point source and ideal gas behaviour; they could relax the strong shock assumption implicit in the similarity solution. In the same year, Brode (1955) published numerically computed results for the point source problem and found that the calculated distributions of the flow parameters closely followed the results from similarity theory for shock overpressure decay up to 20 atmospheres. He used the method of artificial viscosity. Both the studies could predict blast wave history down to shock overpressures as low as 0.1 atmosphere. Brode (1959) relaxed the ideal gas assumption. He used the thermal equation of state for approximating the thermodynamic behaviour of real gas at high temperatures, encountered in strong explosions. The computed results showed that the blast wave overpressure at any radius is lower than the corresponding value predicted by the assumption of ideal gas behaviour, since a part of the available energy is absorbed by the ionization and dissociation processes occurring at high temperatures.

It was already clear that the numerical approaches need not be limited to a point source or to other simplifying assumptions implicit in the analytic treatment of this problem. Finite compressed gas ball explosions are characterized by the multiple wave phenomena that ensue after the rupture of the diaphragm separating the compressed gas and the surrounding medium. Away from solid boundaries, the flow that develops can be assumed to be well-described by the inviscid equations of gas dynamics with almost arbitrary but admissible thermodynamic behaviour. The wave system that develops in the flow is almost completely described by the eigenstructure of the above mentioned equations which, though nonlinear, are strictly hyperbolic in time, subject to some necessary but very liberal assumptions about the thermodynamic behaviour of the material medium through which the blast wave propagation occurs; solution to the Riemann problem with arbitrary data exists and is computable. The last observation provides the necessary underpinnings for the success and reliability of most of the numerical schemes that have been developed during the last three decades, as we shall elaborate in the following.

The point source approximation, no longer required by a numerical approach, was dispensed with quite early by Brode (1959) who described numerical results for two cases of considerable practical importance, namely, the explosion of a spherical charge of TNT and the sudden release of initially static high pressure gas from spherical enclosures. Using a Lagrangian approach, Brode succeeded in capturing the complete post-explosion wave structure and the birth and evolution of the secondary shock which originates at the tail of the inward facing rarefaction wave. The calculations further revealed that the contact front moves initially outwards following the blast wave but decelerates and subsequently reverses direction to move inward. Brode's (1959) calculations were carried out using a real gas equation of state for both air and helium; his results were subsequently compared with the experimental data by Boyer (1960). These early calculations, using

finite sources, confirmed that the late evolution of the blast wave history
is independent of the early time behaviour and is closely approximated by
the point source calculations as long as basic assumptions in the analytic
solutions are not violated.

The first finite-difference-based numerical algorithm for the solution of
the Euler equations for the finite source problem appears to have been described by Payne (1957). He applied a numerical approach, later called
Lax-Friedrich (1954) scheme, to the quasi-conservative form of the one-
dimensional equations of gas-dynamics to obtain solutions for the imploding cylindrical shocks. The geometric singularity occurring at the axis was
addressed by some ad hoc extrapolation formulae which ensured that the
calculations did not fail at the instant when the imploding shock reflects
off the axis. The monotone nature of the Lax-Friedrich scheme made sure
that the computed shock transition was oscillation-free, even as it smeared
the contact discontinuity. Payne (1957) found very good agreement for the
computed strength of the converging cylindrical shock with Chisnell's (1957)
analytic results. The problem of the singularity at the axis was overcome
in a subsequent work by Lapidus (1971) who solved the same problem in
cartesian co-ordinates in two space dimensions using a two-step variation of
the Lax and Wendroff (1960) scheme in conservative form. Being second
order in space and time the scheme is not monotone and exhibits post-shock
oscillations in computed solutions; however, the contact discontinuity can
be clearly discerned. Abarbanel and Goldberg (1972) addressed the problem of the converging cylindrical shock using a quasi-conservative approach
in cylindrical co-ordinates. They derived a second order difference scheme
based on the classical Lax-Wendroff technique and established the linear
stability of the scheme under slightly relaxed Courant-Friedrichs-Lewy condition typical of single-step explicit schemes (see Richtmyer and Morton
(1967)). Their computed solutions disagreed somewhat with those of Payne
(1957) and Lapidus (1971) with respect to shock arrival times at the centre.

A numerical algorithm based on a Reimann solver for the above problem appears to have been first advocated by Sod (1977). Glimm's (1965)
random choice method, together with an exact Riemann solver, was used to
integrate the homogeneous part of the equations of one-dimensional gas dynamics. Operator-splitting was used to integrate in time the singular source
terms. The cell-centred nature of the scheme ensured that the boundary
conditions at the axis could be implemented by simple reflection to fix the
values for a ghost cell. The inherent virtue of Glimm's method enabled Sod
(1977) to capture sharp shock and contact discontinuities; the rarefaction
fan, however, showed a jagged transition.

All the methods mentioned above were based on either first order schemes
or nonmonotone higher order schemes and assumed ideal gas thermodynam-
ics. Glimm's scheme, despite its ability to produce sharp shock and contact
transitions in an Eulerian framework, could not be satisfactorily extended

to multiple dimensions to ensure similar performance. The high resolution schemes developed by Harten (1983) and the essentially nonoscillatory (ENO) schemes of Shu and Osher (1988) made possible robust and reliable computation of discontinuous flows using high order accurate schemes. Special techniques such as sub-cell resolution (Harten (1989)) have been developed to sharpen the captured discontinuous fronts. Liu et al. (1999) used a high resolution total variation diminishing (TVD) scheme of Harten (1983) with suitable modifications for sharp capture of contact surfaces in the flow fields caused by cylindrical and spherical explosions. Here, the perfect gas equation was assumed. Computed results for explosions in air were compared with those of Brode (1955) and found to be in close agreement. Comparison with experimental data from Boyer (1960), however, showed some notable disagreements such as the arrival time of the secondary shock at the centre. Liu et al. (1999) also solved the problem of cylindrical implosion. Comparisons of relevant results with Sod's (1977) computations showed agreement within 10 percent. In a related work, Liu et al. (1999a) extended the scope of the method to explosions in water.

8.2   A Brief Review of Difference Schemes for Hyperbolic Systems

It is clear from the analysis of flows with shock waves reported in earlier chapters that very few realistic problems can be solved exactly. The exact solutions that have been found such as Taylor-Sedov for the point explosion or Guderley's solution for the converging shocks are asymptotic in nature, holding under specific limiting conditions. Thus, there is a need to develop numerical schemes which can effectively reproduce solutions of the governing system of partial differential equations with appropriate boundary conditions across discontinuities such as shocks, as they develop; these solutions must satisfy given initial conditions. The locus of these surfaces of discontinuity which evolve with the flow must also be found as part of the solution. This would require ticklish recursive procedures such as those employed in the classical method of characteristics.

Von Neumann and Richtmyer (1950) proposed an approach, called the method of artificial viscosity, which eliminates the need to apply such boundary conditions explicitly. Using this method, the solutions can be found as accurately as desired by a suitable choice of mesh sizes and other parameters occurring in the problem. The shock discontinuities are treated correctly and automatically whenever and wherever they may arise.

The purpose of this additional term is to introduce a dissipative mechanism in the shock layer such as viscosity (an artificial term, not 'real' viscosity, with the dimension of pressure) which smears the shock so that the mathematical surfaces of discontinuity are replaced by thin layers in which temperature, density, pressure and velocity vary rapidly but continuously. More specifically, the artificial viscosity term is chosen to meet the following specific requirements: (1) equations of motion with the introduction of the extra term possess solutions without discontinuities; (2) the thickness of shock layers must be of the order of the mesh size, Δx, chosen for the numerical scheme, independent of the strength of the shock and of the conditions prevailing ahead of the shock; (3) the effect of the artificial term is negligible outside the shock layer(s); and (4) the Rankine-Hugoniot conditions must hold when all other dimensions characterising the flow are large in comparison with the shock thickness. Von Neumann and Richtmyer (1950) chose the expression

$$q = -\frac{(c\Delta x)^2}{V}\frac{\partial U}{\partial x}\left|\frac{\partial U}{\partial x}\right|, \qquad (8.2.1)$$

for the one-dimensional case and showed by seeking a travelling wave solution of the governing system of equations in plane symmetry, including q, that this term meets the requirements (1)–(4) set out above. Here U is the fluid velocity, V is its specific volume, and c is a (dimensionless) constant of order unity. It was found that the travelling wave solution of the plane gasdynamic system is a half sine wave, which may be pieced together with two other constant solutions. The half sine is of order Δx provided c in (8.2.1) is a constant close to unity. q is found to be negligible in comparison with p everywhere because of the factor $(\Delta x)^2$, except in the shock layer where the derivative $\partial U/\partial x$ is very large. The finite difference scheme used to discretise the governing system of PDEs is detailed in section 8.3, where we discuss the application of the present method to the study of spherical explosion in air (Brode (1957)). As we note in that section, the dissipative term introduces its own stability requirement which is more stringent than the familiar Courant, Friedrich and Lewy condition; however, this condition is not too severe if the amount of dissipation introduced is enough to produce a shock thickness comparable with the spatial mesh size. One may refer to the book of Richtmyer and Morton (1967) for a discussion of the stability conditions in the present context. In spite of the spherical symmetry of the problem studied by Brode (1955), the q term in (8.2.1) chosen for the planar symmetry was found to serve quite adequately.

Sachdev and Prasad (1966), following the work of Von Neumann and Richtmyer (1950), investigated the effect of artificial heat conduction term in lieu of the viscosity term. The qualitative features of the solution were found to be essentially the same for both the dissipative mechanisms.

It is of some interest to study the effect of dissipation on the level of the difference equations rather than that of differential equations. We discuss this matter in the context of the Lax-Wendroff scheme. Here again an additional artificial viscosity term is added to the hyperbolic equation, simulating a diffusive term proportional to u_{xx}. This term must also meet the usual requirements, referred to earlier. It must have a coefficient that vanishes as the mesh sizes tend to zero so that it remains consistent with the hyperbolic equation. Besides, this coefficient must vanish sufficiently quickly so that the order of accuracy of the high order methods on smooth solutions is unaffected. This term must be large near the discontinuities and small in the smooth regions. To illustrate these ideas we look at the Lax-Wendroff difference form of the linear equation $u_t + au_x = 0$ (here a is a constant) with the addition of an artificial difference form of viscosity:

$$
\begin{aligned}
U_j^{n+1} &= U_j^n - \frac{\nu}{2}(U_{j+1}^n - U_{j-1}^n) + \frac{1}{2}\nu^2(U_{j+1}^n - 2U_j^n + U_{j-1}^n) \\
&\quad + kQ(U_{j+1}^n - 2U_j^n + U_{j-1}^n),
\end{aligned} \tag{8.2.2}
$$

where $\nu = ak/h$ is the so-called Courant number and Q is the coefficient of artificial viscosity. It is known (see LeVeque (1992)) that the Lax-Wendroff scheme itself is a third order accurate approximation to the solution of the PDE

$$
u_t + au_x = \frac{h^2}{6}a(\nu^2 - 1)u_{xxx}. \tag{8.2.3}
$$

The modified Lax-Wendroff scheme (8.2.2) with artificial viscosity produces a third order approximation to the solution of the PDE

$$
u_t + au_x = \frac{h^2}{6}a(\nu^2 - 1)u_{xxx} + h^2Qu_{xx}. \tag{8.2.4}
$$

The dispersive term u_{xxx}, which causes oscillations in the Lax-Wendroff scheme, must now compete with the dissipative term involving u_{xx} and, for Q sufficiently large, should yield nonoscillatory solution. This, unfortunately, is not true since, with constant Q, it is still a linear method and is second-order accurate. Q must be made to depend on the data U^n. The method then becomes nonlinear in the manner of the method of artificial viscosity due to Von Neumann and Richtmyer (1950). As in the latter approach, it is hard to determine an appropriate form for Q that introduces just enough dissipation to preserve monotonicity (nonoscillatory character) without causing unnecessary smearing.

The local truncation error $L(x, t)$ for the scheme (8.2.2) can be written as

$$
\begin{aligned}
u(x, t) &= u(x, t + k) + \frac{k}{2h}\{u(x + h, t) - u(x - h, t)\} \\
&\quad - \frac{k^2}{2h^2}\{u(x + h, t) - 2u(x, t) + u(x - h, t)\} \\
&\quad - Qk\{u(x + h, t) - 2u(x, t) + u(x - h, t)\} \\
&= u(x, t + k) + \frac{k}{2h}\{u(x + h, t) - u(x - h, t)\}
\end{aligned}
$$

$$-\frac{k^2}{2h^2}\{u(x+h,t)-2u(x,t)+u(x-h,t)\}$$
$$-Qh^2 u_{xx}(x,t)+O(h^4)$$
$$=\;O(k^2)\quad\text{as}\quad k\to 0,\tag{8.2.5}$$

since $h^2=O(k^2)$ as $k\to 0$.

The Lax-Wendroff method remains second order accurate for any choice of $Q=$ constant.

Before turning to more recent high resolution methods wherein the nonoscillatory requirement can be imposed more directly, we briefly discuss the Lax-Friedrichs scheme which was used by Payne (1957) in the context of converging shock waves (see section 8.4). If we again consider the equation $u_t+au_x=0$ and replace u_t by a forward (in time) approximation and u_x by a spatially centered approximation, we obtain

$$\frac{U_j^{n+1}-U_j^n}{k}+a\frac{(U_{j+1}^n-U_{j-1}^n)}{2h}=0,\tag{8.2.6}$$

or

$$U_j^{n+1}=U_j^n-\frac{k}{2h}a(U_{j+1}^n-U_{j-1}^n)=0.\tag{8.2.7}$$

This scheme, though natural and simple, suffers from severe stability problems and is of little practical use. If, in (8.2.7), we replace U_j^n by $\frac{1}{2}(U_{j-1}^n+U_{j+1}^n)$, it is found that this scheme is stable provided k/h is sufficiently small. This changed scheme may be directly incorporated in $u_t+au_x=0$. Thus, we have

$$\frac{1}{k}\left[U_j^{n+1}-\frac{1}{2}(U_{j-1}^n+U_{j+1}^n)\right]+\frac{1}{2h}a[U_{j+1}^n-U_{j-1}^n]=0.\tag{8.2.8}$$

The local truncation error for this scheme is

$$L_k(x,t)\;=\;\frac{1}{h}\left[u(x,t+k)-\frac{1}{2}(u(x-h,t)+u(x+h,t))\right]$$
$$+\frac{1}{2h}a\left[u(x+h,t)-u(x-h,t)\right].\tag{8.2.9}$$

Assuming the solution to be smooth, we may expand the right hand side of (8.2.9) in a Taylor series about (x,t) and obtain

$$L_k(x,t)\;=\;\frac{1}{k}\left[\left(u+ku_t+\frac{1}{2}k^2u_{tt}+\cdots\cdots\right)-\left(u+\frac{1}{2}h^2u_{xx}+\cdots\cdots\right)\right]$$
$$+\frac{1}{2h}a\left[2hu_x+\frac{1}{3}h^3u_{xxx}+\cdots\cdots\right].$$
$$=\;u_t+au_x+\frac{1}{2}\left(ku_{tt}-\frac{h^2}{k}u_{xx}\right)+O(h^2),\tag{8.2.10}$$

and since $u(x,t)$ is the exact solution of $u_t + au_x = 0$, we may write (8.2.10) as

$$
\begin{aligned}
L_k(x,t) &= \frac{1}{2}k \left(a^2 - \frac{h^2}{k^2} \right) u_{xx}(x,t) + O(k^2) \\
&= O(k) \quad \text{as} \quad k \to 0,
\end{aligned}
\tag{8.2.11}
$$

provided we assume that $k/h = $ constant as the mesh size is refined. This explains why the truncation is defined as $L_k(x,t)$ rather than $L_{k,h}(x,t)$. By a careful analysis of the remainder in the Taylor's theorem, assuming uniform boundedness of the appropriate derivatives of $u(x,t)$, one may prove a sharp bound of the form

$$
|L_k(x,t)| \leq Ck \quad \text{for all} \quad k < k_0,
\tag{8.2.12}
$$

where C depends only on the initial data u_0. The Lax-Friedrichs method, summarised above, is first order accurate since the local error $L_k(x,t)$ depends linearly on k.

As we mentioned earlier, Lax-Friedrichs scheme is only first order accurate on smooth data and gives unacceptably smeared shock profiles. To get over these deficiencies, the so called "high resolution" methods were developed. These methods are second order accurate in the smooth regions and yield much sharper discontinuities. We shall briefly describe one of these, namely, the Godunov (1959) scheme. This scheme uses information via characteristics within the framework of a conservation method. The basic idea is to solve the Riemann problem forward in time for piecewise constant initial data. Since these are exact solutions of the conservation laws, they lead to 'conservative numerical methods' (see LeVeque (1992)). Choose $\tilde{u}^n(x,t_n)$, the solution at t_n, as the initial data for the conservation law

$$
u_t + f(u)u_x = 0.
\tag{8.2.13}
$$

This problem must be solved to obtain $\tilde{u}^n(x,t_n)$ for $t_n \leq t \leq t_{n+1}$. Equation (8.2.13) is solved exactly for a short time by choosing the initial data $\tilde{u}^n(x,t)$ in a piecewise constant manner. The solution is obtained simply by putting together these Riemann solutions; it holds till waves from the neighbouring Riemann problems begin to interact. For details of constructing these solutions we refer the reader to LeVeque (1992). After the exact solution is obtained over the interval $[t_n, t_{n+1}]$, the approximate solution U_j^{n+1} at t_{n+1} is defined by averaging it from $x_{j-1/2}$ to $x_{j+1/2}$:

$$
U_j^{n+1} = \frac{1}{h} \int_{x_{j-1/2}}^{x_{j+1/2}} \tilde{u}^n(x,t_{n+1})dx.
\tag{8.2.14}
$$

These values are then used to define new piecewise constant data $\tilde{u}^{n+1}(x,t_{n+1})$ and the process of solution is repeated. Making use of the fact that the cell average (8.2.14) may be easily computed by using the integral form of the conservation law, one may write (8.2.14) as

$$U_j^{n+1} = U_j^n - \frac{k}{h}[F(U_j^n, U_{j+1}^n) - F(U_{j-1}^n, U_j^n)], \qquad (8.2.15)$$

where the numerical flux function F is given by

$$F(U_j^n, U_{j+1}^n) = \frac{1}{k} \int_{t_n}^{t_{n+1}} f(\tilde{u}^n(x_{j+1/2}, t))dt, \qquad (8.2.16)$$

showing further that the Godunov scheme may be written in 'conservation form'. We may observe that the value $\tilde{u}^n$ along the line $x = x_{j+1/2}$ depends only on points U_j^n and U_{j+1}^n of this Riemann problem. Denoting this value by $u^*(U_j^n, U_{j+1}^n)$, we may write (8.2.16) as

$$F(U_j^n, U_{j+1}^n) = f(u^*(U_j^n, U_{j+1}^n)), \qquad (8.2.17)$$

and hence the Godunov scheme (8.2.15) becomes

$$U_j^{n+1} = U_j^n - \frac{k}{h}[f(u^*(U_j^n, U_{j+1}^n)) - f(u^*(U_{j-1}^n, U_j^n))]. \qquad (8.2.18)$$

The stability condition for the present scheme may be appropriately derived (see LeVeque (1992)). We may observe that the Godunov scheme under the Courant-Friedrich-Lewy condition is total variation diminishing (see section 8.5).

8.3 Blast Wave Computations via Artificial Viscosity

One of the earliest attempts to numerically simulate a spherical blast wave is due to Brode (1955) who considered two models for initial conditions: (i) strong shock, point source solution due to Taylor (1950), Sedov (1946) and Von Neumann (1941) (see sections 3.1–3.4); and (ii) hot high pressure isothermal spheres. We shall discuss the solution subject to initial conditions (i) in some detail and summarize the results for the latter. Brode (1955) set before himself the following practical conditions for accomplishing the numerical solution: the difference scheme must be stable, it must yield reasonably accurate results, it must conserve numerical significance (must tend to the solution of the original system of PDEs as mesh sizes tend to zero), and must be fast enough to give the desired solution with a sensible expenditure of machine time.

To meet the above goals, Brode (1955) employed the method of artificial viscosity proposed earlier by Von Neumann and Richtmyer (1950). This artificial viscosity term has the dimension of pressure (see (8.3.4), (8.3.5), and

(8.3.9)). The introduction of this term must meet the following conditions: (1) the governing equations must possess solutions without (shock) discontinuities; (2) the thickness of the shock layer must be everywhere of the same order as the interval length chosen for the numerical solution, independent of the strength of the shock and the conditions of the material into which it propagates; (3) the effect of the artificial viscosity term must be negligible outside the shock layer(s), and (4) the Rankine-Hugoniot conditions must hold when all other dimensions characterising the flow are large compared to the shock thickness. These conditions were met by the term first suggested by Von Neumann and Richtmyer (1950) for plane symmetry. The term chosen by Brode (1955) (see (8.3.9)) satisfies all the requirements laid down above except that there is no convenient steady state solution in the spherically symmetric system by which one may show that the Rankine-Hugoniot conditions are satisfied. Thus, the form (8.3.9) chosen by Brode (1955) is asymptotically the same as the verified form of artificial viscosity in plane symmetry. We have already discussed in section 3.1 the similarity solution of the gasdynamic equations including an artificial viscosity term; this solution describes the strong blast wave, has the correct nonviscous behaviour in the regions away from the shock, and reasonable transition in the shock layer (Latter (1955)).

We may observe that Brode (1955) used Von Neumann's (1941) point explosion solution in Lagrangian co-ordinates for the purpose of initial conditions. This solution, detailed in section 3.3, is explicit in terms of the parameter θ, the ratio of kinetic energy to internal energy, and is therefore convenient to use. Brode (1955) also employed the Lagrangian form of the basic equations of motion.

We denote by p, ρ, u, and c the pressure, density, particle velocity and speed of sound, respectively. The corresponding undisturbed quantities will be denoted by the subscript '0'. The overpressure will be denoted by $\Delta p = p - p_0$, where p_0 is the atmospheric pressure. The pressure will, in general, be measured in atmospheres, that is, in units of p_0. Thus, the excess pressure will be written as $\Delta p = p - 1$. The Lagrangian co-ordinate and time are r_0 and t, respectively, while the Eulerian co-ordinate is denoted by $r = r(r_0, t)$.

To render the variables nondimensional, we choose the typical length ϵ arising from the total energy, E_{tot}, of the blast wave and the ambient pressure p_0:

$$\epsilon^3 = \frac{E_{tot}}{p_0} = \frac{4\pi}{p_0} \int_0^R \rho \left(E_{int} + \frac{u^2}{2} \right) r^2 dr - \frac{4\pi R^3}{3(\gamma - 1)}, \qquad (8.3.1)$$

where E_{int} is the specific internal energy. The term $4\pi R^3/3(\gamma - 1)$ in (8.3.1) arises from the pre-shock internal energy of the gas, engulfed by the shock. R is the shock radius. The Eulerian co-ordinate r, Lagrangian co-ordinate

r_0, and the time t are rendered nondimensional with the help of ϵ and c_0 where c_0 is the sound speed in the undisturbed medium:

$$\lambda = r/\epsilon, \quad \lambda_0 = (r_0/\epsilon), \quad \tau = tc_0/\epsilon. \tag{8.3.2}$$

The nondimensional form of equations of motion in Lagrangian co-ordinates is

$$\frac{\partial \lambda}{\partial x} = \frac{1}{\rho \lambda^2} \quad \text{or} \quad \frac{\partial \rho}{\partial \tau} = -\rho \left(\frac{2u}{\lambda} + \frac{\partial u/\partial x}{\partial \lambda/\partial x} \right), \tag{8.3.3}$$

$$\frac{\partial u}{\partial t} = -\frac{\lambda^2}{\gamma} \frac{\partial}{\partial x}(p+q), \tag{8.3.4}$$

$$\frac{\partial p}{\partial \tau} = \frac{1}{\rho} \frac{\partial \rho}{\partial \tau}[\gamma p + (\gamma - 1)q], \tag{8.3.5}$$

$$u = \frac{\partial \lambda}{\partial \tau}. \tag{8.3.6}$$

Here the Lagrangian space co-ordinate has been redefined as

$$x = \frac{1}{3}(r_0/\epsilon)^3. \tag{8.3.7}$$

The equation for internal energy for an ideal gas is assumed in the form

$$E = \frac{p}{\rho(\gamma - 1)} \frac{\rho_0}{p_0}. \tag{8.3.8}$$

The artificial viscosity term q in (8.3.4)–(8.3.5) acts like a pressure term and was chosen by Brode (1955) as

$$q = \frac{9\gamma(\gamma + 1)}{4} \left(\frac{M}{3\pi} \right)^2 \rho(\Delta x)^2 \left(\frac{\partial u}{\partial x} \right) \left(\frac{\partial u}{\partial x} - \left| \frac{\partial u}{\partial x} \right| \right), \tag{8.3.9}$$

where Δx is the grid size and M is the number of grid zones in the shock front. We have already discussed the nature of this term earlier in this section. Latter (1955) verified that, with this form of q, the nonlinear ODEs that result from the reduction via similarity transformations give shock thickness for the spherically symmetric case quite close to that for the plane symmetry (see section 3.1). It is clear from the form (8.3.9) that q is zero in the expansion region where $\partial u/\partial x > 0$ and is nonzero only in the compression phase of the shock. In Lagrangian co-ordinates it has the additional advantage that it eliminates a spurious contribution near the region where the positive velocity gradient is large.

Brode (1955) wrote the following difference form for the system (8.3.3)–(8.3.6):

$$u_l^{n+1/2} = u_l^{n-1/2} - \frac{\Delta \tau (\lambda_l^n)^2}{(\Delta x)_l \gamma} \left[p_{l+1/2}^n - p_{l-1/2}^n + q_{l+1/2}^{n-1/2} - q_{l-1/2}^{n-1/2} \right],$$

$$\tag{8.3.10}$$

$$\lambda_l^{n+1} = \lambda_l^n + u_l^{n+1/2}\Delta\tau, \tag{8.3.11}$$

$$\rho_{l-1/2}^{n+1} = \rho_{l-1/2}^n \left(\frac{1-W}{1+W}\right), \tag{8.3.12}$$

where

$$W = \Delta\tau \left(\frac{2(u_l^{n+1/2} + u_{l-1}^{n+1/2})}{\lambda_l^{n+1} + \lambda_l^n + \lambda_{l-1}^{n+1} + \lambda_{l-1}^n} + \frac{u_l^{n+1/2} - u_{l-1}^{n+1/2}}{\lambda_l^{n+1} + \lambda_l^n - \lambda_{l-1}^{n+1} - \lambda_{l-1}^n}\right), \tag{8.3.13}$$

$$q_{l-1/2}^{n+1/2} = 9\frac{\gamma(\gamma+1)}{2}\left(\frac{M}{3\pi}\right)^3$$
$$\times \rho_{l-1/2}^{n+1}\left[u_{l-1}^{n+1/2} - u_l^{n+1/2}\right]^2 \text{ for } u_{l-1}^{n+1/2} > u_l^{n+1/2}, \tag{8.3.14}$$

$$q_{l-1/2}^{n+1/2} = 0 \text{ for } u_{l-1}^{n+1/2} \leq u_l^{n+1/2},$$

$$p_{l-1/2}^{n+1} = \frac{\left[\frac{\gamma+1}{\gamma-1}\rho_{l-1/2}^{n+1} - \rho_{l-1/2}^n\right]p_{l-1/2}^n + 2\left(\rho_{l-1/2}^{n+1} - \rho_{l-1/2}^n\right)q_{l-1/2}^{n+1/2}}{\frac{\gamma+1}{\gamma-1}\rho_{l-1/2}^n - \rho_{l-1/2}^{n+1}}. \tag{8.3.15}$$

Two stability conditions are required by the above difference scheme. One is the usual Courant-Friedrich-Lewy condition, namely,

$$\Delta\tau \leq \Delta x/\lambda^2(p/\rho)_{max}^{1/2}, \tag{8.3.16}$$

and the other arises from the parabolic nature of the equation in the shock layer,

$$\Delta\tau \leq \frac{\gamma}{4}(\Delta x)^2 \left[\frac{1}{\lambda^2 q}\left|\frac{\partial u}{\partial x}\right|\right]_{min}. \tag{8.3.17}$$

The space mesh sizes were chosen to be unequal—smaller in the shock layer and larger outside. This resulted in a sharp shock at very little cost in computing time. The time mesh size was doubled as soon as the stability conditions would allow it. The artificial viscosity method was thus found to be quite general in nature, easy to apply, and (asymptotically) reproduced the Rankine-Hugoniot conditions. Brode (1955) attempted some other finite difference schemes such as that due to Du Fort and Frankel (1953) for diffusion type of equations but found them less fruitful.

To check the veracity of results, Brode (1955) ran the computer program with different zone spacing, different viscosity terms, different time increments and, occasionally, different forms of differencing. This helped to ensure that the results were reliable. The conservation of total energy of the blast did not prove to be a sensitive test of the accuracy of the computations.

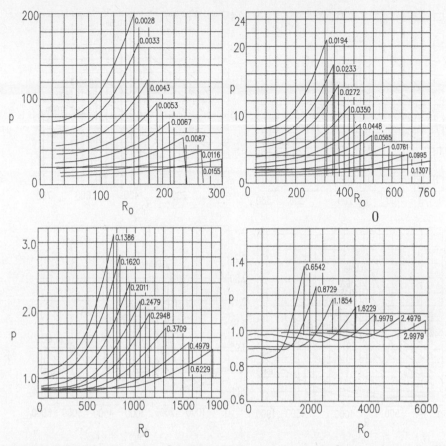

Figure 8.1 Pressure in units of undisturbed pressure versus Lagrangian co-ordinate R_0 for point source solution at different times (Brode, 1955).

Figures 8.1–8.3 show pressure, particle velocity and density (in nondimensional form) versus Lagrangian co-ordinate, $R_0 = r_0 \left\{ (E_{tot}/p_0)^{1/3} \right\}^{-1}$, at different times; the initial conditions were chosen from the point source solution with $\gamma = 1.4$. The flow immediately behind the shock according to this solution is given by the fitted curve $u_s = 0.30(\lambda_s)^{-3/2}$, $\rho_s = (6p + 1)/(p + 6)$ for this value of γ; here p is the pressure ratio at the shock. It is curious that the strong shock behaviour is predicted to quite low pressures. For example, the ratio of central pressure to shock pressure remains 37 percent down to 20 atmospheres and decreases slowly to 33 percent by 3 atmospheres. Beyond this value a negative phase ensues, the pressure falling as low as 0.8 atmosphere near the center.

The variation of particle velocity and density with the Lagrangian co-ordinate follows the strong shock form until the shock overpressure is as low as 3 atmospheres. As the shock wave decays to become relatively weak, the particle velocity profile transforms gradually from its almost linear form in the early stages to much like the overpressure at large distances. The density at the center remains zero for all times since there is no heat conduction or

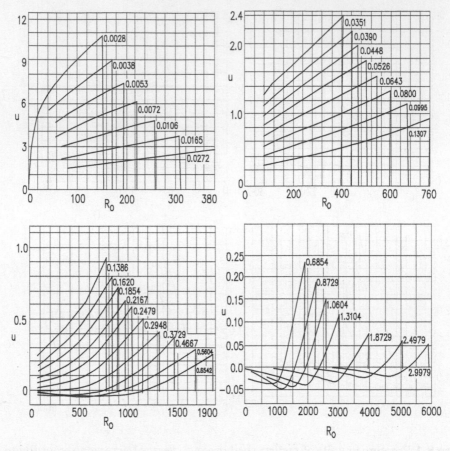

Figure 8.2 Particle velocity in units of undisturbed speed of sound versus Lagrangian co-ordinate R_0 at different times (Brode, 1955).

radiation term in this model to remove the temperature singularity there.

In these figures, the first two sets correspond to strong shocks while the latter ones refer to finite pressure ratio across the shock. Here, $(6p + 1)/(p + 6) = 5.83$ for $p = 200$. This is different from the value 6 for $p \to \infty$. Curiously, in this case, the temperature at the shock is raised to a higher value than that for the infinitely strong shock: "a finite shock is hotter than would be predicted by the strong shock theory".

Figure 8.4 shows overpressure, particle velocity, density and compression in units of their peak values at the shock, versus the Eulerian co-ordinate at different times. The strong shock form prevails in the first two figures at the early times $t = 0.00147, 0.0166$. At later times the characteristic positive phase is followed by a larger, weaker negative phase, and an eventual return to near pre-shock values at the origin. Brode (1955) gave approximate expressions for the variation of overpressure and dynamic pressure at the shock with time, which seem to agree with the numerical results within 10 percent. These expressions were obtained by suitably altering the analytic

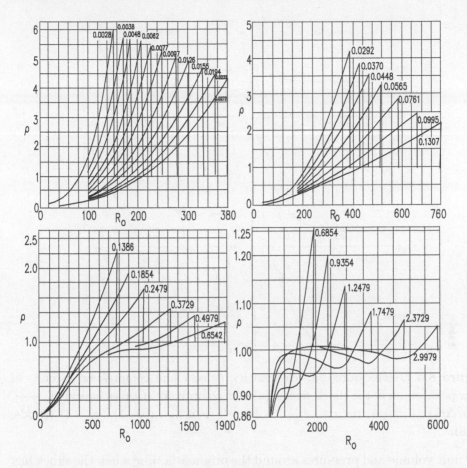

Figure 8.3 Density in units of undisturbed density versus Lagrangian co-ordinate R_0 for point source solution at different times (Brode, 1955).

solution for the point source model in the light of the numerical solution.

For the initial isothermal spheres of gas at rest, the main result is that the flow, starting with these initial conditions, will assume the general shape and value of the point source solution (to within 10 percent) after the shock wave has engulfed a mass of air 10 times the initial mass of the sphere. At the earlier times (before the inward travelling rarefaction reaches the center), the shock strength is less than that predicted by the point source solution.

Comparing the solutions arising from different initial conditions, Brode (1955) observed that a point source should leave a higher temperature and therefore a longer percentage of energy near the origin. This energy, no longer available to the shock wave, would therefore lead to a much faster decay of the shock in comparison to that for the initially isothermal sphere. Actually, no appreciable difference in the low end of the shock overpressure radius relation is observed between the point source and isothermal sphere solutions. For the latter, there is a multiple shocking of the inner regions. The result is that there are nearly identical distributions of residual energies

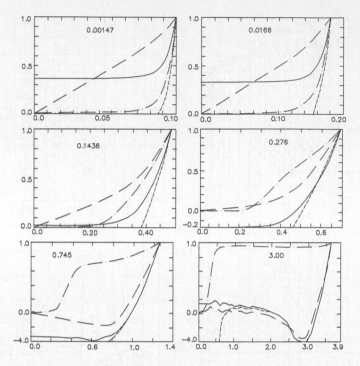

Figure 8.4 Overpressure, particle velocity, density and compression in units of their peak values at the shock versus Eulerian co-ordinate at times indicated. ———, $(\Delta p/\Delta p_s)$; – – – –, (u/u_s); –·–·–·–·, (ρ/ρ_s); ·····, $(\rho - 1)/(\rho_s - 1)$ (Brode, 1955).

per unit volume and pressures around the origin at a time when the shock has progressed to 6 times the initial radius. Moreover, the average temperature (or density) of the gas initially inside the isothermal sphere approaches, to within 10 percent, the average temperature for a corresponding mass around the point source.

This is in spite of the fact that the temperature at the center for the latter model is infinitely large. We have discussed in some detail the analytic solution describing the sudden expansion of a high pressure gas into the ambient atmosphere in sections 7.2 and 7.3; flows with or without secondary shocks were analytically examined.

Brode (1959) followed up his earlier study by a more realistic model for a blast wave from a spherical charge. The initial conditions for this model were approximately those of a centered detonation of a bare sphere of TNT of loading density 1.5 g/cm^3 as specified by the detonation wave descriptions of Taylor (1950). The equation of state of this TNT was realistically modeled after that of Jones and Miller (1948) while the equation of state of air was obtained by a fit to computed data of several previous authors. The problem was solved numerically, using the technique developed earlier, Brode (1955). All the flow variables were depicted as functions of time and distance. The main contribution of this numerical study is the detection of a secondary shock which originates as an imploding shock following the inward rarefaction into the high pressure explosive gases (see section 7.3). A series of subsequent minor shocks were also seen to appear between the origin and the contact surface. The influence of the assumptions about equations of state of the high explosive and of air was not found to be as great as might be predicted on purely thermodynamic grounds. The pressures and velocities are found to be less sensitive to variations of equations of state than are temperatures and densities. This is particularly true in the strong shock high temperature regions; the secondary shock is significantly influenced by the equation of state for the high explosive.

The work of Brode (1955) was extended by Plooster (1970) to cylindrical symmetry. It was envisioned that a cylindrical pressure wave results from instantaneous energy release along a line in a quiescent atmosphere. The applications of this model include exploding wires, long explosive charges, electric sparks, and supersonic aircraft or projectiles. The important natural phenomenon that this model describes is the lightning discharge. The computations reported by Plooster (1970) cover a wide range of initial conditions and use both the ideal gas equation of state and a more realistic equation of state for air and, as in the earlier work of Brode (1955), extend well into the weak shock region. The artificial viscosity term was used in the manner of Brode (1955). Five different sets of initial conditions were chosen:

1. Line source, ideal gas. This model simply extends the solution of Brode (1955) for the point source initial conditions to cylindrical symmetry. The initial conditions were chosen from the analytic solution of Lin (1954) for this geometry.

2. Isothermal cylinder, constant density, ideal gas. Here the energy is supplied to a cylinder of finite radius whose density is equal to the ambient air density; this represents a very rapid heating of a column of air in a time so short that it cannot expand appreciably during the period of heating. This is analogous to Brode's (1955) initial isothermal sphere.

3. Isothermal cylinder, constant density, real gas equation of state.

4. Isothermal cylinder, low density, ideal gas.

5. Isothermal cylinder, high density, ideal gas. This initial condition simulates very roughly the flow field resulting from the detonation of a line charge of high explosive.

The results were presented in different formats: shock wave overpressures versus radius and gas pressure, density, and flow velocity versus time at different radii. All variables were expressed in dimensionless form. It was shown that experimental measurements of shock strength from detonation of long high explosive charges were in good agreeement with the numerical solutions.

An investigation closely related to that of Brode (1955) was concurrently published by Goldstine and Von Neumann (1955). Here, only the point explosion model was studied. The governing system of gasdynamic equations was again expressed in Lagrangian co-ordinates; the initial conditions were obtained by solving numerically the system of ODEs resulting from self-similar form of the solution rather than the analytic form of Von Neumann (1941). The major departure from Brode's (1955) work is in the choice of the finite difference scheme and noninclusion of any artificial viscosity term. The shocks that are fitted are sharp, requiring an iterative procedure to satisfy the Rankine-Hugoniot conditions exactly. Since the computational scheme is rather involved, we skip the details and refer the reader to the original work of Goldstine and Von Neumann (1955). The results of their computations were depicted in a large number of graphs. As in the work of Brode (1955), the shock overpressure versus shock radius was shown to follow the law $p - 1 = AR^{-n}$, where n is a slowly varying function of R and A is constant (see section 8.4). The pressure in the region behind the shock was found both as a function of radial distance for fixed times and as a function of time for fixed distances.

8.4   Converging Cylindrical Shock Waves

We have discussed in sections 6.2 and 6.3 the analytic character of the converging shock wave solutions, referring to them as self-similar solutions of the second kind. Their importance in the study of shock waves compares that of Taylor-Sedov solution for the point explosion. The first numerical investigation of these solutions was carried out by Payne (1957) and merits a detailed discussion (cf. Brode (1955) for the explosion problem). The numerical approach adopted by Payne (1957) is the so-called Lax scheme (1954) which requires that the governing equations be expressible in a conservation form. It also has an artificial viscosity term built into the scheme of differencing of the conservation laws (see Richtmyer and Morton (1967)). Since for the converging shock waves it is not possible to write the system of gasdynamic equations in a conservation form, Payne (1957) had to modify Lax's scheme to appropriately difference the pressure term in the momentum equation (which is in a nonconservative form). The second difficulty arises from the singular nature of equations at the axis of the cylindrical flow. Here, again, special numerical treatment of the governing system of equations in this neighbourhood had to be devised. Sod (1977) could eliminate both these difficulties by a judicious combination of Glimm's method and operator splitting as we shall discuss in some detail later in this section.

The converging shock originates due to the sudden rupture of a cylindrical diaphragm, separating two uniform regions of gas at rest with a higher pressure in the outside region. We denote the ratios of pressures and densities on two sides of the diaphragm by p^* and $\rho^*(p^* > 1)$. The situation here is again analogous to the shock tube problem (see sections 7.2 and 7.3). Here, if $p^* = \rho^*$, that is, if the uniform temperatures on two sides of the diaphragm are the same, then a shock travels into the low pressure region, followed by a contact surface and an expansion wave. The latter moves into the high pressure region. By a suitable choice of p^* and ρ^* it is possible to obtain a flow which has a shock wave and an expansion wave, but no contact discontinuity. This is what was arranged by Payne (1957) for the initial conditions in cylindrical flow; for this purpose he assumed that the gas outside the cylindrical diaphragm was initially at a higher temperature than that inside. It was found that the contact surface does not affect the converging shock; the presence of the contact surface, however, leads to numerical inaccuracy in the flow. Any inaccuracy in the expansion wave also affects the region of the converging shock wave.

If r_0, ρ_0, p_0, and a_0 are the initial radius of the diaphragm, density, pressure and sound speed in the undisturbed medium, respectively, we may normalise the corresponding variables by these quantities. The time t may be scaled by $t_0 = r_0/a_0$. The total energy of the medium per unit volume may be defined as

$$\widehat{E} = \frac{p}{\gamma - 1} + \frac{1}{2}\rho u^2, \tag{8.4.1}$$

and may be rendered nondimensional by p_0. We may thus write the equations of motion in normalised variables as

$$(r\rho)_t + (r\rho u)_r = 0, \tag{8.4.2}$$

$$(r\rho u)_t + (r\rho u^2)_r + \gamma^{-1/2}rp_r = 0, \tag{8.4.3}$$

$$(rE)_t + (rEu + rpu)_r = 0. \tag{8.4.4}$$

Payne (1957) found it convenient to introduce the variables

$$a = r\rho, \quad b = r\rho u, \quad c = rE \tag{8.4.5}$$

in (8.4.2)–(8.4.4) and rewrite this system as

$$a_t + b_r = 0, \tag{8.4.6}$$

$$b_t + (bu)_r + \gamma^{-1/2}rp_r = 0, \tag{8.4.7}$$

$$c_t + (cu + rpu)_r = 0. \tag{8.4.8}$$

The original (nondimensional) variables are therefore given by

$$\rho = \frac{a}{r}, \quad u = \frac{a}{b}, \quad p = \frac{(\gamma - 1)c}{r} - \frac{\gamma(\gamma - 1)}{2}bu. \tag{8.4.9}$$

It may be observed that equation (8.4.7) is not in a conservation form and the term $\gamma^{-1/2} r p_r$ must be handled separately in the context of Lax's scheme.

Consider the mesh points as $r_k = k\Delta r$, $t = t_n$, where k and n are nonnegative integers and Δr is such that $K = (\Delta r)^{-1}$ is an integer. We write $\Delta t_n = t_{n+1} - t_n$. Denoting any function $\phi(r, t)$ at $r_k = k\Delta r$, $t = t_n$ by $\phi_{k,n}$, Lax's scheme replaces the derivatives as follows:

$$\frac{\partial \phi}{\partial t} = \frac{\phi_{k,n+1} - \frac{1}{2}(\phi_{k+1,n} + \phi_{k-1,n})}{\Delta t_n}, \tag{8.4.10}$$

$$\frac{\partial \phi}{\partial r} = \frac{\phi_{k+1,n} - \phi_{k-1,n}}{2\Delta r}. \tag{8.4.11}$$

We may thus replace the derivatives in (8.4.6)–(8.4.8) via (8.4.10) and (8.4.11) and solve for a, b, and c at time t_{n+1} in terms of the quantities at t_n:

$$a_{k,n+1} = \frac{1}{2}(a_{k-1,n} + a_{k+1,n}) + \frac{\Delta t_n}{2\Delta r}(b_{k-1,n} - b_{k+1,n}), \tag{8.4.12}$$

$$b_{k,n+1} = \frac{1}{2}(b_{k-1,n} + b_{k+1,n}) + \frac{\Delta t_n}{2\Delta r}\{b_{k-1,n}u_{k-1,n} - b_{k+1,n}u_{k+1,n}$$
$$+ \gamma^{-1/2} r_k(p_{k-1,n} - p_{k+1,n})\}, \tag{8.4.13}$$

$$c_{k,n+1} = \frac{1}{2}(c_{k-1,n} + c_{k+1,n}) + \frac{\Delta t_n}{2\Delta r}\{c_{k-1,n}u_{k-1,n} - c_{k+1,n}u_{k+1,n}$$
$$+ r_{k-1}p_{k-1,n}u_{k-1,n} - r_{k+1}p_{k+1,n}u_{k+1,n}\}. \tag{8.4.14}$$

The physical quantities u, ρ, and p may now be obtained with the help of (8.4.9). It is clear that equations (8.4.12)–(8.4.14) do not apply at $k = 0$. It is also observed that the values of the variables at points on the two staggered lattices, $k + n$ even and $k + n$ odd, are independent of each other.

Payne (1957) replaced the pressure term in (8.4.7) by

$$\frac{\Delta t_n}{2\Delta r}\gamma^{-1/2}\left(r_{k-1}p_{k-1,n} - r_{k+1}p_{k+1,n} + \int_{(k-1)\Delta r}^{(k+1)\Delta r} p\, dr\right), \tag{8.4.15}$$

where the integral in turn is replaced by

$$(p_{k-1,n} + p_{k+1,n})\Delta r. \tag{8.4.16}$$

This choice still retains the property of the two staggered lattices being independent when $k + n$ is even or when it is odd.

At the axis of the cylindrical flow, we have $u = 0$. Therefore, $a = r\rho = 0$, $b = r\rho u = 0$, $c = rE = 0$. The difference scheme (8.4.12)–(8.4.14) is not applicable here. Payne (1957) derived an alternative form by applying basic conservation laws of mass and energy in a cylinder of radius Δr. Writing the conservation of mass equation in this mesh, we have

$$\frac{\partial}{\partial t}\left\{\int_{r=0}^{r=\Delta r} \rho(r, t)d(r^2)\right\} + 2\Delta r\rho(\Delta r, t)u(\Delta r, t) = 0. \tag{8.4.17}$$

We may now approximate the integral in (8.4.17) by $\frac{1}{2}\{\rho(0,t)+\rho(\Delta r,t)\}\Delta r^2$, take the derivative of this expression with respect to t and use the difference scheme (8.4.10) for $\partial\rho/\partial t$ etc. We then repeat this process for the interval $2\Delta r$ and obtain

$$\rho_{0,n+1} = \rho_{1,n} + \frac{\Delta t_n}{\Delta r}\left(\frac{1}{4}\rho_{3,n}u_{3,n} - \frac{11}{4}\rho_{1,n}u_{1,n}\right). \tag{8.4.18}$$

By a similar argument applied to the conservation of energy one may obtain

$$\begin{aligned}E_{0,n+1} =\ & E_{1,n} + \frac{\Delta t_n}{\Delta r}\left(\frac{1}{4}E_{3,n}u_{3,n} + \frac{1}{4}p_{3,n}u_{3,n}\right.\\ & \left. -\frac{11}{4}E_{1,n}u_{1,n} - \frac{11}{4}p_{1,n}u_{1,n}\right).\end{aligned} \tag{8.4.19}$$

Therefore, the pressure at the center is

$$p_{0,n+1} = (\gamma-1)E_{0,n+1}, \tag{8.4.20}$$

(see (8.4.1) with $u = 0$). Equations (8.4.19)–(8.4.20) also use a staggered mesh.

With the above difference scheme some oscillations in pressure and density were observed near the axis which grew in their amplitude. This was remedied by using $\frac{1}{3}(p_{k-1,n} + 4p_{k,n} + p_{k+1,n})\Delta r$ instead of (8.4.16) at the point $k = 1$. Thus, for $k = 1$ the difference scheme (8.4.13) was replaced by

$$\begin{aligned}b_{1,n+1} =\ & \frac{1}{2}b_{2,n} + \frac{\Delta t_n}{2\Delta r}\Big\{ -b_{2,n}u_{2,n}\\ & +\gamma^{-1/2}\Delta r\left(\frac{1}{3}p_{0,n} + \frac{4}{3}p_{1,n} - \frac{5}{3}p_{2,n}\right)\Big\}.\end{aligned} \tag{8.4.21}$$

This, however, led to the nonindependence of the two staggered lattices since the term $p_{1,n}$ appears in (8.4.21).

The initial conditions between the axis $k = 0$ and the diaphragm $k = K$ were chosen in the following way. At the diaphragm $k = K$, these were chosen to be the average of the values of $a_{k,0}$, $b_{k,0}$ and $c_{k,0}$ at $k = K+1$ and $k = K-1$. This is due to the sensitivity of the data near the diaphragm. In view of the special treatment of the flow near the axis, all the points of the network corresponding to $k = 0,1,2,\ldots$ at each time step had to be used. Thus, the initial conditions are taken to be

$$\begin{aligned}u_{k,0} &= 0 \quad \text{for all } k,\\ \rho_{k,0} &= 1, p_{k,0} = 1 \quad \text{for } k < K,\\ \rho_{K,0} &= \frac{1}{2}(\rho^* + 1) + \frac{\Delta r}{2}(\rho^* - 1),\\ p_{K,0} &= \frac{1}{2}(p^* + 1) + \frac{\Delta r}{2}(p^* - 1),\\ \rho_{k,0} &= \rho^*, \quad p_{k,0} = p^* \quad \text{for } k < K,\end{aligned} \tag{8.4.22}$$

where ρ^* and p^* are (dimensionless) constants.

Payne (1957) carried out the calculations with a uniformly spaced set of mesh points from the axis to twice the radius of the cylindrical diaphragm. The mesh size was taken to be 1/64th of the radius. For 140 time steps of the integration over these 128 points the time on the Manchester University Mark 1 computer was 5 hours—very large compared to what the modern computers would take. The computations were checked by repeating the calculation with twice the mesh size or half the mesh size. The Courant-Friedrichs-Lewy criterion for the stability of the numerical scheme was taken in the form

$$\Delta t_n/\Delta r \leq A/(\text{velocity of the shock}), \qquad (8.4.23)$$

since the velocity of the shock is highest in the flow. The constant A was varied between 0.75 and 0.85 to ensure stability; for stronger shocks, the value of A was about 0.75, requiring shorter time intervals. The velocity of the shock at the previous time was used to get Δt_n from (8.4.23). The shock location was found by identifying the point where the pressure p is the average of the pressure behind and in front of the shock. For the converging shock, the pressure behind the shock was chosen to be the local maximum of $p_{k,n}$ while that ahead was the undisturbed value $p_{0,n}$ at the axis. This agrees roughly with the prescription of Lighthill (1956) for the shock location. For diverging shocks, the pressure ahead was chosen in an ad hoc manner at the point where $p(r - \Delta r, t_n) - p(r, t_n) = B$; suitable values for the constant B were found to be 0.1 for a weak shock of initial strength 2, 0.2 for a shock of initial strength 4, and 0.4 for a strong shock of initial strength 8.

When the shock approached the axis, the time Δt_n was taken to be constant, as obtained earlier in the calculations. For this (small) value of Δt_n, the reflected shock was so diffused that it was impossible to identify it. Therefore, the time interval at this point could be chosen to be relatively large and constant. It was possible to adjust its value with reference to the velocity of the outgoing shock. A judicious choice helped to keep the solution stable with a smooth shock which was not too diffused.

We discuss in some detail the numerical results for the initial conditions $p^* = \rho^* = 4$. The initial distribution for the cylindrical diaphragm was chosen to be the solution of the (plane) shock tube problem with these conditions. They give rise to a converging shock of strength 1.93, a contact surface, and an expansion wave. The mesh size Δr was chosen to be 1/128. The results are shown in Figures 8.5–8.8, which give pressure, particle velocity, density and temperature at 0.2 time intervals. Figure 8.5 shows that the shock strength increases with time leading to increase in pressure at any point behind it.

The shock itself is about six mesh points wide. It reaches the center at $t = 0.66$, attaining a high but finite value of the pressure there and is,

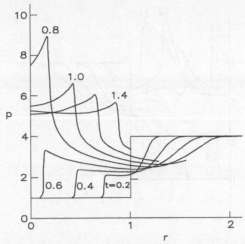

Figure 8.5 Pressure versus radius at 0.2 time intervals for a flow initiated by a cylindrical diaphragm with initial pressure and density ratios 4 (Payne, 1957).

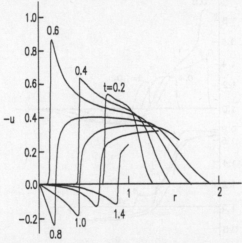

Figure 8.6 Particle velocity versus radius at 0.2 time intervals for a flow initiated by a cylindrical diaphragm with initial pressure and density ratios 4 (Payne, 1957).

then reflected. As the reflected shock engulfs the disturbed gas ahead of it, the pressure behind it at any fixed point decreases with time. Figure 8.6 shows the velocity of the gas as it is overtaken by the converging shock. The latter imparts it a negative (inward) value. The reflected shock subsequently increases the gas velocity so that it becomes positive though small. At any given point behind the converging shock the velocity increases with time; after it has been passed by the diverging shock it decreases.

The behaviour of density at different times (Figure 8.7) is similar to that for pressure, except that its rise across the shock is smaller, corresponding to an increase in temperature. A contact surface—with a gradual change over an increasing number of mesh points—appears in density and temperature

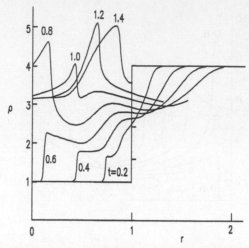

Figure 8.7 Density versus radius at 0.2 time intervals for a flow initiated by a cylindrical diaphragm with initial pressure and density ratios 4 (Payne, 1957).

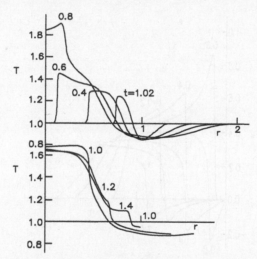

Figure 8.8 Temperature versus radius at 0.2 time intervals for a flow initiated by a cylindrical diaphragm with initial pressure and density ratios 4 (Payne, 1957).

(see Figures 8.7 and 8.8). The contact surface moves inward behind the converging shock and is later traversed by the diverging shock. This traversal is completed at $t = 1.4$ leaving behind a region of high temperature between the axis and the contact surface.

Figure 8.9 shows variation of pressure, density and particle velocity with time at $r = 0$ and $r = 0.375$. The solution qualitatively resembles that of Guderley (1942) but is different in magnitude. This is because the shock here is not assumed to be infinitely strong.

The initial conditions $p^* = \rho^* = 4$ gave rise to a strong contact discontinuity behind the shock which Lax's scheme was unable to handle. Payne (1957), therefore, assumed another set of initial conditions $p^* = 3.52$,

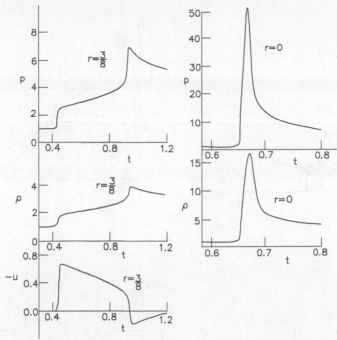

Figure 8.9 Variation of pressure, density and particle velocity with time at $r = 0$ and $r = 0.375$ (Payne, 1957).

$\rho^* = 2.44$ which led to a shock of the same strength, namely, 1.93. In this case the shock tube solution gives a contact surface of zero strength. The converging cylindrical shocks in both these instances behave identically, but the reflected divergent shocks behave differently as they move out. For the case $p^* = \rho^* = 4$, the reflected shock propagates slower as it moves into the colder region. The result of eliminating the contact surface for the case with $p^* = 3.52$ or $\rho^* = 2.44$, is that it is possible now to obtain a stronger converging shock. Such results were obtained for shocks of initial strengths 4 and 8.

Figure 8.10 shows the distribution of parameters for initial shock strength 8 at 0.1 time intervals. The general behaviour of various quantities is the same as for the shock of initial strength 1.93, discussed earlier. The main difference is that the stronger shocks increase in their magnitude more rapidly as they approach the axis. Moreover, the expansion fan significantly increases the diffusion effect for the stronger shocks.

Figures 8.11 and 8.12 give shock strength, $z = \frac{p - p_0}{p_0}$, versus radius for converging shocks of initial strengths 1.93 and 8, respectively. The agreement with the results by Chisnell's method (which we have often referred to as the Chester (1954), Chisnell (1957) and Whitham (1958) approach) is found to be remarkable. Comparison with other analytic results may be found in chapter 6. Payne (1957) also studied the effect of varying γ on the propagation of converging shocks. There was no major qualitative change.

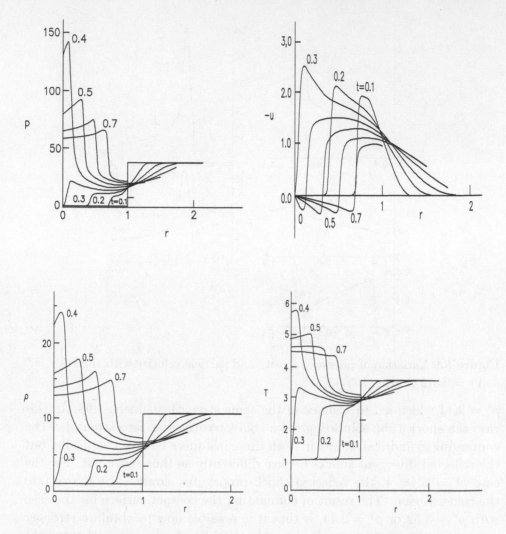

Figure 8.10 Pressure, velocity, density and temperature versus radius at intervals 0.1 of time for a flow with a converging cylindrical shock of initial strength 8 (Payne, 1957).

However, when γ was changed from 1.4 to 5/3, there was a larger density change across the contact surface. The production of entropy by the traversal of the shocks (converging and diverging) was also briefly discussed by Payne (1957).

It is clear from Payne's application of the Lax (1954) scheme to the converging shock problem that there are difficulties near the singular point $r = 0$. Moreover, the momentum equation, not being in a conservation form, must also be treated separately. The numerical results suggest that the discontinuities—the contact surface and the shock—are not precisely located; considerable manipulations are needed to get a reasonable shock. The mesh sizes must be appropriately changed to avoid oscillations and obtain somewhat sharp shocks.

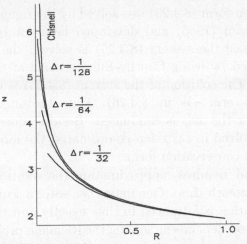

Figure 8.11 Shock strength versus radius for a converging cylindrical shock of initial strength 1.93 (Payne, 1957).

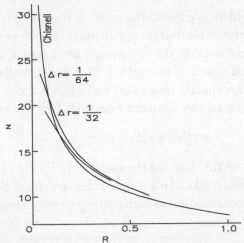

Figure 8.12 Shock strength versus radius for a converging cylindrical shock of initial strength 8 (Payne, 1957).

To get over these difficulties, Sod (1977) investigated converging spherical and cylindrical shocks by using a judicious combination of Glimm's (1965) random choice method and operator splitting. The system of gasdynamic equations in the vector form

$$\mathbf{U}_t + \mathbf{F}(\mathbf{U})_r = -\mathbf{W}(\mathbf{U}), \qquad (8.4.24)$$

was treated by operator-splitting. It was written as the system

$$\mathbf{U}_t + \mathbf{F}(\mathbf{U})_r = 0, \qquad (8.4.25)$$

which represents one-dimensional equations of gas dynamics in cartesian co-ordinates, and

$$\mathbf{U}_t = -\mathbf{W}(\mathbf{U}). \qquad (8.4.26)$$

The conservation form (8.4.25) was solved by the random-choice method introduced by Glimm (1965) and developed later for hydrodynamics by Chorin (1976). Once the system (8.4.25) is solved, the system of ODEs (8.4.26) is integrated by using Cauchy-Euler scheme at the interior points for one time step. The solution of the system (8.4.25) is used to determine the inhomogeneous term $-W$ in (8.4.26). Thus, the singular nature of the original system near the axis is eliminated. Besides, since the equations of gas dynamics are solved in cartesian co-ordinates, the momentum equation can be written in a conservation form.

Glimm's method requires, approximating the solution by a piecewise constant function at each time. One must then solve a sequence of Riemann problems. The solution is advanced in time exactly and the new values are sampled. The method depends on solving the Riemann problems exactly and inexpensively. We refer the reader to Sod (1976, 1977) for further details. Here we summarize the results of Sod (1977) for the converging cylindrical shock.

The physical problem is exactly the same as that treated by Payne (1957) and is solved with the same initial conditions. The (normalised) pressure and density inside and outside the diaphragm are 1 and 4, respectively. This gives rise to an initial shock of strength 1.93, a contact discontinuity, and a rarefaction wave. The spatial mesh size was chosen to be $\Delta r = 0.01$ while Δt was chosen subject to the Courant-Friedrich-Lewy condition

$$\max(|u| + c)\Delta t/\Delta r \leq 1.$$

The general features of the flow are the same as in Payne (1957). The results are shown in Figures 8.5–8.12. One major achievement of Sod's study is that the shock and the contact discontinuity are perfectly sharp. However, due

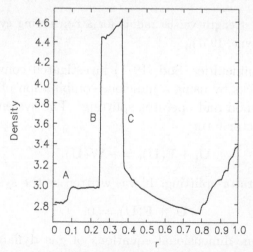

Figure 8.13 Density profile after interaction of diverging shock and contact discontinuity at time $t = 0.6$ (Sod, 1977).

to the randomness of Glimm's method, at a given time the position of the shock or contact discontinuity may not be exact; on the average it is.

The interaction of the reflected shock with the contact discontinuity results in a reflected (converging) shock A, and a contact discontinuity B propagating towards the axis and a transmitted (diverging) shock C (see Figure 8.13). This reflected shock was not perceived by earlier investigators (Payne (1957), Abarbanel and Goldberg (1972), and Lapidus (1971)).

We conclude the discussion of converging shocks by summarizing recent numerical results of Liu, Khoo and Yeo (1999). They used a modified total variation (TVD) scheme due originally to Harten (1977, 1978) (see section 8.5). They chose the initial conditions the same as in the study of Sod (1977), namely, $p_H = 4$, $\rho_H = 4$, $u_H = 0$, $p_0 = 1$, $\rho_0 = 1$, $u_0 = 0$, where H denotes conditions inside the high pressure isothermal cylinder. The cylindrical diaphragm was located initially at 0.25. The major physical features observed were the same as in the study of Sod (1977). The converging shock with increasing strength implodes at the axis at about $t = 0.15$, in agreement with the figure obtained by Sod (1977). The temperature and density attain their maximum values there at that time. The observed contact discontinuity was always found to be sharp.

The results of Liu et al. (1999) show a good general agreement with those of Sod (1976) except for a minor divergence of the locus of the main shock after its divergence from the axis—the difference, however, never exceeding 10%. Sod's (1976) results, being first order, are probably less accurate as was pointed out by Liu et al. (1999). Unlike in some previous studies such as Payne (1957), both the main shock, the secondary shock and the contact discontinuity are successfully captured by the modified TVD scheme employed by Liu et al. (1999). However, the third shock, generated by the interaction of the reflected shock with the contact discontinuity, is too weak and is not detected precisely even by the modified TVD scheme of Liu et al. (1999). This shock did not attract much mention in earlier literature.

8.5 Numerical Simulation of Explosions Using Total Variation Diminishing Scheme

In sections 8.3 and 8.4 we dealt with the early numerical investigations in the context of explosion and implosion phenomena. These studies gave a reasonable qualitative picture but were not sophisticated enough to yield high accuracy or precise description of discontinuities. In this section we describe a recent study due to Liu, Khoo and Yeo (1999), which uses the total variation diminishing (TVD) scheme of Harten (1977). It was appropriately modified to give a high resolution of contact discontinuities which, in the scheme used by Payne (1957), were rather vaguely found. The model we discuss here is the release of a high pressure gas into a quiescent medium, due originally to McFadden (1952) and Friedman (1961), which we have discussed in much detail in sections 7.2 and 7.3. This release leads to a main shock rushing out through the quiescent low pressure gas and to a rarefaction wave moving inward into the high pressure gas. This flow also results in the formation of a contact discontinuity and, subsequently, a secondary shock. This shock is weak initially and propagates outwards with the expanding gas. It grows in strength and becomes fairly strong in a short time. Soon after, this secondary shock stops propagating outward, attains zero velocity, and then begins to implode on the center.

The contact discontinuity also moves outward initially behind the main shock. After a certain time it ceases its journey outward and begins to move inwards to the origin. This is due to inward moving flow by the converging secondary shock. After reflection from the center, the secondary shock moves out, interacts with the inward moving contact surface and continues to propagate outward. An inward rarefaction is produced as a result of this interaction, leading to the formation of a third shock. This shock wave is very weak and is not easy to detect. At the point of formation of the third shock all the pre-existing discontinuities are rather weak and the whole flow region is nearly uniform. These detailed features were numerically observed by Liu et al. (1999) and make for a fascinating study.

In the present section we detail the work of Liu et al. (1999). The scheme used by these investigators is a modified form of TVD, referred to as the artificial compression method (ACM). It was found particularly useful in increasing resolution of the contact discontinuities.

We write equations of motion for an inviscid, non-heat-conducting, radially symmetric flow in the form

$$U_t + (F(U))_r = W(U), \qquad (8.5.1)$$

where

$$U = \begin{bmatrix} \rho \\ \rho u \\ E \end{bmatrix}, \quad F(U) = \begin{bmatrix} \rho u \\ \rho u^2 + p \\ (E + p)u \end{bmatrix},$$

$$W(U) = -\frac{\alpha - 1}{r} \begin{bmatrix} \rho u \\ \rho u^2 \\ (E + p)u \end{bmatrix}. \qquad (8.5.2)$$

The variables here are expressed in terms of ρ_0 and a_0, the undisturbed density and speed of sound, respectively. The independent variables t and r have been rendered nondimensional by $4r_0/a_0$ and $4r_0$, where r_0 is the initial radius of the compressed gas. Here, $\alpha = 2, 3$ for cylindrical and spherical symmetry, respectively. The energy of the gas (in nondimensional variables) is given by

$$E = \frac{p}{\gamma - 1} + \frac{1}{2}\rho u^2, \tag{8.5.3}$$

where $\gamma = c_p/c_v$. The system (8.5.1)–(8.5.2) is recast in nearly conservative form as

$$\frac{\partial \tilde{U}}{\partial t} + \frac{\partial F(\tilde{U})}{\partial r} = S(\tilde{U}), \tag{8.5.4}$$

where

$$\tilde{U} = r^{\alpha-1}U, \quad S(\tilde{U}) = \frac{\alpha - 1}{r}\left\{\begin{array}{c} 0 \\ \tilde{p} \\ 0 \end{array}\right\}, \quad \tilde{p} = r^{\alpha-1}p. \tag{8.5.5}$$

For spherical symmetry, we have

$$\tilde{U} = \left\{\begin{array}{c} r^2\rho \\ r^2\rho u \\ r^2 E \end{array}\right\}, \quad F(U) = \left\{\begin{array}{c} r^2\rho u \\ r^2(\rho u^2 + p) \\ r^2(E + p)u \end{array}\right\}, \quad S(\tilde{U}) = \frac{2}{r}\left\{\begin{array}{c} 0 \\ r^2 p \\ 0 \end{array}\right\}. \tag{8.5.6}$$

It may be observed that the RHS of (8.5.4) is always positive; there is considerable evidence to suggest that this fact enhances numerical stability. Besides, since $S(\tilde{U})$ is continuous through the contact surface, this form may be convenient for flows involving these discontinuities.

The system (8.5.4) is still singular at the origin; it is first written there in the conservation form

$$\rho_t + \alpha(\rho u)_r = 0, \tag{8.5.7}$$

$$E_t + \alpha[(E + p)u]_r = 0. \tag{8.5.8}$$

The finite difference scheme applied to (8.5.7) and (8.5.8) would, therefore, be compatible with the original system at the origin.

We must also have

$$u(0, t) = 0. \tag{8.5.9}$$

We now discuss the total variation diminishing (TVD) (see (8.5.19) for definition) scheme in some detail. When a hyperbolic equation is linear and the numerical scheme simulating it is also linear, the convergence of the numerical approximation is implied by the consistency and stability of the scheme. This is not the case for nonlinear problems. Conventional shock-capturing schemes such as Lax-Wendroff (see Richtmyer and Morton (1967)) for the solution of nonlinear hyperbolic conservation laws produce overshoots and undershoots near the discontinuity; these schemes may also select a nonphysical solution. To overcome these difficulties one may add a large amount of dissipation but that results in the smearing of the discontinuity on many grids.

To obviate these difficulties a new class of schemes was introduced by Harten (1983). These schemes must satisfy the following requirements : (i) they must be total variation diminishing (see (8.5.19)); (ii) they must be consistent with the conservation law and satisfy entropy inequality, (see (8.5.13) below); and (iii) they must be second order accurate away from shocks. The condition (i) guarantees that the scheme does not generate spurious oscillations while (ii) ensures that the numerical solution converges to the entropy solution.

Consider the scalar hyperbolic consevation law

$$\frac{\partial u}{\partial t} + \frac{\partial}{\partial x} f(u) = 0, \quad t > 0, \tag{8.5.10}$$

with the initial condition

$$u(x, 0) = u_0(x), \tag{8.5.11}$$

where the flux function f is smooth. We now define the entropy condition for the scalar equation (8.5.10). A solution of (8.5.10) with a shock propagating with speed s and satisfying the Rankine-Hugoniot condition

$$f(u_l) - f(u_r) = s(u_l - u_r), \tag{8.5.12}$$

is said to satisfy the entropy condition if

$$\frac{f(u_l) - f(u_r)}{u_l - u_r} \geq s \geq \frac{f(u_r) - f(u_l)}{u_r - u_l}, \tag{8.5.13}$$

for all u between u_l and u_r where u_l and u_r are values of u on the left and right of the shock.

It is clear from (8.5.12) that $f'(u)$ is the characteristic speed. If the function f is convex, that is, if $f''(u)$ is positive, then (8.5.13) implies that s must lie between $f'(u_l)$ and $f'(u_r)$ and $u_l > u_r$.

A conservative and consistent method will yield a unique weak solution if it satisfies a discrete version of the entropy condition (LeVeque (1992), p. 142).

We now define some terms required for the discussion of TVD schemes. A difference scheme for (8.5.10) is said to be in conservation form if there exists a continuous function F such that

$$u_i^{n+1} = u_i^n - \lambda(F_{i+1/2}^n - F_{i-1/2}^n). \tag{8.5.14}$$

where $F_{i+1/2} = F(u_{-k+1,\cdots,u_k})$. F is thus the numerical flux function and λ is the ratio of time step to space step.

The difference scheme (8.5.14) is consistent with the conservation law (8.5.10) if

$$F(u, u, \cdots, u) = f(u).$$

The scheme (8.5.14) may be put in a viscous form if there exists a function Q of $2k$ variables, called the coefficient of numerical viscosity, such that

$$Q_{i+1/2} = Q(u_{i-k+1}, \cdots, u_{i+k}),$$

and (8.5.14) may hence be written as

$$
\begin{aligned}
u_i^{n+1} &= u_i^n - \frac{\lambda}{2}(f(u_{i+1}^n) - f(u_{i-1}^n)) \\
&\quad + \frac{1}{2}\{Q_{i+1/2}^n(u_{i+1}^n - u_i^n) - Q_{i-1/2}^n(u_i^n - u_{i-1}^n)\}, \quad (8.5.15)
\end{aligned}
$$

The numerical flux therefore is

$$F_{i+1/2}^n = \frac{1}{2}\left\{f(u_{i+1}^n) - f(u_i^n) - \frac{Q_{i+1/2}^n}{\lambda}(u_{i+1}^n - u_i^n)\right\}. \quad (8.5.16)$$

The numerical scheme (8.5.14) is said to be L^∞-stable if there exists a constant $c > 0$, independent of n and Δt, such that $\|U^{n+1}\|_{L^\infty} \leq c\|U^0\|_{L^\infty}$, where $\|U^n\|_{L^\infty} = \sup_i |u_i^n|$. The scheme (8.5.14) is said to be in incremental form if there exist two functions C and D,

$$C_{i+1/2}^n = C(u_{i-k+1}^n, \cdots u_{i+k}^n), \quad D_{i+1/2}^n = D(u_{i-k+1}^n, \cdots u_{i+k}^n), \quad (8.5.17)$$

such that we may write

$$u_i^{n+1} = u_i^n + D_{i+1/2}^n(u_{i+1}^n - u_i^n) - C_{i-1/2}^n(u_i^n - u_{i-1}^n). \quad (8.5.18)$$

A common feature of all TVD schemes is that second (or higher) order accuracy is surrendered at the extrema since one cannot have both second order accuracy everywhere and the TVD property. Second order TVD schemes are second order accurate away from extrema. These schemes are said to enjoy second order resolution (SOR).

The scheme (8.5.14) is called total variation diminishing (TVD) if

$$\sum_{i=-\infty}^{\infty} |u_i^{n+1} - u_{i-1}^{n+1}| \leq \sum_{i=-\infty}^{\infty} |u_i^n - u_{i-1}^n|. \quad (8.5.19)$$

The scheme (8.5.18) is TVD if

$$C_{i+1/2}^n \geq 0, \quad D_{i+1/2}^n \geq 0 \quad \text{and} \quad C_{i+1/2}^n + D_{i+1/2}^n \leq 1 \quad \text{for all} \quad i, n; \quad (8.5.20)$$

Any numerical scheme which can be put in the viscous form (8.5.15) is TVD if the coefficients in the viscous term satisfy

$$\lambda |a_{i+1/2}^n| \leq Q_{i+1/2}^n \leq 1, \quad (8.5.21)$$

where

$$a_{i+1/2}^n = \begin{cases} \frac{f(u_{i+1}^n) - f(u_i^n)}{u_{i+1}^n - u_i^n} & \text{if } u_{i+1}^n \neq u_i^n \\ f'(u_i^n) & \text{otherwise.} \end{cases} \tag{8.5.22}$$

For a three-point scheme to be TVD, the conditions (8.5.20) and (8.5.21) are both necessary and sufficient.

We may mention that the Lax-Wendroff scheme is not TVD since the coefficient of viscosity $Q_{i+1/2}^{LW} = (\lambda a_{i+1/2})^2$ does not satisfy the condition (8.5.21).

The construction of the SOR scheme changes the three-point first order accurate TVD scheme into a five point second order accurate TVD scheme. For this purpose we consider a three-point TVD scheme which is in a conservative form and is consistent with the entropy condition (Lax-Friedrichs and Godunov schemes, for example). The basic idea here is to use a modified flux $f + (1/\lambda)g$ in place of f where the function g is discretised as

$$g_i = \frac{s_i}{2} \min\{(Q(\lambda a_{i+1/2}^n) - (\lambda a_{i+1/2})^2)|u_{i+1}^n - u_i^n|,$$
$$(Q(\lambda a_{i-1/2}^n) - (\lambda a_{i+1/2})^2)|u_i^n - u_{i-1}^n|\}. \tag{8.5.23}$$

Here,

$$s_i = \begin{cases} \text{sgn}(u_{i+1}^n - u_i^n) & \text{if } (u_{i+1}^n - u_i^n)(u_i^n - u_{i-1}^n) > 0 \\ 0 & \text{otherwise,} \end{cases} \tag{8.5.24}$$

and $Q(\lambda a_{i+1/2}^n)$ is the coefficient of numerical viscosity satisfying (8.5.21). The numerical flux $\tilde{F}$ is now defined by

$$\begin{aligned} \tilde{F}_{i+1/2} &= \frac{1}{2}\{f(u_{i+1}^n) + f(u_i^n) + \frac{1}{\lambda}(g_i + g_{i+1}) \\ &\quad - \frac{1}{\lambda}Q(\lambda a_{i+1/2} + \nu_{i+1/2})(u_{i+1}^n - u_i^n)\} \\ &= \frac{1}{2}\{\tilde{f}(u_i^n) + \tilde{f}(u_{i+1}^n) - \frac{1}{\lambda}\tilde{Q}_{i+1/2}(u_{i+1}^n - u_i^n)\}, \end{aligned} \tag{8.5.25}$$

where $\tilde{f}(u_i^n) = f(u_i^n) + (1/\lambda)g_i$, $\tilde{Q}_{i+1/2}^n = Q(\lambda a_{i+1/2} + \nu_{i+1/2})$ and

$$\nu_{i+1/2}^n = \begin{cases} (g_{i+1} - g_i)/(u_{i+1}^n - u_i^n) & \text{if } u_{i+1}^n \neq u_i^n, \\ 0 & \text{otherwise.} \end{cases}$$

The new difference scheme for (8.5.10) becomes

$$u_i^{n+1} = u_i^n - \lambda\{\tilde{F}_{i+1/2}^n - \tilde{F}_{i-1/2}^n\}. \tag{8.5.26}$$

The scheme (8.5.26) is five-point, is in conservation form, and is consistent with the conservation law (8.5.10). This scheme is TVD. This follows from

(8.5.21) if we replace f by $\tilde{f}$, $Q(\lambda a_{i+1/2})$ by $Q(\lambda a_{i+1/2} + \nu_{i+1/2})$ and $\lambda a_{i+1/2}$ by $\lambda a_{i+1/2} + \nu_{i+1/2}$.

Harten (1983) extended the scalar TVD scheme to systems of nonlinear hyperbolic equations to ensure the stability of the numerical scheme over a long time as well as to enhance the resolution of contact discontinuity. The main difficulty in extending the TVD scheme to nonlinear systems is that the total variation of the solution may increase when there is a two-wave interaction. To overcome this difficulty Harten (1983) extended the TVD scheme to systems in such a manner that the resulting scheme is TVD for the "locally frozen" constant coefficient systems.

Liu et al. (1999) modified Harten's scheme by incorporating an artificial compression in the so called artificial compression method (ACM) to increase the resolution of contact discontinuities. In general, a numerical scheme coupled with the use of an artificial compression will considerably increase the resolution near a contact discontinuity and yet not change the main features of the original scheme. With this end in view, Liu et al. (1999) combined Harten's TVD scheme with ACM for application to the explosion problem. Let $\tilde{U}_i^{n+1} = r_{i+1/2}^{\alpha-1} U_i^{n+1}$, $\Delta_{i+1/2} U^n = U_{i+1}^n - U_i^n$. Furthermore, let $A_{i+1/2}^n = A(U_i^n, U_{i+1}^n)$ be the mean Jacobian matrix such that

$$F(U_{i+1}^n) - F(U_i^n) = A_{i+1/2}^n(U_{i+1}^n - U_i^n). \qquad (8.5.27)$$

Let $a_{i+1/2}^l$ and $R_{i+1/2}^l$ ($l = 1, 2, 3$) be the left eigenvalues and the corresponding eigenvectors of $A_{i+1/2}^n$ and let $(\alpha_{i+1/2}^1, \alpha_{i+1/2}^2, \alpha_{i+1/2}^3)^T = R_{i+1/2}^{-1} \Delta_{i+1/2} U^n$, where $R_{i+1/2}$ is a 3×3 matrix whose columns are the right eigenvectors of $A_{i+1/2}^n$.

Corresponding to the function g_i defined by (8.5.23) for the scalar case, we write $g_i^{H,l}$ for the system (8.5.4) as

$$g_i^{H,l} = \min \text{mod}(\sigma(\lambda_r a_{i-1/2}^l)\alpha_{i-1/2}^l, \sigma(\lambda_r a_{i+1/2}^l)\alpha_{i+1/2}^l)), \quad l = 1, 2, 3, \qquad (8.5.28)$$

where

$$\min \text{mod}(x, y) = \begin{cases} \text{sgn}(x)\min(|x|, |y|) & \text{if} \quad xy > 0, \\ 0 & \text{otherwise}, \end{cases} \qquad (8.5.29)$$

$$\sigma(x) = \frac{1}{2}\phi(x) - \frac{1}{2}x^2, \qquad (8.5.30)$$

and $\lambda_r = \Delta t / \Delta r$. Δt and Δr are time and space step sizes. $\phi(x)$ is the coefficient of viscosity (Roe (1981)), chosen to be

$$\phi(x) = \begin{cases} |x| & \text{if} \quad |x| \geq 2\epsilon \\ \frac{x^2 + 4\epsilon^2}{4\epsilon} & \text{otherwise}. \end{cases} \qquad (8.5.31)$$

Liu et al. (1999) modified the function $g_i^{H,l}$ in (8.5.28) and wrote

$$g_i^l = (1 + q_{i+1/2}^l)g_i^{H,l}, \quad l = 1, 2, 3, \tag{8.5.32}$$

where $q_{i+1/2}^l$ is the artificial compression term added to ensure a sharp contact discontinuity. It is defined as

$$q_{i+1/2}^l = \tilde{s}_{i+1/2} \frac{|\alpha_{i+1/2}^l - \alpha_{i-1/2}^l|}{|\alpha_{i-1/2}^l| + |\alpha_{i+1/2}^l|} \frac{\phi(\lambda_r a_{i+1/2}^l)}{\sigma(\lambda_r a_{i+1/2}^l)}, \quad l = 1, 2, 3, \tag{8.5.33}$$

where

$$\tilde{s}_{i+1/2} = \frac{1}{2}|\operatorname{sgn}(a_{i+1/2}^2 - a_{i-1/2}^2) - \operatorname{sgn}(a_{i+1/2}^3 - a_{i-1/2}^3)|. \tag{8.5.34}$$

When $q_{i+1/2}^l = 0$, the numerical flux reduces to Harten's flux for the system (8.5.4).

Thus, the modified numerical scheme for (8.5.4) is

$$\tilde{U}_i^{n+1} = \tilde{U}_i^n - \lambda_r(H_{i+1/2}^n - H_{i-1/2}^n) + \frac{1}{2}\Delta t S(\tilde{U}_j^{n+1}) + \frac{1}{2}\Delta t S(\tilde{U}_j^n). \tag{8.5.35}$$

The numerical flux $H_{i+1/2}^n$ now is

$$H_{i+1/2}^n = \frac{1}{2}\{F(\tilde{U}_i^n) + F(\tilde{U}_{i+1}^n) + r_{i+1/2}^{\alpha-1}G_{i+1/2}^n\}, \tag{8.5.36}$$

where

$$
\begin{aligned}
G_{i+1/2}^n &= R_{i+1/2}\Phi_{i+1/2}^n, \\
\Phi_{i+1/2}^n &= (\phi_{i+1/2}^1, \phi_{i+1/2}^2, \phi_{i+1/2}^3)^T, \\
\phi_{i+1/2}^l &= \frac{1}{\lambda_r}(g_i^l + g_{i+1}^l - \sigma(\lambda_r a_{i+1/2}^l + \nu_{i+1/2}^l)\alpha_{i+1/2}^l), \quad l = 1, 2, 3 \\
\nu_{i+1/2}^l &= \begin{cases} \frac{g_{i+1}^l - g_i^l}{\alpha_{i+1/2}^l} & \text{if } \alpha_{i+1/2}^l \neq 0 \\ 0 & \text{otherwise.} \end{cases}
\end{aligned}
\tag{8.5.37}
$$

The present scheme is TVD under the CFL condition $\frac{\Delta t}{\Delta r}\max(|u| + c) \leq (5 - \sqrt{17}/2)$. This condition is more severe than that for the TVD scheme of Harten (1983). For the actual computation Liu et al. (1999) required that

$$\frac{\Delta t}{\Delta r}\max(|u| + c) \leq 0.8, \tag{8.5.38}$$

Liu et al. (1999) assumed the same initial conditions for the high pressure gas as were first adopted by Brode (1957) and experimentally verified by Boyer (1960). The initial sphere was taken to be of radius 2 in; it contained a gas at a pressure of 326 psi and temperature 299 K. The outside medium is air at 15 psi and has the same temperature as inside, namely, 299 K. The nondimensional values of these quantities are $p_H = 15.514$, $\rho_H = 21.7333$, $u_H = 0$, $p_0 = 0.715$, $\rho_0 = 1.0$, $u_0 = 0$. Observe that the pressure has been rendered nondimensional by $\rho_0 a_0^2 = \gamma p_0 = 1.4 p_0$; hence $p_0 = 0.715$. Since the initial radius has been rendered nondimensional by $4 r_0$, its value is 0.25; the time is related by $t' = 293 t(\mu s)$ or $t = t'/293$.

Liu et al. (1999) first solved the (plane) shock tube problem by Harten's TVD scheme to confirm its effectiveness. They chose the initial conditions as $p_H = 4, \rho_H = 4, u_H = 0$. They used a modified form of Harten's scheme as well as Harten's original scheme. The number of mesh points in each scheme was chosen to be the same with mesh size $\Delta x = 0.01$. Liu et al. (1999) claim that the definition of contact surface improves while the general results remain the same everywhere else. This improvement was brought about by introducing the ACM technique. As we observe below, the same effect is also brought about in the spherical explosion problem.

The temperature profiles at different times are shown in Figures 8.14(a)–(c). As we have discussed in sections 7.2 and 7.3, the secondary shock is first formed at the point on the tail of the rarefaction wave where pressure and temperature attain their minima. This fact was used to identify the inception of the secondary shock in the numerical solution. It was also observed numerically that the rarefaction wave reaches the center at about 60 μs and brings about a rapid decrease in pressure there. At about 140 μs, the pressure at the center is lower than that at the tail of the rarefaction wave. From this point onwards, velocity, density and pressure between the center and the secondary shock are almost constant, though they continue to diminish with time. At $t' = 140$ μs, the secondary shock ceases its outward motion and begins to move inward with increasing strength. Now the problem in this region corresponds to that of the collapse of a converging spherical shock into a nonuniform region. The following features may be observed from Figures 8.14(a)–(c). The maximum temperature is observed at the contact discontinuity until about 100 μs. Thereafter, it occurs in the immediate neighbourhood of the front shock. As time passes, the converging shock becomes stronger and the maximum of temperature shifts until it is attained at the center at $t' = 360$ μs when it implodes there. Even as the reflected shock moves out through the contact discontinuity, the maximum temperature continues to occur at the center of the explosion.

The secondary shock, after formation, first moves away from the center. At about $t' = 170$ μs, it attains zero velocity and then moves inward until it implodes at the center at about 360 μs. During this motion, the particle velocity behind it is also inward. This affects the motion of the contact discontinuity. The latter ceases its outward motion and begins to move inward at about $t' = 240$ μs. This inward contact discontinuity meets the (outgoing) reflected secondary shock at about 550 μs, resulting in its

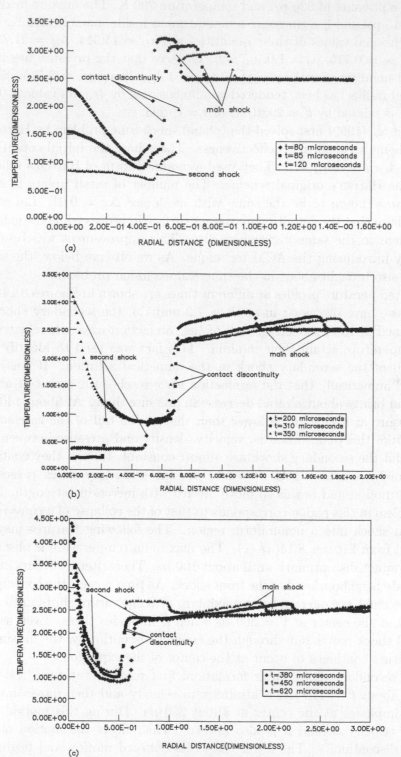

Figure 8.14 (a)–(c) The temperature profiles at different times for the spherical explosion centred at the origin (Liu et al., 1999).

outward motion again. After the transmission of the reflected secondary shock, the contact surface is relatively still. No reflection from the contact discontinuity due to interaction with the secondary shock was observed by Liu et al. (1999). The results of Liu et al. (1999) show close agreement with the analysis of Brode (1959). The numerical solution of Liu et al. (1999) shows good qualitative agreement with the experimental results of Boyer (1960), though there is considerable quantitative divergence. For example, the time of implosion of the secondary shock in the experiment was about 180 μs later (a discrepancy of 50%) than that predicted by the numerical solution. Moreover, the contact discontinuity in the experiment did not indicate any inward radial movement before its interaction with the reflected secondary shock. This divergence, Liu et al. (1999) attribute to the inadequacy of the experiments and the energy losses due to thermal heat conduction etc.

Qualitatively similar numerical results were obtained for the case of cylindrical explosion in air.

References

Abarbanel, S., and Goldberg, M. (1972) Numerical solution of quasi-conservative hyperbolic systems—the cylindrical shock problem, *J. Comp. Phys.*, 10, 1.

Akulichev, V.A., Boguslavskii, Yu.Ya., Ioffe, A.I., and Naugol'nykh, K.A. (1968) Radiation of finite-amplitude spherical waves, *Sov. Phys. Acoustics*, 13, 281.

Andriankin, E.I., Kogan, A.M., Kompaneets, A.S., and Krainov, V.P. (1962) The propagation of a strong explosion in a nonhomogeneous atmosphere, *Zh. Prik. Mech. Tekh. Fiz.*, 6, 3.

Ardavan-Rhad, H. (1970) The decay of a plane shock wave, *J. Fluid Mech.*, 43, 737.

Bach, G.G., and Lee, J.H.S. (1970) An analytical solution for blast waves, *AIAA Journal*, 8, 271.

Bach, G.G., Kuhl, A.L., and Oppenheim, A.K. (1975) On blast waves in exponential atmospheres, *J. Fluid Mech.*, 71, 105.

Barenblatt, G.I. (1979) Similarity, Self-similarity and Intermediate Asymptotics, Consultants Bureau, New York.

Barenblatt, G.I. (1996) Scaling, Self-similarity and Intermediate Asymptotics, Cambridge University Press, New York.

Bethe, H.A. (1942) Theory of shock waves for an arbitrary equation of state, *Off. Sci. Res. Dev. Rep.*, 545.

Boyer, D.W. (1960) An experimental study of the explosion generated by a pressurised sphere, *J. Fluid Mech.*, 9, 401.

Brinkley, S.R., and Kirkwood, J.G. (1947) Theory of propagation of shock waves, *Phys. Rev.*, 71, 606.

Brode, H.L. (1955) Numerical solutions of spherical blast waves, *J. Appl. Phys.*, 26, 766.

Brode, H.L. (1957) Theoretical solutions of spherical shock tube blasts, Rand Corporation, Rep. RM-1974.

Brode, H.L. (1959) Blast wave from a spherical charge, *Phys. Fluids*, 2, 217.

Brushlinskii, K.V., and Kazhdan, Ja.M. (1963) On auto-models in the solution of certain problems of gas dynamics, *Russ. Math. Surv.*, 18, 1.

Butler, D.S. (1954) Converging spherical and cylindrical shocks, Armament Res. Estab. Rep., 54/54.

Chan, B.C., Holt, M., and Welsh, R.L. (1968) Explosions due to pressurized spheres at the ocean surface, *Phys. Fluids*, 11, 714.

Chernii, G.G. (1960) Application of integral relationships in problems of propagation of strong shock waves, *PMM*, 24, 121.

Chester, W. (1954) The quasi-cylindrical shock tube, *Phil. Mag.*, 45, 1293.

Chester, W. (1960) The Propagation of Shock Waves Along Ducts of Varying Cross Section, Advances in Applied Mechanics, VI, Academic Press, New York.

Chisnell, R.F. (1955) The normal motion of a shock wave through a non-uniform one-dimensional medium, *Proc. Roy. Soc. A*, 232, 350.

Chisnell, R.F. (1957) The motion of a shock wave in a channel, with application to cylindrical and spherical shock waves, *J. Fluid Mech.*, 2, 286.

Chisnell, R.F. (1998) An analytic description of converging shock waves, *J. Fluid Mech.*, 354, 357.

Chorin, A.J. (1976) Random choice solution of hyperbolic systems, *J. Comp. Phys.*, 22, 517.

Cole, R.H. (1948) Underwater Explosions, Princeton University Press, New Jersey.

Courant, R., and Friedrichs, K.O. (1948) Supersonic Flow and Shock Waves, Interscience Publishers Inc., New York.

Domb, C., and Sykes, M.F. (1957) On the susceptibility of a ferromagnetic above the Curie point, *Proc. Roy. Soc. A*, 240, 214.

Du Fort, E.C., and Frankel, S.P. (1953) Mathematical tables and other aids to computation, VIII, 135.

Engquist, B., and Osher, S. (1980) Stable and entropy satisfying approximations for transonic flow calculations, *Math. Comp.*, 34, 45.

Ferro Fontan, C., Gratton, J., and Gratton, R. (1975) Self-similar spherical implosion, *Phys. Lett.*, 55A, 35.

Ferro Fontan, C., Gratton, J., and Gratton, R. (1977) Self-similar implosion of shells, *Nucl. Fusion*, 17, 135.

Friedman, M.P. (1961) A simplified analysis of spherical or cylindrical blast waves, *J. Fluid Mech.*, 11, 1.

Fujimoto, Y., and Mishkin, E.A. (1978) Analysis of spherically imploding shocks, *Phys. Fluids*, 21, 1933.

Gelfand, I.M. (1959) Some problems of the theory of quasi-linear equations, *Uspekhi Mat. Nauk.*, 14, 82.

Glimm, J. (1965) Solutions in the large for nonlinear hyperbolic systems of equations, *Comm. Pure Appl. Math.*, 18, 697.

Godlewski, E., and Raviart, P. (1991) Hyperbolic Systems of Conservation Laws, Mathematiques & Applications, Publication trimestrielle numero 3/4:300 F.

Godunov, S. (1959) Finite difference methods for numerical computation of discontinuous solutions of equations in fluid dynamics, *Mat. Sb.*, 47, 271.

Goldman, E.B. (1973) Numerical modeling of laser produced plasmas: the dynamics and neutron production in densely spherical symmetric plasmas, *Plasma Phys.*, 15, 289.

Goldstine, H.H., and Von Neumann, J. (1955) Blast wave calculations, *Comm. Pure Appl. Math.*, 8, 327.

Grigorian, S.S. (1958a) Cauchy's problem and the problem of a piston for one-dimensional, non-steady motions of a gas (automodel motion), *PMM*, 22, 244.

Grigorian, S.S. (1958b) Limiting self-similar, one-dimensional non-steady motions of a gas (Cauchy's problem and the piston problem), *PMM*, 22, 417.

Guderley, G. (1942) Starke kugelige und zylindrische Verdichtungsstösse in der Nähe des Kugelmittelpunktes bzw. der Zylinderachse, *Luftfahrtforschüng*, 19, 302.

Harten, A. (1977) The artificial compression method for computation of shocks and contact discontinuities: I. Single conservation laws, *Comm. Pure Appl. Math.*, 30, 611.

Harten, A. (1978) The artificial compression method for computation of shocks and contact discontinuities: III. Self-adjusting hybrid schemes, *Math. Comput.*, 32, 363.

Harten, A. (1983) High resolution schemes for hyperbolic conservation laws, *J. Comput. Phys.*, 49, 357.

Harten, A. (1989) ENO scheme with subcell resolution, *J. Comput. Phys.*, 83, 148.

Harten, A., and Hyman, J.M. (1983) Self-adjusting grid methods for one-dimensional hyperbolic conservation laws, *J. Comp. Phys.*, 50, 235.

Hayes, W.D. (1968) The propagation upward of the shock wave from a strong explosion in the atmosphere, *J. Fluid Mech.*, 32, 317.

Hayes, W.D. (1968a) Self-similar strong shocks in an exponential medium, *J. Fluid Mech.*, 32, 305.

Hayes, W.D. (1968b) The propagation of the shock wave from a strong explosion in the atmosphere, *J. Fluid Mech.*, 32, 317.

Jones, H., and Miller, A.R. (1948) The detonation of solid explosives: the equilibrium conditions in the detonation wavefront and the adiabatic expansion of the products of detonation. *Proc. Roy. Soc. A*, 194, 480.

Keller, J.B. (1956) Spherical, cylindrical and one-dimensional gas flows, *Quart. Appl. Math.*, 14, 171.

Kidder, R.E. (1974) Theory of homogeneous isentropic compression and its applications to laser fusion, *Nuclear Fusion*, 14, 53.

Kirkwood, J.G., and Bethe H.A. (1942) Progress report on "the pressure wave produced by an underwater explosion, I", OSRD No. 588.

Kochina, N.N., and Mel'nikova, N.S. (1958) On the unsteady motion of gas driven outward by a piston, neglecting the counter pressure, *PMM*, 22, 622.

Kogure, T., and Osaki, T. (1962) Propogation of shock waves in inhomogeneous medium, *Pub. Astr. Soc. Japan*, 14, 254.

Kompaneets, A.S. (1960) A point explosion in an inhomogeneous atmosphere, *Sov. Phys. Dokl.*, 5, 46.

Korobeinikov, V.P., and Riazanov, E.V. (1959) Solutions of singular cases of point explosions in a gas, *PMM*, 23, 539.

Laumbach, D.D., and Probstein, R.F. (1969) A point explosion in a cold exponential atmosphere, *J. Fluid Mech.*, 35, 53.

Landau, L.D. (1945) On shock waves at a large distance from the place of their origin, *Sov. J. Phys.*, 9, 496.

Lapidus, A. (1971) Computation of radially symmetric shocked flows, *J. Comp. Phys.*, 8, 106.

Latter, R. (1955) Similarity solution for a spherical shock wave, *J. App. Phys.*, 26, 954.

Lax, P.D. (1954) Weak solutions of nonlinear hyperbolic equations and their numerical computation, *Comm. Pure. Appl. Math.*, 7, 159.

Lax, P., and Wendroff, B. (1960) Systems of conservation laws, *Comm. Pure Appl. Math.*, 13, 217.

Lazarus, R.B. (1980) Comments on "Analysis of spherical imploding shocks", *Phys. Fluids*, 23, 844.

Lazarus, R.B. (1981) Self-similar solutions for converging shocks and collapsing cavities, *SIAM J. Num. Anal.*, 18, 316.

Lazarus, R.B., and Richtmyer, R.D. (1977) Similarity solutions for converging shocks, Los Alamos Scientific Laboratory Report, LA 6823-MS.

Lee, B.H.K. (1968) The initial phases of collapse of an imploding shock wave and the application to hypersonic internal flow, *CASI Trans.*, 1, 57.

LeVeque, R.J. (1992) Numerical Methods for Conservation Laws, Birkhauser Verlag, Berlin.

Lewis, C.H. (1961) Plane, cylindrical and spherical blast waves based on Oshima's quasi-similarity model, AEDC-TN-61-157, Arnold Engineering Development Center, Air Force Systems Command, U.S. Air Force, Tenn.

Lighthill, M.J. (1948) The position of the shock wave in certain aerodynamic problems, *Quart. J. Mech. Appl. Math.*, 1, 309.

Lighthill, M.J. (1956) Viscosity effects in sound waves of finite amplitude *in* Surveys in Mechanics, ed. G.K. Batchelor and R.M. Davies, pp. 250–351, Cambridge University Press.

Liu, T.G., Khoo, B.C., and Yeo, K.S. (1999) The numerical simulations of explosion and implosion in air: Use of a modified Harten's TVD scheme, *Int. J. Numer. Meth. Fluids*, 31, 661.

Ludford, G.S.S., and Martin, M.H., (1954) One-dimensional anisentropic flows, *Comm. Pure Appl. Math.*, 7, 45.

Lutzky, M., and Lehto, D.L. (1968) Shock propagation in spherically symmetric exponential atmospheres, *Phys. Fluids*, 11, 1466.

Marshak, R.E. (1958) Effect of radiation on shock wave behaviour, *Phys. Fluids*, 1,24.

McFadden, J.A. (1952) Initial behaviour of a spherical blast, *J. Appl. Phys.*, 23, 1269.

McVittie, G.C. (1953) Spherically symmetric solutions of the equations of gas dynamics, *Proc. Roy. Soc. A*, 220, 339.

Meyer-ter-Vehn, J., and Schalk, C. (1982) Self-similar spherical compression waves in gas dynamics, *Z. Naturforsch.*, 37a, 955.

Mishkin, E.A., and Fujimoto,Y. (1978) Analysis of a cylindrical imploding shock wave, *J. Fluid Mech.*, 89, 61.

Mishkin, E.A. (1980) Reply to the comments by Roger B. Lazarus, *Phys. Fluids*, 23, 844.

Nadezhin, D.K., and Frank-Kamenetskii, D.A. (1965) The propagation of shock waves in the outer layers of a star, *Sov. Astr.*, 9, 226.

Nageswara Yogi, A.M.N. (1995) On the Analytic Theory of Explosions. Ph.D. Thesis, I.I.Sc., Bangalore.

Osher, S. (1981) The use of the Riemann solvers, the entropy condition and difference approximatations, *SIAM J. Num. Anal.*, 21, 217.

Oshima, K. (1960) Blast Waves Produced by Exploding Wires, Rept. 358, Aeronautical Research Institute, Univ. of Tokyo, Tokyo, Japan.

Payne, R.B. (1957) A numerical method for a converging cylindrical shock, *J. Fluid Mech.*, 2, 185.

Plooster, M.N. (1970) Shock waves from line sources: numerical solutions and experimental measurements, *Phys. Fluids*, 13, 2665.

Raizer, Yu.P. (1964) Motion produced in an inhomogeneous atmosphere by a plane shock of short duration, *Sov. Phys. Dokl.*, 8, 1056.

Rankine, W.J.M. (1870) On the thermodynamic theory of waves of finite longitudinal disturbance, *Phil. Trans.*, 160, 277.

Reinicke, P., and Meyer-ter-Vehn, J. (1991) The point explosion with heat conduction, *Phys. Fluids A*, 3, 1807.

Richtmyer, R.D., and Morton, K.W. (1967) Difference methods for initial value problems,Wiley-Interscience, New York.

Richtmyer, R.D., and Von Neumann, J. (1950) A method for the numerical calculation of hydrodynamic shocks, *J. Appl. Phys.*, 21, 232.

Roe, P.L. (1981) The use of the Riemann problem in finite difference schemes *in* Seventh International Conference on Numerical Methods in Fluid Dynamics, Lecture Notes in Physics, 141, 354.

Rogers, M.H. (1958) Similarity flows behind strong shock waves, *Quart. J. Mech. Appl. Math.*, 11, 411.

Sachdev, P.L. (1970) The Brinkley-Kirkwood theory and Whitham's Rule, *ZAMP*, 21, 481.

Sachdev, P.L. (1971) Blast wave propagation in an inhomogeneous atmosphere, *J. Fluid Mech.*, 50, 669.

Sachdev, P.L. (1972) Propagation of a blast wave in uniform or non-uniform media: a uniformly valid analytic solution, *J. Fluid Mech.*, 52, 369.

Sachdev, P.L. (1976) On the theory of weak spherical shocks, *Ind. J. Pure Appl. Math.*, 7, 1261.

Sachdev, P.L. (2000) Self-similarity and Beyond—Exact Solutions of Nonlinear Problems, Chapman & Hall/ CRC Press, New York.

Sachdev, P.L., Dowerah, S., Mayil Vaganan, B. and Philip, V. (1997) Exact analysis of a nonlinear partial differential equation of gas dynamics, *Quart. Appl. Math.*, 55, 201.

Sachdev, P.L., Neelam Gupta and Ahluwalia, D.S. (1992) Exact analytic solutions describing unsteady plane gas flows with shocks of arbitrary strength, *Quart. Appl. Math.*, 50, 677.

Sachdev, P.L., and Prasad, P. (1966) A numerical procedure for shock problems using artificial heat conduction. *J.Phys. Soc. Japan*, 21, 2715.

Sachdev, P.L., and Venkataswamy Reddy, A. (1982) Some exact solutions describing unsteady plane gas flows with shocks, *Quart. Appl. Math.*, 40, 249.

Saito, T., and Glass, I.I. (1979) Applications of random-choice method to problems in shock and detonation-wave dynamics, UTIAS Report, No. 240.

Sakurai, A. (1953) On the propagation and structure of the blast wave I, *J. Phys. Soc. Japan*, 8, 662.

Sakurai, A. (1954) On the propagation and structure of the blast wave II, *J. Phys. Soc. Japan*, 9, 256.

Sakurai, A. (1959) Exploding wires based on a conf. exploding wire phenomena, (W.G. Chace and H.K. Moore, eds.), I, Plenum Press, New York.

Sakurai, A. (1960) On the problem of a shock wave arriving at the edge of a gas, *Comm. Pure. Appl. Math.*, 13, 353.

Sakurai, A. (1965) Blast Wave Theory *in* Basic Developments in Fluid Dynamics (ed. M. Holt), Academic Press, New York.

Schatzman, E. (1949) The heating of the solar corona and chromosphere, *Ann. d'Astr.*, 12, 203.

Sedov, L.I. (1946) Propagation of intense blast waves, *PMM*, 10, 241.

Sedov, L.I. (1959) Similarity and Dimensional Methods in Mechanics, Academic Press, New York.

Shu, C.W., and Osher, S. (1988) Efficient implementation of essentially non-oscillatory shock-capturing schemes. I, *J. Comp. Phys.*, 77, 439.

Sod, G.A (1976) The computer implementation of Glimm's method, Lawrence Livermore Lab. Rep., UCID-17252.

Sod, G.A. (1977) A numerical study of a converging cylindrical shock, *J. Fluid Mech.*, 83, 785.

Stanyukovich, K.P. (1960) Unsteady Motion of Continuous Media, Pergamon Press, New York.

Stekette, J.A. (1972) An explosion wave in the non-homentropic flows of Martin and Ludford, *Quart. Appl. Math.*, 30, 167.

Stekette, J.A. (1976) Transformations of the equations of motion for the unsteady rectilinear flow of a perfect gas, *J. Engg. Math.*, 10, 69.

Stekette, J.A. (1977) Unsteady rectilinear flows of a non-homentropic gas, *Acta Astronautica*, 6, 413.

Stoner, R.G., and Bleakney, W. (1948) The attenuation of spherical shock waves in air, *J. Appl. Phys.*, 19, 670.

Sweby, P.K. (1984) High resolution schemes using flux limiters for hyperbolic conservation laws, *SIAM J. Num. Anal.*, 21, 995.

Taylor, G.I. (1946) The air wave surrounding an expanding sphere, *Proc. Roy. Soc. A*, 186, 273.

Taylor, G.I. (1950) The formation of a blast wave by a very intense explosion, I, *Proc. Roy. Soc. A*, 201, 159.

Taylor, G.I. (1950a) The formation of a blast wave by a very intense explosion, II. The atomic explosion of 1945, *Proc. Roy. Soc. A*, 201, 175.

Taylor, J.L. (1955) An exact solution of the spherical blast wave problem, *Phil. Mag.*, 46, 317.

Theilheimer, F. (1950) The determination of the time constant of a blast wave from the pressure-distance relation, Navord Report No. 1734, U.S. Naval Ord. Lab.,White Oak, Md.

Troutman, W.W., and Davis, C.W. (1965) The two-dimensional behaviour of shocks in the atmosphere, Air Force Weapons Lab., Rep. AFWL-TR-65-151.

Ustinov, M.D. (1967) Ideal gas flow behind a finite-amplitude shock wave, *Izv. Akad. Nauk. SSSR. Mekh. Zhid. Gaza.*, 2, 88.

Ustinov, M.D. (1986) Some one-dimensional unsteady adiabatic gas flows with plane symmetry, *Izv. Akad. Nauk. SSSR, Mekh. Zhid. Gaza.*, 5, 96.

Van Dyke, M., and Guttmann, A.J. (1982) The converging shock wave from a spherical or cylindrical piston, *J. Fluid Mech.*, 120, 451.

Van Leer, B. (1974) Towards the ultimate conservative difference scheme II: monotonicity and conservation combined in a second order scheme, *J. Comp. Phys.*, 14, 361.

Von Mises, R. (1958) Mathematical Theory of Compressible Fluid Flow, Academic Press, New York.

Von Neumann, J. (1941) The point source solution, National Defence Research Committee, Div. B Report AM-9; see Collected Works of J. Von Neumann, Vol. VI, Pergamon Press, Oxford, p. 219.

Von Neumann, J., and Richtmyer, R.D. (1950) A method for the numerical calculation of hydrodynamic shocks, *J. Appl. Phys.*, 21, 232.

Walton, T.S., (1952) Numerical solution of the equations for a discrete model of a spherical blast, *Phys. Rev. A*, 87, 910.

Waxman, E., and Shvarts, D. (1993) Second type self-similar solutions to the strong explosion problem, *Phys. Fluids A*, 5, 1035.

Wecken, F. (1950) Expansion einer gaskugel hohen druckes, *Z. Angew. Math. Mech.*, 30, 270.

Welsh, R.L. (1967) Imploding shocks and detonations, *J. Fluid Mech.*, 29, 61.

Whitham, G.B. (1950) The propagation of spherical blast, *Proc. Roy. Soc. A*, 203, 571.

Whitham, G.B. (1952) The flow pattern of a supersonic projectile, *Comm. Pure Appl. Math.*, 5, 301.

Whitham, G.B. (1956) On the propagation of weak shock waves, *J. Fluid Mech.*, 1, 290.

Whitham, G.B. (1958) On the propagation of shock waves through regions of non-uniform area or flow, *J. Fluid Mech.*, 4, 337.

Whitham, G.B. (1974) Linear and Nonlinear Waves, John Wiley & Sons, New York.

Zel'dovich, Ya. B., and Raizer, Yu. P. (1967) Physics of Shock Waves and High-Temperature Hydrodynamic Phenomena, Vol. 2, Academic Press, New York.

Index